NEW DIRECTIONS IN AFRICA–CHINA STUDIES

Interest in China and Africa is growing exponentially. Taking a step back from the 'events-driven' reactions characterizing much coverage, this timely book reflects more deeply on questions concerning how this subject has been, is being and can be studied.

It offers a comprehensive, multi-disciplinary and authoritative contribution to Africa–China studies. Its diverse chapters explore key current research themes and debates, such as agency, media, race, ivory, development or security, using a variety of case studies from Benin, Kenya and Tanzania, to Angola, Mozambique and Mauritius. Looking back, it explores the evolution of studies about Africa and China. Looking forward, it explores alternative, future possibilities for a complex and constantly evolving subject.

Showcasing a range of perspectives by leading and emerging scholars, *New Directions in Africa–China Studies* is an essential resource for students and scholars of Africa and China relations.

Chris Alden is Professor of IR at the LSE, UK, Senior Research Associate at the South African Institute for International Affairs and Research Associate, University of Pretoria.

Daniel Large is Assistant Professor at the School of Public Policy, Central European University, Hungary.

NEW DIRECTIONS IN AFRICA–CHINA STUDIES

Edited by Chris Alden and Daniel Large

LONDON AND NEW YORK

First published 2019
by Routledge
2 Park Square, Milton Park, Abingdon, Oxon OX14 4RN

and by Routledge
711 Third Avenue, New York, NY 10017

Routledge is an imprint of the Taylor & Francis Group, an informa business

British Library Cataloguing in Publication Data
A catalogue record for this book is available from the British Library

Library of Congress Cataloging in Publication Data
Names: Alden, Chris, editor, contributor. | Large, Daniel, editor, contributor.
Title: New directions in Africa-China studies / edited by Chris Alden and Daniel Large.
Description: New York, NY : Routledge, 2018. | Includes bibliographical references and index.
Identifiers: LCCN 2017054624| ISBN 9781138714632 (hardback) | ISBN 1138714631 (hardback) | ISBN 9781138714670 (paperback) | ISBN 1138714674 (paperback) | ISBN 9781315162461 (ebook)
Subjects: LCSH: Africa–Relations–China. | China–Relations–Africa.
Classification: LCC DT38.9.C5 N49 2018 | DDC 327.6051–dc23
LC record available at https://lccn.loc.gov/2017054624

ISBN: 9781138714632 (hbk)
ISBN: 9781138714670 (pbk)
ISBN: 9781315162461 (ebk)

Typeset in Bembo
by Taylor & Francis Books

MIX
Paper from
responsible sources
FSC **FSC® C013604**
www.fsc.org

Printed and bound by CPI Group (UK) Ltd, Croydon, CR0 4YY

CONTENTS

Views from downstairs: ethnography, identity, and agency   143

Views from upstairs: elites, policy and political economy   241

Conclusion   327

ILLUSTRATIONS

Figures

Tables

CONTRIBUTORS

Chris Alden is Professor of IR at the LSE UK, Senior Research Associate at the South African Institute for International Affairs and Research Associate, University of Pretoria. His publications include *China in Africa* (London: Zed, 2007), *Mozambique and the Construction of the New African State* (Basingstoke, UK: Palgrave, 2001) and the co-edited *China and Mozambique: From Comrades to Capitalists* (Auckland Park: Fanele, 2014), with Sérgio Chichava, and *China Returns to Africa: A Continent and a Rising Power Embrace* (London and New York: Hurst Publishers and Columbia University Press, 2008), with Daniel Large and Ricardo Soares de Oliveira.

Ana Cristina Alves is Assistant Professor at Nanyang Technical University, Singapore. She was Senior Researcher at the South African Institute for International Affairs between 2010 and 2014 and co-edited, with Marcus Power, *China and Angola: A Marriage of Convenience?* (Oxford: Pambazuka Press, 2012).

Kweku Ampiah is Associate Professor of Japanese Studies at the University of Leeds. His publications include *The Political and Moral Imperatives of the Bandung Conference of 1955: The Reactions of US, UK and Japan* (London: Global Oriental, 2007); *Crouching Tiger, Hidden Dragon? China and Africa* (Scottsville: University of KwaZulu-Natal Press, 2007), co-edited with Sanusha Naidu; and *The Dynamics of Japan's Relations with Africa: South Africa, Tanzania and Nigeria* (London: Routledge, 1997).

Li Anshan is Professor at the School of International Studies at Peking University and Vice President of the Chinese Society of African Historical Studies. His publications include *A History of Overseas Chinese in Africa to 1911* (English edn., New York: Diasporic Africa Press, 2013), *British Rule and Rural Protest in Southern Ghana* (New York: Peter Lang Publishing, 2002), *FOCAC 2015: A New Beginning of*

China-Africa Relations (Pretoria: Africa Institute of South Africa, 2015) with Garth Shelton and Funeka Yazini April.

Ross Anthony is Director of the Centre for Chinese Studies, Stellenbosch University, South Africa.

Gabriel Bamana is Associate Researcher at the Centre for East Asian Studies at the University of Groningen, and Researcher at the National University of Mongolia. A native of the D.R. Congo, he holds a PhD in Anthropology (University of Wales, UK). His recent publications include a manuscript on *Tea Practices in Mongolia*.

Lina Benabdallah is Assistant Professor of Politics and International Affairs at Wake Forest University. Her current book project examines China's multilateral foreign policy in continental Africa and seeks to theorise the power dynamics within seemingly equal Global South diplomatic relations.

Alvin Camba is a doctoral candidate in the Department of Sociology at Johns Hopkins University and a non-resident fellow at the Stratbase ADR Institute.

Tatiana Carayannis is Program Director of the SSRC's Understanding Violent Conflict Program. Her publications include *Making Sense of the Central African Republic* (London: Zed Books, 2015), co-edited with Louisa Lombard.

Sérgio Chichava is Senior Researcher at the Institute of Social and Economic Studies (IESE) in Mozambique. He co-edited *China and Mozambique: from Comrades to Capitalists* (Auckland Park: Fanele, 2014), with Chris Alden; and *Mozambique and Brazil: Forging New Partnerships or Developing Dependency?* (Auckland Park: Fanele, 2017), with Chris Alden and Ana Alves.

Ho-Fung Hong is Associate Professor in the Department of Sociology at Johns Hopkins University. His publications include *The China Boom: Why China Will Not Rule the World* (New York: Columbia University Press, 2015), and *Protest with Chinese Characteristics: Demonstrations, Riots, and Petitions in the Mid-Qing Dynasty* (New York: Columbia University Press, 2011).

Honita Cowaloosur is Head of the Africa Division at Enterprise Mauritius, the export promotion agency of the Government of Mauritius. She has a PhD in International Relations from the University of St Andrews, UK. She was previously a Phandulwazi Nge Fellow at the Centre of Chinese Studies, University of Stellenbosch.

T. Tu Huynh is Acting Associate Professor at Jinan University, Guangzhou, and Co-Founder of the Chinese in Africa/Africans in China Research Network.

Daniel Large is Assistant Professor at the School of Public Policy, Central European University and a Non-Resident Senior Fellow at the China Policy Institute, University of Nottingham.

Jamie Monson is Professor of African History in the Department of History, and Director of African Studies, at Michigan State University. She is also a visiting Professor of African Studies at Zhejiang Normal University, China. She is the author of *Africa's Freedom Railway* (Bloomington: Indiana University Press, 2009).

Yoon Jung Park is Executive Director of the Chinese in Africa/Africans in China Research Network, Adjunct Associate Professor at Georgetown University, and Senior Research Associate (non-resident), Sociology Department, Rhodes University. She is author of *A Matter of Honour: Being Chinese in South Africa* (Johannesburg: Jacana Media Pty; Lanham, MD: Lexington Books, 2009), and is completing her second book on Chinese migrants in Africa.

Garth le Pere teaches international political economy at the University of Pretoria and has published widely on Africa–China relations and South African foreign policy. His publications include the edited volume *China in Africa: Mercantilist Predator, or Partner in Development* (Midrand, ZA: Institute for Global Dialogue; Johannesburg: SAIIA 2007), and *China, Africa and South Africa: South-South Cooperation in Global Era* (Midrand, ZA: Institute for Global Dialogue, 2007) co-authored with Garth Shelton. Dr le Pere is the former executive director of the Institute of Global Dialogue.

Maddalena Procopio completed her doctorate at the LSE and was previously affiliated to the University of Nairobi. She is an Associate Research Fellow in the Africa Programme at the Institute for International Political Studies (ISPI), Milan, Italy.

Mzukisi Qobo teaches international political economy at the University of Johannesburg, and is the former head of the Global Powers Programme, South African Institute of International Affairs, and former head of research at the South African Department of Trade and Industry. He is co-author, with Prince Mashele, of *The Fall of the ANC: What Next?* (Johannesburg: Picador Africa, 2014).

Stephanie Rupp is Assistant Professor of Anthropology at Lehman College, City University of New York. She is the author of *Forests of Belonging: Identities, Ethnicities, and Stereotypes in the Congo River Basin* (Seattle: University of Washington Press, 2011).

Derek Sheridan received his PhD in Anthropology from Brown University in 2017. His dissertation 'The Ambivalence of Ascendance: Chinese Migrant Entrepreneurs and the Interpersonal Ethics of Global Inequality in Dar es Salaam, Tanzania', examines how Chinese expatriates and ordinary Tanzanians in the urban economy negotiate the emerging interdependencies and inequalities of South–South connections through the interpersonal ethics of social interactions. He was Visiting Assistant Professor with the Department of Anthropology at Brandeis University, 2017–2018.

Tang Xiaoyang is the Deputy Director at the Carnegie-Tsinghua Center for Global Policy and Associate Professor in the Department of International Relations at Tsinghua University. He is the author of *Zhongfei Jingji Waijiao Jiqi Dui Quanqiu Chanyelian De Qishi* [*China–Africa Economic Diplomacy and Its Implication to Global Value Chain*] (Beijing: World Knowledge Publishers, 2014), and has published extensively on Asia–Africa relations.

Folashadé Soulé-Kohndou is a postdoctoral researcher in International Relations at the University of Oxford with the Oxford–Princeton Global Leaders fellowship programme.

Cobus van Staden is a Senior Researcher: China–Africa, at the South African Institute of International Affairs, and a visiting lecturer at the Department of Media Studies, University of the Witwatersrand, Johannesburg. He is also a co-founder of the online multimedia platform, the China Africa Project.

Ian Taylor is Professor in IR and African Political Economy at St Andrews. His other positions include Chair Professor at the School of International Studies, Renmin University, and Professor Extraordinary in Political Science at the University of Stellenbosch. His publications include *Global Governance and Transnationalising Capitalist Hegemony: The Myth of the 'Emerging Powers'* (London: Routledge, 2017), *Africa Rising? Diversifying Dependency* (Oxford: James Currey, 2014), and *China's New Role in Africa* (Boulder: Lynne Rienner, 2009).

Yu-Shan Wu is a PhD Candidate (International Relations) at the University of Pretoria, and a Research Associate at the University of the Witwatersrand's Africa–China Reporting Project, and the South African Institute of International Affairs.

George T. Yu is Professor Emeritus of Political Science at the University of Illinois at Urbana-Champaign. His books include *China and Tanzania: A Study in Cooperative Interaction* (Berkeley: Center for Chinese Studies, University of California, 1970), and *China's African Policy: A Study of Tanzania* (New York: Praeger, 1975).

ACKNOWLEDGEMENTS

Thanks are extended to all the participants at the SSRC China–Africa Working Group meeting in June 2015, especially Mamadou Diouf and Seifudein Adem; to those participating in the SSRC-Yale Conference 'Making Sense of the China-Africa Relationship' in November 2013; those at the SSRC who assisted with these, including Mignonne Fowlis, Ciara Aucoin and Aaron Pangburn; and Ilaria Carrozza of the LSE. Thanks to William Mangimela, Edwin Riga, Magdi El-Ghazouli, Jok Madut Jok, and Ana Cristina Alves for helping with front cover translations. Thanks to two anonymous reviewers, Leanne Hinves and especially Leela Vathenen at Routledge for their interest, patience and support. Particular thanks are due to Yoon Park and the CAAC, and to Tatiana Carayannis at the SSRC for enabling and guiding the project. We would like to offer special thanks to the Henry Luce Foundation, including Helena Kolenda, who generously funded the SSRC's China Africa Knowledge Project that this book came out of.

PREFACE: NEW DIRECTIONS IN AFRICA–CHINA STUDIES

The twenty-first century has witnessed China's emergence as Africa's largest trading partner and the leading contributor to United Nations' peacekeeping personnel of any of the five permanent members of the Security Council. The first decade of the new millennium saw the emergence of a small cottage industry of publications on China's role in Africa, as well as a growing number of graduate students across disciplinary fields choosing to write doctoral dissertations on some aspect of the China–Africa relationship. The scale and speed of China's growing engagement in Africa and Africa's engagement with China in terms of human and capital flows, and the growing convergence of China's economic and security interests in Africa marks a moment of unprecedented global change. These global transformations are particularly manifest in China's growing trade, investment, aid, and security engagement with the African region both bilaterally and multilaterally, as it is here that China is testing out its new 'going out' policy of global investment and its new role as a global leader. In the last decade, Africa has become an important laboratory for Chinese public policy, which suggests that understanding China's engagement in Africa is a key case for, and window on to, understanding China in the world. It is also important to understanding the transformations underway in the global south.

A preliminary mapping study of China–Africa research and knowledge networks undertaken by the Social Science Research Council (SSRC) in January 2012[1] identified three general trends and recurring themes in academic and policy research activities around China and Africa: 1) the impact of Chinese economic engagement in Africa, especially in the resource sector but also large-scale infrastructure projects, on the economic development of African states; 2) China's interaction with African security, including its growing involvement in international peace operations as well as the impact of Chinese economic interests on ongoing and emerging conflicts in the region; and 3) historic as well as more recent migration patterns, their resulting diasporas, including both African trading communities in China and Chinese labour and independent migrants in Africa, and questions of identity. A subsequent SSRC mapping study published in 2017[2] identified two additional, emerging thematic concentrations of research: 4) environment and public health and 5) media studies, including media representations of the Africa–China relationship. Despite the vast amount of research produced in recent years on Africa–China relations, the work has remained largely under-theorised and fragmented, and mostly driven by events of the day.

This changing global context has challenged researchers to build new intellectual partnerships and cross-regional knowledge networks, and to develop new frameworks, theories, and approaches with which to understand and explain these dynamic social processes and inform public conversations. Within China itself there are ongoing efforts to build the country's own knowledge base and approach to the African region as part of its 'going out' and more recent 'responsible stakeholder' policy. African universities, on the other hand, have well-established departments for the study of Europe and the West, but relatively little capacity for the study of contemporary China or other regions of Asia.

This book grows out of the China–Africa Knowledge Project, a multi-year initiative of the SSRC supported partially by two generous grants from the Henry Luce Foundation. In 2013, as the volume of scholarly and public policy work on Africa and China reached a critical mass, the SSRC saw both a window of opportunity and an urgent need to provide the rigorous theoretical groundwork necessary for future research, mobilise collaborations across scholars and institutions in China, Africa, North America and Europe, and help set priorities for research and teaching. The SSRC China–Africa Knowledge Project thus responded to the need to build connections and coherence across this work, and to the need to critically integrate work on China and Africa (and, more broadly, China and the world) with scholarly discourses and theories about global and transnational structures and flows. It also began to address the teaching and learning needs of graduate students interested in the study of China, Africa, and their international relations, and raised critical questions about the politics of knowledge production on this topic. By studying how China's emergent role in the world is itself being studied and analysed, we actively built more generative connections between scholars across disciplines and regions, while organising a growing and fragmented body of knowledge and connecting it to important trends in the social sciences relevant for understanding Africa's and China's new international relations. The Project reflected two central components of the SSRC mission: bringing social science knowledge and thinking to bear on important global issues, and catalysing innovation in social science research through interdisciplinary, comparative, and cross-regional intellectual encounters. We are certain that along with the Project's many activities, this book will inject a measure of useful reflection and direction into an emerging body of inter-disciplinary and trans-regional work that has the potential to inform critical research and policy across multiple regions.

Dr. Tatiana Carayannis
Program Director, SSRC Understanding Violent Conflict Initiative & Project
Director, SSRC China–Africa Knowledge Project

Notes

1 Tatiana Carayannis and Nathaniel Olin, *A Preliminary Mapping of China-Africa Knowledge Networks*, New York: Social Science Research Council, January 2012 (www.ssrc.org/p ublications/view/392EA92D-FF5E-E111-B2A8-001CC477EC84/).
2 Tatiana Carayannis and Lucas Niewenhuis, *China-Africa: State of the Literature 2012–2017*. New York: Social Science Research Council, December 2017 (www.ssrc.org/publica tions/view/china-africa-state-of-the-literature-2012%E2%80%932017/).

LIST OF ABBREVIATIONS

AFRASO	Africa's Asian Options project, Goethe University, Frankfurt
AIIB	Asian Infrastructure Investment Bank
AGETIP	National Agency of Works of Public Interest (Benin)
AUPSC	African Union Peace and Security Council
ARPONE	*Associação dos Agricultores e Regantes do Bloco de Ponela para o Desenvolvimento Agro-Pecuário e Mecanização Agrícola de Xai-Xai*
ASA	Africa's Security Architecture
ASI	African Studies Initiative, University of Minnesota
BRI	Belt and Road Initiative
BRICS	Brazil, Russia, India, China and South Africa
CAAC	Chinese in Africa-Africans in China Research Network
CAAS	Chinese Association of African Studies
CARI	Johns Hopkins China-Africa Research Initiative
CASS	Chinese Academy of Social Sciences
CATTF	China-Africa Think Tanks Forum
CCTV	China Central Television
CDB	China Development Bank
CFA	West African or Central African franc
CI	Confucius Institute
CIF	China International Fund
CITES	Convention against Illegal Trade in Endangered Species
CNMG	China Nonferrous Metal Mining Group
CNOOC	China National Offshore Oil Corporation
CNPC	China National Petroleum Corporation
CPC	Communist Party of China
CRI	China Radio International
CSEZA	Chinese Special Economic Zones in Africa

CSAHS	Chinese Society of African Historical Studies
DAC	OECD Development Assistance Committee
DTAA	Double Tax Avoidance Agreement
EIA	Environmental Investigation Agency
EPZ	Export Processing Zone
ECOWAS	Economic Community of West African States
ECTZ	Economic Cooperation and Trade Zones
EU	European Union
FDI	Foreign Direct Investment
FNLA	Frente Nacional de Libertação de Angola
FOCAC	Forum on China-Africa Cooperation
FONGZA	Zambézia Forum of Non-Governmental Organisations (Mozambique)
FRELIMO	Frente para a Libertação de Moçambique
HIPC	heavily indebted poor countries
ha	hectare
HS	human security
ICBC	Industrial and Commercial Bank of China
ICC	International Criminal Court
IUCN	International Union for the Conservation of Nature
IMF	International Monetary Fund
IPSS	Institute for Peace and Security Studies, Addis Ababa University
IR	International Relations
ISI	import substitution industrialisation
IWAAS	Institute of West Asian and African Studies, CASS
JFET	JinFei Economic Trade and Cooperation Zone Co. Ltd (Mauritius)
LDCs	Least Developed Countries
LPA	Lagos Plan of Action
MOFA	Ministry of Foreign Affairs
MPLA	Movimento Popular de Libertação de Angola
NEPAD	New Economic Partnership for African Development
NGO	non-governmental organisation
NTS	non-traditional security
NOC	national oil company
OAU	Organization of African Unity
OECD	Organisation for Economic Cooperation and Development
PKU	Peking University
PLA	People's Liberation Army
PRB	Popular Republic of Benin
UNEC	United Nations Economic Commission
UNEP	United Nations Environment Programme
R2P	Responsibility to Protect
SAP	Structural Adjustment Programme

SEZ	Special Economic Zones
SOAS	School of Oriental and African Studies (University of London)
SOE	state-owned enterprise
SPV	Special Purpose Vehicle
SSRC	Social Science Research Council
SYNTRA-TTP	Syndicat des Travailleurs de l'Administration des Transports et des Travaux Publics (Benin)
TAZARA	Tanzanian–Zambian Railway
TISCO	Taiyuan Iron and Steel (Group) Co. Ltd
TRAFFIC	The Wildlife Trade Monitoring Network
UCLA	University of California, Los Angeles
UNECA	United Nations Economic Commission for Africa
UNEP	United Nations Environment Program
UNITA	National Union for the Total Independence of Angola
WTO	World Trade Organization

Introduction

1

STUDYING AFRICA AND CHINA

Chris Alden and Daniel Large

The exponential growth of studies concerning Africa and China relations has only relatively recently begun to more systematically consider questions about the nature of scholarship on these themes and their relation to established academic disciplines. With scholarship on Africa–China now more seriously reflecting on questions of theory, method, and, with these, epistemology and the politics of knowledge, this book seeks to contribute towards this end.

This introductory chapter contextualises how China–Africa relations have been studied to date, and seeks to open up questions about the study of the study of Africa and China. It begins by tracing, in descriptive and chronological terms, the evolution of studies about China and Africa. More analytically, it then problematises this. Is China–Africa a field of study? How has this been changing? Such methodological, epistemological, and ultimately ontological questions can advance different types of inquiry, and open up important questions concerning the politics of knowledge. China–Africa studies is a field of power embedded in deeper, historically produced questions about African Studies, and is also at the forefront of ongoing global shifts in the nature and future direction of these. The chapter ends by outlining the book's aims and contents.

The evolution of China–Africa studies: one account

The Forum on China–Africa Cooperation III (FOCAC) Beijing summit in November 2006 was a catalyst behind the widespread current interest in China's relations with Africa. A decade earlier, in 1996, Chinese President Jiang Zemin's tour of Egypt, Kenya, Ethiopia, Mali, Namibia, and Zimbabwe, and his address to the Organisation of African Unity, attracted far less attention. FOCAC III, however, was the first heads of state summit and marked a very public culmination of China's Year of Africa that had begun with the release of Beijing's first Africa

Policy. FOCAC III was notable as a political spectacle showcasing the nature, ambition and mounting power of attraction China had in its relations with all but a handful of Taiwan-recognizing African states. It also showcased the centrality of business ties, and efforts not just to celebrate historic relations but selectively harness these to the pursuit of enhancing political and economic relations. More substantively, it was notable for rendering connections that had been growing after 1989 visibly manifest and, following the Africa-focused G8 Gleneagles summit of July 2005, for declaring an ambitious set of economic, political and other goals that contrasted with the aid-based Gleneagles process.

FOCAC III stimulated greater interest and scholarly engagement concerning China and Africa relations (e.g. Gaye 2006; Alden 2007; Taylor 2009; Bräutigam 2009). While gathering pace with expanding interest in China's Africa links from 2006, this phase followed earlier periods of scholarly interest (Ogunsanwo 1974 and see George T. Yu chapter in this volume). Indeed, many of the challenges scholarship on China–Africa relations is grappling with today have antecedents in these previous engagements.

While the circumstances were markedly different, some of the underlying questions about how to research, frame and attempt to understand China–Africa relations that surfaced in the 1960s remain pertinent. Knowledge of the different layers of scholarship across time remains necessary and highly instructive, even if the context, scale and magnitude of Chinese relations with Africa now represent a marked departure from late and post-colonial African politics. Research on China's post-colonial and more recent Africa relations has, in different ways, grappled with questions about the politics of formal claims and self-presentation contrasted with actual political realities and conduct. For instance, much like recent efforts to de-sensationalise crude narratives of China's role in Africa, and rebut myths through verified empirical analysis, Emmanuel Hevi attempted to tell "Africa what Communist China is really like" and, while warning Africa about Chinese communism, also criticised the "holier-than-thou air" through which Western countries "arrogated to themselves the sacred duty of protecting Africa from the encroachments of the East's ideological invasion" (Hevi 1963, 9; 1967, 66). In an effort to address variations on the "red menace" or "yellow peril" in Africa (see Cooley 1965), the journalist Alan Hutchinson attempted to investigate actual Chinese conduct, rather than imagined, assumed or projected to Western audiences (Hutchinson 1975). In more scholarly fashion, Bruce Larkin offered a theoretically informed, "more complex understanding" of China's Africa relations (Larkin 1971, 8).

Scholarship expanded and contracted in tandem with the changing dynamics of China's relations with African states from the mid-1950s. Integral to these were relations with Taiwan, the US, and the Soviet Union. In this way, George Yu's in-depth case study of China's multidimensional "policy of selective interaction" with Tanzania helped not just understanding of China's relations with a single, important case but also China's wider policy in other parts of Africa (Yu 1975, xv; 10). The methodological problems of studying relations between China, a country of continental scale, and the African continent were very present at this time, but the context was different in being dominated by other concerns and geopolitical rivalry. Pointing to underlying continuities, as well as notable departures concerning the study of China and Africa, Yu's chapter in this volume historicises questions of method and theory.

Despite multiple post-colonial African trajectories, the single most influential underlying factor in China–Africa relations – and efforts to research these – was Chinese domestic politics. Following the Maoist periods of China's Africa relations, a transitional phase unfolded reflecting and externalizing aspects of the reforms inaugurated under Deng Xiaoping. These combined, significantly, with changes in Africa during the 1980s, a comparatively quiet, if important decade from China's perspective due to the Communist Party of China's (CPC) primary focus on domestic reform. Its engagement in Africa underwent reform, reflecting the economic changes underway in China. Scholarship waned, with notable exceptions (Snow 1988; Taylor 2006; Monson 2009). Deborah Bräutigam's seminal work on Chinese aid in Sierra Leone, Liberia, and the Gambia, was thus published at a time when interest in China, but not China–Africa per se, was rising (Bräutigam 1998).

FOCAC III did much to change this, but its impact should not be overstated. It was preceded by more than a decade of research engagement, notably in South Africa, which was indicative of deeper processes of change afoot. For South Africa in transition from minority rule to democracy in the early 1990s, the contestation over diplomatic recognition of Taipei or Beijing generated a heated public debate amongst scholars, journalists, and policy makers. Nelson Mandela's quixotic promotion of "dual recognition" extended this early academic interest in China and lay the foundation for what was to become an active local research community (Alden 2001).

Underscoring the important African contribution to scholarship and policy-related analysis concerning China and East Asia, this also provides the basis for revising conventional accounts of the development of Asian studies in Africa. For some time, and not uniquely, Japan and East Asia had been leading poles of attraction, including around questions of industrial policy and development (see Hart 2002). Sectoral studies of China and Taiwan's economy explored the content of their development experiences and relevance to South Africa (Lin and Shu 1994; Shelton and Alden 1994; Liu and Hong 1996; Davies 1998). Alert observers in southern Africa detected various kinds of growing connections between, and direct and indirect impacts of China as an increasing economic influence in the region (e.g. Kaplinsky and Posthuma 1994; Cornelissen 2000; Kaplinsky, 2008). More dedicated analysis of China's evolving role and its implications beyond southern Africa followed (e.g. Muekalia 2004; Abraham 2005; Haugen and Carling 2005; Ali 2006; Draper and Le Pere 2006; Sidiropoulos 2006; Gaye 2006; Alden and Davies 2006; Alden 2007; Davies and Corkin 2007; Le Pere and Shelton 2007; Le Pere 2007; Ampiah and Naidu 2008). The growth of interest in China as more than a subject of scholarly interest, but one regarded as important to South Africa's international re-engagement, economic policy and development strategy, continued in an elevated, expanded sense when interest in China's wider continental role became the subject of more global research, analysis and media commentary.

In contrast, despite a number of scholars being active in the field within China, African studies in China languished until Africa became more prominent in foreign policy.

The high profile staging of FOCAC III in Beijing, and obvious importance attached by the Chinese government to Africa, stimulated wider media attention, and engagement with the policy implications of China's expanding engagement in the continent for development policy, Western interests and research (e.g. IPPR 2006; Tjonneland 2006; Gill et al., 2007; Mawdsley 2007). The starting point for many of the waves of scholarship on China and Africa that followed was embedded within the terrain of classic IR, with its concern for great power politics, hierarchies of states, and systemic drivers of change. The realist thrust of much work dominated lines of enquiry but there emerged efforts to explore the relationship along alternative theoretical lines. In the larger context of China's rise in global affairs, Africa therefore became the subject of debate over Chinese foreign policy, and a case-study in the "China threat" (e.g. Naim 2007). Argument over discerning China's real motivation became a staple and tended to employ economic reductionism and monolithic explanation (resource exploitation by an authoritarian unitary state). Counter-arguments stressed benevolence, benign or positive intentions and the possibilities for win-win outcomes (e.g. Li 2007).

Increasing publications on China–Africa meant a diversification of themes, countries, and macro analysis (see Strauss and Saavedra 2009).[1] New initiatives undertook more rigorous research, often using collaborative partnerships and fieldwork.[2] The broadening and deepening of scholarship gained momentum (Strauss 2013). The subject of China–Africa grew to involve a large, expanding and diverse range of scholarly publications and activity, as well as dynamic and diverse policy engagements. Country case studies ploughed new ground, exposing through detailed examination of Chinese ties in Sudan, Angola, Mozambique, Namibia, and beyond, greater insight into everything from the particulars of the political economy to the history of local Chinese migrant communities (Ali 2006; Large and Patey 2011; Power and Alves 2012; Alden and Chichava 2014; April and Shelton 2015; Dobler 2017). It partly contributed to – and joined – mounting interest in "emerging powers" like India, Brazil or Turkey (e.g. Chaturvedi et al. 2012). This contrasted – quite conspicuously, for the most part – with the relative lack of comparable interest in, and attention to, Africa's relations with former colonial powers like France or the UK, and the US (see Engel and Olsen 2005; Gallagher 2011). China in Africa was branded a "burgeoning cottage industry" (Cornelissen et al. 2012, 2). The ever more prominent attention in subsequent years only elevated interest further.

New momentum: FOCAC VI and Chinese foreign policy under Xi Jinping

South Africa's staging of FOCAC VI in December 2015, and use of the summit in which President Xi Jinping participated as a means to advance South Africa's place within Africa and on the global stage (Alden and Wu 2016), injected new impetus. Assisted by an increasing volume of multi-lingual publications (e.g. Adel et al. 2017), scholarship, in turn, endeavoured to keep pace with developments and the changing nature of relations, as evident in a number of thematic areas.

First, economics and trade continue to attract wider interest from global corporate quarters, in the context of a new phase of slowing growth rates within China and concern about the implications of this for Africa (Chen et al. 2016; Sun et al. 2017). One theme attracting scholarly and policy interest concerns efforts to establish Chinese industry in Ethiopia, and other countries, as part of renewed interest in industrialisation (see Tang in this volume). Supplementing macro-economic studies were the growth of micro-processes sensitive to fluctuations in trade. Analysis of formal economic sectors and official data has also moved into various aspects of informal African economies (Xiao 2015; Makungu 2013; Hilson et al. 2014). Other themes rooted in economics, such as labour relations (Giese and Thiel 2015; Rounds and Huang 2017), urban development (Benazeraf 2014; Huynh 2015; Dittgen 2015), or agriculture (Qi et al. 2012; Bräutigam and Zhang 2013), have also been the object of further studies. With expanding economic relations, questions of environmental impact have inevitably become prominent in a range of ways. International NGOs have expanded advocacy efforts on China's role in Africa, testifying to the ongoing importance of a range of policy engagements, in which ivory stands out together with a range of other issues involving illicit trade (e.g. Greenpeace 2015).

The use of global frames and methodological approaches has taken scholarship beyond the limiting confines of methodological nationalism in diverse ways. One concerns the human processes bound up in and contributing toward different, embodied forces of "low-end globalisation" (Mathews 2015). The connected themes of human movement, Africans in China (Bodomo 2012; Bodomo 2015; Castillo 2016; Zhou et al. 2016) and Chinese in Africa (Park 2009) have been amongst the most interesting. One aspect, raising issues extending beyond migration, has been more attention to the hitherto neglected, overlooked, or marginalised theme of gender (Huynh 2015; Giese and Thiel 2015).

Media studies is perhaps the most prominent growing sub-field (e.g. Wu 2012; Li 2016; Puppin 2017) and demonstrates, as van Staden and Wu put it in this volume, how intensely "mediated" China–Africa relations are. This has involved the creation of new research centers and media developments.[3] Debates about China's "soft power" ambitions and spread of culture, including through Confucius Institutes, continue (King 2013; Hartig 2015). One aspect connected to discussions about the propagation of values concerns ethics and culture (Metz 2015).

Efforts to better locate and examine Chinese engagement in African politics, as a number of chapters in this volume demonstrate, form an important trend. Scholarship has moved beyond conceptions of China as a unitary actor to explore the diverse Party-state and non-state entities previously subsumed under the label China. This reflects a move from when China was relatively, if not entirely, new to Africanist circles, and when Africa was the same for scholars of Chinese politics and IR. The opening up of scholarship to more disaggregated Chinese engagements, such as at the provincial level (Shen and Fan 2014) or interest in Chinese social organisation and NGOs in Africa (Hsu et al. 2016), has been fruitful. Another part of this involves the better location of Chinese engagements within African politics to the point of rendering this, not China, the starting point for analysis. Notwithstanding the wide variations in African regime types, China was previously held in some quarters to be undermining democratic African politics in a zero-sum manner. The incorporation of the Chinese role as one part of globally connected political economy structures and processes and, crucially, as used by governing African regimes, has meant a more realistic analysis (e.g. Burgis 2015). This begins with African politics and extraverted techniques of managing and benefiting from China as one of a number of external partnerships (e.g. Soares de Oliveira 2015).

The shifting political dynamics of relations, occurring at multiple levels, reflect processes of evolution. The linear teleology informing assumptions about a somehow inexorable Chinese ascendancy have been disturbed by political and economic changes, taking relations into more uncertain, less determined areas. At the same time, as Alves and Chichava argue here, from the perspective of African politics this is less a story of change and more a historically familiar pattern. On the back of an extended period of more concerted Chinese government engagement in Africa since 2000, the theme of Chinese "learning" and the evolution of its engagement has also become the object of more attention (Giese 2015; Patey 2017). One part of this, China's expanding security role, is another emerging area of inquiry (see Benabdallah and Large in this volume). Debates about the shrinking or expansion of African "policy space" has also developed in more nuanced ways. Whereas after 2006 much Western analysis posited Chinese competition with established OECD engagements, more recent analyses suggest China's engagement has in fact been converging not diverging with Western interests and practice (Kragelund 2015; Swedlund 2017). How what is called China fits into different levels and types and processes of African politics is a theme of wider relevance to many questions concerning the status of China and Africa studies as a field of study.

Another, related and overarching theme concerns geopolitical changes. With Taiwan firmly relegated to a minor, marginal position (Cabestan 2016), Africa in China's changing global politics, through and outside of such forums as the BRICS or the G20, has become more significant. Most saliently, China's Belt and Road Initiative (BRI) is stimulating new waves of attention featuring different kinds of detached and involved scholarship (e.g. Ehizuelen and Abdi 2017). China under President Xi has already seen an ever stronger contrast with the Hu Jintao era of "peaceful development (rise)" in which China–Africa relations took off. China's growing power and determination to enhance its foreign policy capacity to conduct "big power diplomacy" (Hu 2016) have taken its Africa relations, and scholarship in turn, into a new era. China's changing foreign policy and global role under President Xi involves confronting the challenges of establishing the human and institutional resources required to better understand and deal with overseas affairs beyond its neighbourhood. Equipping itself to be a global power necessitates a role for knowledge, including academic. If the US had to build its intellectual and policy capacity to engage the world as a superpower, of which Area Studies was notoriously a part, then China now appears to be facing the same challenge (Ferchen 2016).

The descriptive, chronological summary of China and Africa studies over time presented here is one, more obvious way to set the scene for this book. It should, however, be questioned. It is broadly based on the main trends in China's domestic politics as the determinative influence in Africa relations, meaning that China–Africa relations are organised according to different generations of Chinese leaders or, broadly, revolution, reform and opening, post-1989 China and the "go global" policy, and, more recently, the China Dream and national rejuvenation. As Li Anshan explores in this volume, and Jamie Monson also notes, Chinese academic institutions have their own histories and traditions in African Studies. Inverting such a framework, however, and beginning instead from 54 African country histories (or other sub-national or regional points of departure) would naturally give very different results. African–China historical periodization might resonate with this broad historical pattern of Chinese encounters and the scholarship these have produced. Privileging the respective political trajectories of any given African country as the basis for tracing the evolution of China relations would render any such analysis quite different, partly because China would be one of a number of external partnerships.

Various alternative tellings of the evolution of studies of Africa and China relations are conceivable. Discipline-based accounts, such as from economics, anthropology or history, would offer different perspectives on the ways through which scholarship on China–Africa has evolved. Or, as Li Anshan and Kweku Ampiah do here, different starting perspective from, broadly, China and Africa would form the basis of different accounts. These would likely also raise questions about the evolving status of Africa–China as a field of study. Given the expansion of work in this area, and its continued growth, however, it seems almost impossible to adequately cover China–Africa. The difficulties inherent in such an endeavour suggest that the days of totalizing China–Africa studies, even quantitative studies for which data continues to pose challenges, are highly questionable given the growing complexity of relations.

Themes and sub-fields

The process by which "China–Africa" has grown as a meta-organizing subject for scholarship and policy related research has involved a gradual and informal organization into subthemes and clusters of interest (Monson and Rupp 2013). Some of these involve the social sciences or humanities. Many concern continuing, unresolved questions about methodology and theory. The spread of China–Africa research has become more global and multilingual, despite the dominance of English, French and Mandarin Chinese. The expansion of Africa studies in China, and the expansion of China studies in parts of the African continent, has proceeded in tandem with an expansion of global interest in Africa and China. Impetus has derived not merely from the stimulus China provides but also the prominence and associated support given by the BRICS at a time when Africa Rising narratives were ascendant. The result is a more globally diffused geography of studies amidst evolving patterns of transnational research collaboration.

A number of more salient sub-themes can be identified. The first concerns the effort to ensure accurate data through positivist methodologies in the pursuit of an enhanced, empirical understanding of the nature of China's engagement. With media attention and narratives about China's conduct in Africa after FOCAC III, it became apparent that the quality of information and analysis was lacking. The proliferation of speculative commentary in a relative vacuum of hard data was frequently characterised by binary tropes (China as development competitor or partner, good or bad, win–win or unequal, and, in turn, the Chinese government promoting or defending its Africa engagement). Against these, efforts were made to provide more considered analysis and accurate data. Investigating problems of data and interpretation in China–Africa studies, Bräutigam (2015) harnessed forensic research methods in the pursuit of empirical accuracy. Such work was particularly notable given the uneven, problematic and at times sensationalist Western media coverage of China's role in Africa. While bringing important issues into focus concerning the basis of knowledge claims, some of the issues raised by the "myth-busting" turn in China–Africa studies (Hirono and Suzuki 2014), such as data about investment or growth, are symptomatic of deeper, prior challenges concerning, for example, the study of African economics (Jerven 2016).

Development is a second, diverse area attracting wide interest, due to its policy relevance and political role in foreign policy and domestic politics. Much attention has been directed toward economic development, development aid and cooperation (Dent 2011; Zhao 2014; Li et al. 2014; Kragelund 2015). Africa's position as the epicenter of development, at least until 2006 when FDI flows exceeded aid flows from the external world to the continent, contributed to this debate. Development engagement has involved policy research collaborations of various kinds (e.g. China-DAC Study Group 2011) as well as more theoretically attuned academic interest (Bräutigam and Tang 2012). In both economic and political areas of development, the notion of a – or *the* – "China model" has been prominent. At one level, this resurrected historical interest in China as a post-colonial socialist alternative for Africa, which gained purchase in some African quarters during periods of the Cold War when multiple Chinese, Soviet or Yugoslav Communist models were contending in the continent, also contributing to hybrid, indigenised forms of "Afrocommunism" (Hamrell and Widstrand 1964; Ottaway and Ottaway 1986).

In other ways, recent interest in "the China model" occurred in radically different circumstances, stimulated by the demonstrable achievements of China's post-1978 economic reform and opening, not claims about the efficacy of ideological fervor and faith. As China has become more involved in global development policy, amidst increasing efforts to consider whether and how aspects of China's domestic development could be selectively applied in other parts of the world, the issue of the China model became the subject of wide, diverse interest. This raised questions about whether the China model was to be understood narrowly, as a set of development policies and practices to achieve principally sectoral aims, or rather, as composites of a whole that included a political system dominated by a single party.[4] It was also advanced by the CPC's efforts to promote its approach to political and economic management as a means to mobilise soft power. However, where early considerations tended to implicitly assume, or more explicitly call for, exporting the China model, more recent analysis has begun to consider an arguably more important dynamic: the role of ideas and selective appropriation and application by African ruling elites as part of processes involving importing and customizing ideas according to political circumstances and needs (see Fourie 2015).

Migration and the flows of peoples within a changing geography of global trade has been a further significant area of academic research and related forms of media-driven attention (e.g. Park 2009; Cisse 2013; Mohan et al., 2013). This has been one part of growing interest in China as an immigrant country (Pieke 2012), of which African immigrants have been one part (e.g. Bodomo 2012; Haugen 2012; Ho 2017). Migration, and the ethnographic methods often utilised to study flows of people, have been especially illuminating in terms of diverse and fine-grained human dimensions of Africa and China relations. This has contributed toward balancing the state-dominated ontology of much IR coverage of China and Africa. As seen in debate generated by Howard French's book, *China's Second Continent*, there has been robust argument about how to understand migration, its implications, and logic of development (French 2014). If migration has been expanding as a sphere of academic interest, it has also been the subject of a number of news items, documentaries, and films.[5]

The move toward comparative analysis – approaching China as one of a number of external powers in Africa – has been a further trend (Adebajo 2010; Cheru and Obi 2010). Partly out of recognition of the limitations of any singular focus on China, and the significance of growing relations between other parts of Asia and Africa, there has been a broadening into various forms of comparative research. One direction has concerned the emerging theme of Africa–Asia relations, in which India's relations with Africa have been prominent (Mawdsley and McCann 2011; Beri 2014; Dubey and Biswas 2016; Modi 2017). Japan has received more attention (e.g. Lumumba-Kasongo 2010; Kato 2017), as have South Korea–Africa relations (e.g. Kim and Gray 2016; Asongu 2015).

Political economy approaches seek to better locate and explore China's role (e.g. Mohan 2013), beyond forms of methodological statism that ignore wider structural forces (Ayers 2013). The tendency to isolate and magnify China's role has been criticised, with Sautman and Yan, for instance, arguing that studies of China–Africa engagements throw broader processes like neoliberalism in Africa into starker relief (Sautman and Yan 2008). Studies utilizing a global political economy framework have thus offered reflections on the extent to which economic investment in Zambia, for example, is or can be considered "Chinese", or reflects qualities and political relations and patterns of exploitation generically familiar to the global behaviour of capital (Lee 2014). Rather than approaching corporate engagement through the prism of its nation-state origins, this perspective recasts the Chinese role as a new chapter in global capitalist relations (Li and Farah 2013). Such an approach, taken too far, runs the risk of downplaying Chinese characteristics and connections. In the attempt to demonstrate conformity to historical extraversion or the logic of capital, the Chinese qualities of these dynamics, mediated though these can be by hybrid global dynamics, can be stripped away and questions about forms of Chinese power avoided. At the same time, China's role risks being overstated. To overcome such analytical challenges, Ian Taylor situates China's engagement within a historically structured global political economy of resource extraction and trade. Instead of emancipation through market-means and economic growth, he argues that China's engagement is one part of the entrenchment of African economic dependency in context of vulnerability to and dependence on global commodity prices, rendered more important due to failure to successfully diversify economics (Taylor 2014).

China–Africa: a field of study?

"What is 'China–Africa' studies?" Jamie Monson's question, explored in her chapter here, gets to the heart of the issue. What is a field of study? is a related, higher-order question about just how China–Africa studies can be understood to date – and might advance in the future.

Questions start with the very choices of description. The choices and preferences implicit in the shorthand language used to distil the subject is evident in the use of even basic vocabulary. The various permutations used to consider China and Africa can convey meanings that inhere to their formulation: "China–Africa" (connections between); "China in Africa" (an interventionary presence); and "China and Africa" (a partnership). In turn, some advocate a re-ordering into "Africa–China", "Africa in China", or "Africa and China" to signify and advance a substantive re-orientation of scholarship. The difference between "in" and "and" has generated debate, amidst critique of "the China in Africa discourse." In turn, this shows how language plays an important role. The dominance of English-language sources outside of China itself is revealing. Publications in French on China and Africa have increased, and Lusophone Africa has its own sub-theme that partly involves Portuguese language sources, but English predominates. The barriers imposed by language, within the context of China's domestic approach to managing controlling information, means a risk of self-referential coverage.

Although the limitations and limits of "China–Africa" as a combination of singular, shorthand abstractions is invariably the first qualification to be made, it remains, unavoidably perhaps, part of the basic framing in coverage and studies. While problematic, at the same time this continues to be routinely employed as a form of necessary shorthand. Emanating from the prior problem of the abstraction of ideas about and representations of Africa, this framing question is hard to avoid and constitutes "a familiar paradox: we cannot generalise about Africa, and yet we must do so" (Cooper and Morrell 2014, 2). One challenge facing analysis stems from this binary combination.[6] For all the interest in finding a vocabulary of description and analysis better able to capture and engage complexity, however, there is also a sense that the basic framing of China–Africa will endure, if only because it frames official interaction and the media coverage flowing from this, thereby serving as a gateway. In this regard, scholarship has a duty to engage in more critical analysis and not merely replicate and amplify state-sponsored discourse.

Area Studies and beyond

The shift in China–Africa studies from being under-researched to under-theorised has been noted (A'Zami 2015). A number of studies have sought to integrate different types of theory into China–Africa work, including efforts to reconceptualise these as indicative of new emerging formations in global politics or economics (Mason 2017). In general, efforts to engage in overt, intentional and systematic theorizing followed the initial empirical focus of many studies. Some efforts to engage theoretical frameworks based these around empirical cases (Power et al. 2012). Perhaps the subject that stimulated most interest and debate of a scholarly kind thus involved a salient "turn" in China–Africa studies toward "African agency," becoming perhaps the main possible exception to the general lack of systematic treatment of theory and method in China–Africa studies. The agency turn rested upon a critique of China–Africa for variously neglecting, ignoring, or erasing African agency (Mohan and Lampert 2012). Followed by other works exploring the theme (Corkin 2013; Gadzala 2015; Lopes 2016), such studies were bound up in a wider conversation about African agency in international politics (Brown and Harman 2013).

Efforts to address and go beyond China–Africa have featured frameworks such as the "Asian drivers", framing China in global political economy terms, and differentiating myriad direct and indirect connections and effects of economic links across multiple different economic cases (Kaplinsky 2008). Other approaches include those framing dynamic human mobility and trade flows in terms of transnational connectedness according to competitive, complementary, and cooperative dynamics (Haugen 2011). Further efforts, including comparative work, and interest in global transformations in terms of interactions between Africa and Asia within a changing geography of the global South, have produced notions such as the "Afrasian imagination" (Desai 2013), "Afrasia" (Mazrui and Adem 2013), and efforts to reconceptualise Area Studies from transregional perspectives.[7] The founding of Association of Asian Studies in Africa, in Accra, September 2015, gave further impetus to new scholarship.

Jamie Monson and Stephanie Rupp underlined the potential of historically grounded, ethnographically derived and theoretically attuned work. They moved the focus from national-level macro analysis to multiple levels of dynamic connection, from community and individual to transnational and transregional, and identified a connected need to factor in a multiplicity of actors. The additional combination of a temporal dimension, that is, approaching and locating "engagements between Africa and China within both historical processes and newly emerging realities," offers further benefits, notably the sense in which China is part of wider questions and issues involving African spaces and protagonists, befitting changing patterns, modes and the critical function of disciplinary theory in analyzing this data. Such a use of carefully grounded ethnographic and historical methods has the potential to enhance understanding not only of area processes but also of global transformations in Africa, in China, and beyond (Monson and Rupp 2013, 41).

Africa in China studies, China in African studies

African studies has undergone radical changes in research methods in the last three decades. While Walter Rodney, Giovanni Arrighi and Edwards Alpers, amongst others, provided a strongly theorised critical and historisised understanding of the political economy of Africa and its international manifestations in the aftermath of independence, by the late 1970s, the majority of work about the continent in fields such as history, politics, or anthropology was predominantly qualitative and characterised by wide theoretical differences.

Qualitative IR research on Africa has tended to become more explicitly theoretical and draw upon insights from, amongst other areas, critical security studies, feminist theory, or postcolonial literatures. Indeed, explorations of discursive power relations using textual analysis have become a feature of articles on African IR (Abrahamsen 2003). Such innovations and departures from the set menu of disciplinary concerns in African studies, broadly conceived, appear conducive to integrating and exploring aspects related to China and Chinese in a global setting. More recently, in the context of an identified decline in long-term immersive field work (see Duffield 2014), the notion "that new ways of analysing the continent have created a greater distance between the researcher and the people they are researching" has become apparent (Cheeseman et al. 2016, 4). In contrast, one problem facing research on Africa within China noted by Li Anshan in his chapter here, namely the initial lack of in-depth fieldwork-based research and immersion leading to direct first-hand knowledge of context, was identified and is being addressed through a new generation of younger Chinese scholars from Chinese, African and Western universities able to conduct longer-term fieldwork.

If China–Africa has become a loose, de facto field of study, at best this process has been driven by accidental and unbounded interest in China–Africa. It is now perhaps less a field, in the sense of a dedicated, organised area of inquiry, and more a dynamic starting point and intersection featuring multiple possible avenues of inquiry and emerging combinations of various kinds of approaches and fields of study. The unifying meta-theme of China's relations with Africa belies such messiness and the diversity of different kinds of academic treatments and media coverage. As this book seeks to begin to do, it is clearly a subject that demands more engaged debate and reflection about the study of the study of China–Africa or Africa and China. In this, and amidst critique of Western scholarship on China in

Africa, questions concerning knowledge and power inevitably and necessarily arise (Foucault 1980). What reasons are there to think that the power-knowledge dynamics familiar to Western scholarship will not apply in China's case?

Power-knowledge

The study of China–Africa is a field of power relations connected to, but going beyond, those already present in African studies. This has been an enduring characteristic of China–Africa as a subject of study, policy and politics, as post-colonial relations showed. The issue of how knowledge is constructed, by whom and to what ends is crucial (see Monson, in this volume). Underpinning questions about knowledge are questions about method, which is fundamental to epistemology: "*how* we know determines *what* we know" (Ashcroft 2014, 65). Looking forward, as the Conclusion to this book discusses, epistemological questions about China and Africa require further interrogation.

Much interest in China–Africa relations before and certainly after 2006 reflected an asymmetrical privileging of China, whose ascendancy in the African continent and the world was the big story. Interest in African agency marks a counter-reaction to this, with arguments favouring a reframing of China–Africa studies into Africa–China studies part of an effort to overhaul and advance a wave of new scholarship from – and for – the continent. Within Africa and African studies, however, China has attained a more prominent position. It is now no longer possible for textbooks on African politics and IR to omit China (see Large 2008). In comparative terms, Africa is not as important in Chinese foreign policy as the other way round. This is perhaps one indirect reason accounting for mounting African interest in China on the one hand, and the growing, significant but still comparatively small (in relation to, for example, the US) Chinese interest in Africa on the other. This is not to diminish African studies in China, or broader interest and engagement; far from it, both have undoubtedly grown in recent years. It is, however, to point out a macro-contrast applicable at different levels to a related combination of scholarship, policy interest, and relative importance structured in asymmetrical ways. While China's role in African politics is receiving more serious attention than previously, there is little sense that African dynamics have serious political impacts in China, beyond certain foreign policy episodes or issues such as migration and race, which themselves mostly fail to attain sustained national political significance.

Several contributors in this book critique "the 'China in Africa' discourse," situating this in the context of the deeper power relations in which scholarship on Africa – and African studies – has historically developed within. This ultimately concerns colonialism and is partly why debate about decolonizing knowledge is topical (see Alden and Large, Chapter 21, in this volume). One contention founded in a longer historical perspective is that China in Africa as a scholarly subject was "invented in the West" due to the way in which China approached Africa "through the West," particularly in the form of former colonial powers and knowledge.[8] That knowledge has been one aspect of African studies in China is undoubted, reflecting more global trends. Whereas China's African studies has in part used knowledge produced in the West, the more recent official push toward less mediated forms of knowledge production is significant.

The emergence of China as a research funder, if sustained and expanded, has the potential to be of great consequence for African studies and knowledge production. A notable development in this regard came in March 2010 when the China–Africa Joint Research and Exchange Program was launched in Beijing. The FOCAC IV Action Plan (2010–2012) contained provisions to enhance cooperation and exchanges between scholars and think tanks. Addressing a seminar on "Development Strategy for the Originality of African Studies in China" in 2017, and couched in terms of President Xi's diplomatic thinking, China's Vice Foreign Minister Zhang Ming called for China's research on Africa to be strengthened. He urged participants:

> to stick to China's own road, do more original thinking and research with theoretical and in-depth meaning, and practical and operable significance as well, strive to mark 'Chinese Brand' on the international research on Africa, hold the discourse right of China-Africa relations in China's own hands, and make the research on Africa more active, in-depth and solid.[9]

Such directions, if continued, can be viewed as part of a process of eroding the dominance of Western funding. At the same time, amidst China's emergence as an attractive destination for higher education students, new generations of African university students educated in China are already beginning to impact changing trends in scholarship and knowledge production (Haugen 2013). China's domestic context, however, is such as to mean it is not just scholarship that is bound up in political trends but other forms of knowledge and control. Moreover, one telling measure of focus of scholarship emanating from China is that, according to one Chinese academic, it too often aimed at "China–Africa" topics rather than the study of Africa itself; meaning, rather than disinterested scholarship, too often there can be political directives behind research (see Strauss 2009. This is clearly not unique to China, e.g. see Martin and McQuade 2014).

The context of African studies in China is unavoidably conditioned by political circumstances, and has wider ramifications. President Xi cited China's role in Africa as part of Beijing's commitment to open global order and carrying out international responsibilities, but such rhetorical gestures toward external openness contrast with the domestic crackdown that has intensified during his reign. The CPC's push for ideological control has intensified under President Xi.[10] This has extended into the education sector, with the president calling for Chinese universities and colleges to serve as "strongholds that adhere to party leadership" and a tightening of the CPC's role in education together with the political responsibilities of professors.[11]

Internet control in China, strengthened by enhanced policing and regulation, also has implications for African studies and China–Africa research and the ability to exchange information.[12] At a time when top Chinese universities are pursuing global ambitions, seen for example in 2017 when Peking University announced it would open a branch in Oxford, foreign universities in China face more challenging political circumstances. Academic publishers have also faced dilemmas in the context of China's "authoritarian information order."[13] A notable example was the demands made in 2017 by the Chinese authorities on Cambridge University Press to remove published academic content from its website in China, including articles on Xinjiang, Tibet and human rights published in the flagship China Studies journal *The China Quarterly*.

Such tendencies potentially have implications for academic freedom beyond China, and are also bound up in processes by which information orders are diverging amidst a fragmentation of state/corporate regulation of digital space. The rapid changes brought about by the internet, and different media technologies and forms of communication, now mean that categories defined by the media take on a life of their own. A striking amount of China–Africa coverage is now produced outside of established African or China studies and bypasses (while influencing and being used by) academic routes or publication outlets. Indeed, much of the energy and agency is being exercised via new forms of engagement, including social media. This has implications for the authoritative nature of different forms of knowledge.

Knowledge production

Calls for rigor, accuracy in data and other sources used as the basis of academic analysis are important. These are rendered even more so by the ease with which rumour, conjecture, falsehoods or simple inaccuracies can travel around the internet. There has been progress, part of a wider attempt to reset and improve the knowledge basis upon which, for example, economists study Africa. The empiricist ambition, however, has limits. Most obviously, the ideal of foundational knowledge is problematic, which is where contending approaches to epistemology become important. Alternative critical approaches, such as those exploring the nature, construction, or mobilization of discourses, challenge the empiricist ambition and do so in a markedly different and rapidly evolving context. China–Africa publications from the 1960s or 1970s had a fairly limited circulation, largely within the academy or policy-circles. Today, with media coverage and the internet, there is more dynamic and influential interaction, such that Chinese in African countries, Africans in China, or more globally, are influenced by a range of media and, crucially, are not merely objects but active subjects.

Academic scholarship is but one source of knowledge production amidst a proliferation of knowledge producers. Apart from the issue of existing infrastructure for social science in Africa, this raises the issue of what knowledge is being produced, how, by whom and for what purposes. The singular notion of knowledge production can be questioned in order to identify and separate different forms of knowledge. In exploring "how knowledge is made in African contexts," for example, Cooper and Morrell argue that knowledge must be understood in a plural sense, complementing an understanding that Africa is not homogenous or monolithic (Cooper and Morrell 2014). A historically more familiar theme accompanying African studies concerns tensions of ownership ultimately rooted in power imbalances. Questioning the extent to which "so-called African Studies" are African, Hountondji thus argues that the study of Africa has been part of a Western initiated and controlled "overall project of knowledge accumulation," which is "massively extraverted, i.e. externally oriented, intended first and foremost to meet the theoretical and practical needs of Northern societies" (Hountondji 2009: 1).

Trends in knowledge production concerning China and Africa have seen multiplying initiatives, widening geographical spread, new sources of funding, and new patterns of collaboration and exchange, such as initiatives falling under the BRICS remit or FOCAC. That a key source of information about China–Africa relations, the FOCAC website, is a state controlled information resource demonstrates how knowledge and information connect with political interests. In contrast, and involving a different web of initiative and funding, are Africa-based ventures. Many are African-led with a defined focus on China–Africa, such as the Centre for Chinese Studies at Stellenbosch University, or engage with China as one part of a broader mission. Many reflect activist ends, such as civil society and trade unions (e.g. Marks 2009), and inspired innovative platforms such as Pambazuka News or the Wits China and Africa Reporting Project. The number of new initiatives may or may not be sustainable, and join the established architecture of social science in Africa, including the Council for the Development of Social Science Research in Africa (CODESRIA).

The need to reimagine the global, and approach this from perspectives that transcend meanings deriving from a Western-centric perspective, is partly where Africa and China are situated. In such an endeavour, fresh interest in reclaiming and using African intellectual and practical engagement with China from past eras to illuminate the current period suggest an effort to attain more autonomous engagement (e.g. see Matambo and Mtshali 2016). This matters, partly because debates about changing African studies, Africa's role in China knowledge, or China's role in producing new African knowledge are subsets of larger structural conditions reflecting the subordination of the continent in global terms, including education and scholarship. One example is the under representation of scholars from Africa in publications in most highly ranked African studies journals (see Briggs and Weathers 2016). In this context, there have been recent renewed calls and initiatives aimed at Africa taking "the intellectual lead on the global discourse about Africa."[14] More generally, and connected to but going beyond China, during the March 2015 first African Higher Education Summit in Dakar, 15 African universities drawn from eight countries formed an African Research Universities Alliance. In the context of such efforts, China is a part of changing forms of research collaboration in the continent (Halvorsen and Nossum 2016).

Aims of the book

Overall, this book aims to contribute to the burgeoning area of China–Africa studies. Drawing on various disciplinary perspectives, it reflects on hitherto neglected issues of method and theory, and analyzes different aspects of China–Africa relations. Situated within wider debates in the social sciences, some chapters also reflect upon the significance of this transnational phenomenon for our understanding of the future of traditional forms of scholarship. The collection has four, more defined objectives.

It endeavours, first, to step back from the "events-driven" reactions and analysis characterizing much coverage of China–Africa relations and reflect more deeply on various questions concerning how this has been, is being and can be studied. Second, it aims to showcase perspectives by a range of scholars in order to demonstrate aspects of and potential for the theoretical and methodological pluralism in studies of Africa and China today. Different generations of contributors are featured, from established, well-known scholars, to younger, upcoming post-doctoral researchers, and from different parts of the African continent (Southern, Eastern, Western, Central and Northern Africa), China, Europe, and North America. This conveys a positive diversity of methodological, theoretical, and more empirically grounded perspectives. A third, wide-ranging aim is to raise questions concerning the politics of knowledge. The influence of a strong media dimension more readily shapes the contours – and provides some of the content – of China–Africa studies than found in other areas of academia. This is arguably a phenomenon common to the origins of many new research areas.[15] At the same time, a discernible trend in the academic literature seems intent on re-producing nationalist-oriented discourses founded on epiphenomena and ideological precepts. All of this underscores the new and evolving conditions under which scholarship operates, challenging academia and researchers to reflect upon questions concerning the politics of knowledge and to find new ways of engaging a widening audience consisting of policy constituencies, students and publics.

Finally, this collection aims to help advance scholarly inquiry. It poses and encourages reflection on methodological and theoretical questions concerning Africa–China studies but also wider issues, such as how China–Africa relates to wider scholarship on Africa, higher education in Africa or the global impact of China on social science. Combined, these are intended to help generate debate and contribute toward new, critical scholarship. The starting subject of Africa and China, however, is by no means confined to scholarship, as the constantly expanding diversity of critical, celebratory and more considered interest spanning social, cultural, media, film, art, or literary coverage demonstrates.[16] While these are important, and themselves the subject of new lines of scholarly inquiry, this volume proceeds from the perspective of academic research. It aims to contribute toward defining new frontiers, engaging emerging trends of global significance about the current state and future trends of social science in a world where Africa is no longer as marginalised and China's influence is increasingly felt.

It is important to also be clear what this collection is not trying to do. One point of departure is the very dynamism and complexity of the area. While ethnographic studies have been able to capture the pace of change, much scholarship often has sought to catch up "with a rapidly evolving and changing situation" (Chan 2013b, 143)

The contributions assembled here can only thus mark a time-bound moment in this development. Second, this volume is not trying to be systematic or exhaustive; indeed, one implication arising from the chapters assembled here concerns the limits of any totalizing explanatory ambition regarding China and Africa relations. Finally, despite arguments favouring a reframing of China–Africa studies into Africa–China studies, this collection uses both interchangeably and does not imbue either with too much meaning. It seeks to open up rather than try to resolve such important current debates.

Overview of chapters

The book's three sections organise its contents into general but overlapping parts. Indicative of the fluid nature of the subject matter, these serve as broad categories organised along thematic lines. The chapters in the first section explore aspects concerning the origins, development, and possible futures of Africa–China studies. These encompass the shift from China in Africa as the predominant mode of studies, at least in Western scholarship, to current arguments about global African studies. George Yu provides personal reflections on how he came to first study Tanzania's relations with China, and the evolution of China–Africa academic studies since then. This is a story of how "a very small, peripheral and solitary field" became "a multi-disciplinary mainstream academic field of study." Facing the impossibility of undertaking research in China, he identified east Africa as where it would be possible to "conduct field research directly observing Chinese behaviour and attempt to assess and verify developments first hand." He was not merely undertaking empirical research but also attempting "to construct a conceptual framework to examine and explain China's foreign policy and behaviour" in an effort to move beyond monocausal explanations. Yu has also been active in important collaborations concerning Chinese and African scholarship. From 1982, he directed a Ford Foundation funded program to train and retrain Chinese Africanists in academic and government research institutions. Such collaborations are touched on by Jamie Monson in her chapter and, going forward, will continue to be important.

The distinguished Chinese Africanist Li Anshan surveys the historical evolution of the study of Africa within China and more recent trends in Chinese scholarship. His chapter provides an authoritative summary of the background, nature, and recent trends in Chinese scholarship on Africa, together with applicable key publications. He notes that: "Economics is the most studied subject, and South Africa the most studied state." While concluding that Chinese scholars have "a long way to go," he also concludes that Chinese scholarship on Africa is benefiting from better opportunities to conduct research in African contexts. This shows how much has changed in a relatively short space of time, and points to important changes underway in terms of expanding and deepening scholarly connections.

Kweku Ampiah evaluates the discourse regarding China's role in Africa's recent socio-economic development, and considers how Africanist scholars have discussed and interrogated the unfolding relations between the relevant stakeholders.

Following previous concerns that African perspectives on China's engagement with Africa were marginalised in a discourse largely dominated by American, British, and European researchers and media outlets, he notes that this ignited a counter-discourse, especially involving African scholars. More generally, noting the dominance of interest in economic questions, he exhorts Africanist scholars to "transcend the perennial focus on trade and development-based scholarship that is too often analysed in the discipline of International Relations, and to engage with and interrogate the underlying socio-cultural imperatives of the China-Africa matrix, the circulation of knowledge engendered through that medium and its power implications." Culture and sociality are significant and warrant further study (see Liu, 2008).

In their chapter, Cobus van Staden and Yu-Shan Wu note how "intensely mediated" China–Africa relations are. They examine China–Africa media relations "as a field of study," and examine how this has been delineated up to the present, including different methodologies and disciplines of study. It also considers what they term "the dual role of media": as not just a subject in China–Africa studies, but also "an instrument *for* the very analysis and deliberation of the relationship." Their chapter also explores new directions in media analysis, including the emergent interest in the Internet as a vector of China–Africa relations, and the role of social media. In considering ways China–Africa media scholarship could advance, they note that less is known "about how African respondents and local media players are impacting China's media practices and engagement." They conclude that irrespective of the shape the field of China–Africa media studies takes in the future, "it will be even more mediated than it is now." As such, "it is incumbent on researchers to trace the role of politics and power in this mediation."

Drawing on critical post-structuralist insights, Ross Anthony deconstructs mainstream "Euro-American discourses" on the Chinese engagement in Africa. The African continent is "a site of contestation in which Euro-American ideologies exert influence through championing, in many regards, their difference from China." He critiques "territorially based, binary, value imbued" Eurocentric thinking and discourse. Most importantly, he argues that this "signals a failure to locate an adequate representation of the world in an age of economic globalisation" in the context of the Anthropocene, the present ecological era in which human-based planetary alteration driven by market capitalism is being inscribed in the fossil record. While noting that in China "there exists a very different China–Africa discourse, couched in a positively charged language of historical struggles against Western imperialism and the championing of development assistance," and the complex discourses within Africa on China, these are subjects identified as needing more research. Overall, China's integration into the global economy has accelerated the problem of the Anthropocene and, in turn, underlined the imperative of "re-thinking global political economy in the context of the Anthropocene."

Jamie Monson explores "Global African Studies and Locating China," weaving in aspects of her career as a historian with reflections on the different phases of China–Africa studies. She also asks: "if scholars today were to ask a major foundation like the Ford Foundation to fund a new fifteen-year global research initiative on Africa and China, what would it look like, and why?" In part, her answer is that African scholars, including those in the global African diaspora, and African universities would be central to any such future initiative. Epistemology would be a core primary starting point of inquiry for such a program. The history, current and future iterations of Area Studies would be "a critical element in moving forward to post-Area Studies scholarship." This would go much further than merely reformulating Area Studies into a transregional framework. She concludes by arguing that "African Studies has the potential to become an Africa-led field that is truly global – indeed it can only become truly global – through the collaboration of interdisciplinary and global partners." Not only are "new paradigms for the study of global Africa" needed, but also "new institutional models that are collaborative and consortial across national and regional boundaries."

Gabriel Bamana opens up an intriguing set of dynamics and questions by reflecting on his efforts to find his place as an African ethnographer in Mongolia. His starting point goes beyond the established convention of an "ethnographer and the ethnographic Other in an ethnocentric context defined by the West-Other power relations" by asking what happens when the ethnographer is "an individual traditionally identified as the ethnographic Other, yet he/she uses tools from a western intellectual tradition to process and mediate knowledge in the universe traditionally identified as that of the ethnographic other." By not conforming at first glance to China–Africa studies, and other established traditions, his chapter demonstrates and points to new lines of inquiry and research connected to but going well beyond China (see Pedersen and Nielsen 2013). As well as demonstrating new avenues in anthropological scholarship on Asia, this chapter offers insights into applied methods and discussion about method and epistemology in ethnographic encounters.

The second section assembles chapters from different disciplinary perspectives under the theme "Views from Downstairs: ethnography, identity and agency" (see Alden and Park 2013). Derek Sheridan's chapter explores the use and abuse of a rumour about Chinese peanuts in Dar es Salaam, extending this more widely to ethnographic writing. He examines the social lives of myths about Africa and China in ordinary discourse among those supposedly closest to the "real story," ordinary Africans and Chinese, showing that even the Chinese in Africa sometimes reproduce myths about the Chinese in Africa. What matters for an ethnographic stance toward myths is less empirical verification and more "intelligibility; why particular stories *make sense* to people, and what that *sense* can tell us about the social and cultural realities in which people live." As well as illuminating China–Africa from a different disciplinary perspective, and demonstrating revealing insights deriving from everyday encounters and worlds away from – yet interconnected with – macro grand narratives, he thus offers critical insights and reflections on method transcending China–Africa.

In examining race in China–Africa relations, T Tu Huynh and Yoon Jung Park note that "the racial assumptions embedded in existing 'China–Africa' discourses have yet to be unpacked." Their analysis proceeds to examine how race has been treated in the existing body of work concerning China and Africa, seeking to "disrupt restrictive binary constructions and ahistorical conceptions of race, which appear in many of the available China–Africa studies". Given that race and racism are ultimately socially constructed, they argue "that a focus on Chinese and African bodies and even (nation)states sometimes obscures our view of larger global structures and processes of power which have been, are, and continue to be racialised and racializing." They conclude by observing that "race making, othering, and hierarchies of power are entwined and ongoing processes require rigorous analysis that disrupt and displace simplistic models, binary thinking, and application of existing theories based on experiences in the global North."

If earlier analysis of China's engagement in Africa sought to disaggregate China and go beyond a unitary conception of the Chinese party–state–military (e.g. Gill and Reilly 2007), then both Maddalena Procopio and Folashadé Soulé-Kohndou offer perspectives that can assist a more in-depth, representative and compelling understanding of African political dynamics. Procopio's study of agency in Kenya–China relations critiques "first generation" research on China–Africa relations for paying disproportionate attention to the Chinese side. By shifting the focus to Kenya, and multiple scales of reference, she offers a different approach, exploring governance as the locus where different forms of agency are found and exercised. The Chinese role in Kenya is bound up in different types, and levels, of domestic politics. The unpacking of different Kenyan layers of engagement with China provides a lens through which to approach other African cases. Likewise, Soulé-Kohndou examines bureaucratic agency in the context of what she terms the "power asymmetry" in Benin's relations with China. While exploring the specificities of a particular and less-studied case, her analysis has wider relevance and use in its applicability to other African contexts through the connecting analytical framework of bureaucratic agency, including in the circumstances of asymmetrical power relations.

Honita Cowaloosur and Ian Taylor employ a modified dependency theory prism to the case of the Mauritius special economic zone, examining how Mauritius has been "facilitating its own underdevelopment by employing strategies which incessantly reproduce dependency." While acknowledging that Mauritius is a particular case, with its own specificities, this enables a set of questions to be raised about the broader African tensions in what they deem efforts "to merge cooperation with dependency, socialism with neoliberal capitalism, and Chinese socialist capitalism with African particularism." Cowaloosur and Taylor's theoretically informed research, which draws on fieldwork, raises questions about the nature, status, and implications of dependency dynamics today. Instead of treating SEZs as "empirical events," they argue that the "absence of a theoretical prism" and the resulting dominant understanding of these as insular entities, matters; their analysis "is as much as a suggestion for reform as a critique of it." By extension, their chapter also has implications for arguments concerning African agency.

Stephanie Rupp contributes not just an analysis of ivory and elephants in Africa and Asia but also reflections on how China and Africa can be studied in ways capable of properly engaging and capturing the dynamism and transcaler complexity such relations can have. Drawing on fieldwork in the Congo River basin and Thailand, she examines the overlapping systems and values involved in the international ivory trade. This is not just a case study but a potentially transferable method and framework of analysis. Her analysis points to the limitations of short-term temporal frameworks, bounded geographies, or categories of public debate and discourse. She emphasises the importance of expanding the geographical scope of "China–Africa" studies, examining relationships concerning a particular commodity, such as ivory, in networks of flows between Africa and Asia over the *long durée* and of framing economic flows in terms of overlapping economic spheres, each shaped by distinct values (Achberger 2015).

The final section contains diverse chapters encompassing "View from upstairs: political elites, policy and political economy". As such, it relates to many of the preceding themes but presents a range of perspectives on more macro and emerging themes in the study of Africa and China. Using Angola and Mozambique as cases, Ana Alves and Sergio Chichava draw on the notion of extraversion (Bayart 2000) to re-center questions of agency in African political terms. They problematise the issue of whether China's outward difference, in terms of stated intentions, principles, and methods of conducting relations, matters in the face of appropriation and incorporation by political elites into time-honored strategies of extraversion. They also explore varieties of neo-patrimonialism and how the Chinese engagements fit into these, suggesting an essential continuity in the manner by which both cases incorporate China into their own agendas consistent with historical patterns.

Mzukisi Qobo and Garth le Pere examine China's role as a development partner in the continent through the prism of its involvement in Africa's resource-industrialisation complex, and assess how meaningful this involvement has been from an African perspective in influencing the knowledge-generating landscape of Africa's growth and development. They consider the "enduring paradox in Africa's post-independence development" of the continent's "inability to use its manifold resource endowments as an impetus for generating sustained industrialisation." One theme running through their analysis, signalling both epistemological and ontological deficits afflicting China–Africa studies and, in particular, its path to development, is "the extent to which Africa has been a reactive subject rather than an objective agent in setting the terms and conditions about the discourse and debates about its development." With the ending of "the era where the West has the answers and African countries can only ask how high they should jump," and the reality that "China is not the saviour either," they therefore conclude by asserting that "African countries should be the drivers of their own development programs and policies and the role of other external actors should be supportive and complementary."

Chris Alden offers a critical examination of the history, role and challenges of "models" in China–Africa relations. Models can be best understood as vehicles for policy learning but, as he discusses, can serve purposes beyond the content of policy in terms of cooperation strategies and co-constitution of shared identities. In examining the limits of Chinese solutions to African problems, and what is involved in the identification, transfer, and nature of learning itself, he identifies key problems, including policy transfer and African institutions, and feigned learning rather than a process with meaningful, sustained impact on targeted actors and their policies.

Tang Xiaoyang's chapter explores the new structural economics of Justin Yifu Lin and Wang Yan (Lin and Yan 2016), partly presented as an alternative to Washington Consensus policy prescriptions and mainstream Western economics theory. As well as contributing to debates about development, and presenting a different perspective to that of Qobo and le Pere, Tang also raises questions about the more influential Chinese contribution to the theories and practices of international development. A key part of new structural economics, the notion that structural economic transformation can be propelled through measures like industrial policies and special economic zones, is supported by examples from China's domestic reform and international practices. Tang notes that Africa is a particular focus area of this theory. However, he also discusses the limitations of New Structural Economics, including its applicability to diverse cases and the political and economic challenges of industrialisation in underdeveloped countries. Tang concludes by affirming the potential of theoretically informed economic analyses and the value of multi-disciplinary research. This chimes with Li Anshan's conclusion about the dominance of economics but moves in Chinese scholarship on Africa involving different disciplines.

In a context whereby China is emerging as "an exporter of capital," Alvin Camba and Ho-Fung Hong suggest that the world "might be moving into a new global political economy in which Chinese overseas direct investment, as an example of South–South economic integration, can serve as a new engine of development in Africa." Their chapter offers a theoretically informed alternative research framework for approaching and understanding China's relations with Africa within the political economy of global transformations. Reacting against analysis of China–Africa economic relations at the national or sectoral level, they argue for the importance of China's domestic political economy, role in other developing regions and in the global political economy at large. Rather than obscure China's multiple concurrent engagements in various world regions, they unpack China's role and actions in Africa in connection with the rise and fall of the China boom, offering a comparison of how different regions in the world respond to the active overtures of Chinese state and private actors and considering the intersection between China's geopolitical ambitions and economic goals.

The relationship between economic development and security has attracted theoretical and applied interest, drawing attention to the growing area of security as a subject of study in China's relations with Africa. Lina Benabdallah and Daniel Large explore the relatively more recent theme of security. As a policy issue, security has travelled from a relatively marginal to a more prominent, manifest position in a fairly short space of time. The relative paucity of scholarly analysis thus far in the face of the growing importance of the theme suggests that this is a notable new frontier in China–Africa and Africa–China studies. While China's engagement with Africa's security can't be approached in isolation from other regional engagements and Beijing's evolving global role, the continent is a revealing theater showcasing Chinese foreign policy experimentation in the realm of security.

Notes

1 See David Shinn's periodically updated bibliography at http://davidshinn.blogspot.co.uk.
2 China–Africa Research Initiative, Johns Hopkins School of Advanced International Studies www.sais-cari.org/publications/.
3 The Center for African Film and Television Research was founded in 2015 at the Institute of African Studies, Zhejiang Normal University (http://ias.zjnu.cn/en/). The new, rebranded China Global Television Network was launched on December 31, 2016, with a broadcast center in Nairobi.
4 Daniel Bell is the strongest advocate of this interpretation of the China Model, though one rarely sees a comprehensive call for adoption amongst African elites or scholars. See Bell (2016).
5 For a range of films, see Roberto Castillo's blog https://africansinchina.net.
6 One ambitious study seeking to overcome this challenge by Shinn and Eisenman (2012), for example, combines macro China–Africa analysis with short individual African country entries, thereby reconciling the need for some continental scale but, concurrently, providing multiple levels of analysis.
7 Africa's Asian Options (AFRASO), an interdisciplinary and transregional research project with scholars from two area studies centers on Africa and East Asia at Goethe University, Frankfurt am Main.
8 Mamadou Diouf, speaking at the SSRC China–Africa Working Group meeting, June 4, 2015.
9 "Vice Foreign Minister Zhang Ming Makes New Requests to Chinese Research Institutes for Enhancing Original Research on Africa," September 14, 2017 at www.fmprc.gov.cn/mfa_eng/wjbxw/t1493456.shtml.
10 Notably, Document no. 9, an internal communiqué instructing cadres to stop universities and media from discussing seven taboo topics: Western constitutional democracy, universal values, civil society, neoliberalism, the Western concept of press freedom, historical nihilism, and questioning whether China's system is truly socialist.
11 *Xinhua*, "Xi Calls for Strengthened Ideological Work in Colleges," December 9, 2016.
12 Indeed, unless accessed via other means, the CAAC Research Network Google group is not readily accessible in China. Many are concerned that internet censorship will hamper scientific research within China. Such trends also impact the China campuses of foreign universities. One US government report found universities it engaged with "indicated they have freedoms on campus that do not exist beyond it, suggesting that they operate within a protected sphere in China." The report found that "universities generally emphasise academic freedom at their institutions in China" but "the environment in which these universities operate presents both tangible and intangible challenges. In particular, Internet censorship presents challenges to teaching, conducting research, and completing coursework." "U.S. Universities in China Emphasise Academic Freedom but Face Internet Censorship and Other Challenges," US Government Accountability Office Report August 2016, GAO-16–757.
13 Jonathan Sullivan, "Censorship and China Studies", *CPI Analysis*, August 19, 2017.
14 T20 Africa Conference, Africa and the G20: Building alliances for sustainable development, February 1–3, 2017: Communiqué of the Conference Co-Hosts.
15 For instance, consider the early impact of Thomas Friedman's influential journalism on subsequent scholarly work on globalisation.
16 For example, *China Remix* (dir. Melissa Lefkowitz and Dorian Carli-Jones, 2015); and *Africans in Yiwu* (dir. Zhang Yong, Hodan O. Abdi and Fu Dong, 2017).

References

Abrahamsen, Rita. 2003. "African Studies and the Postcolonial Challenge." *African Affairs* 102(407): 189–210.

Achberger, Jessica. 2015. "Follow the Tofu: Mapping Chinese Agricultural Value Networks in Zambia." Paper presented at Africa-Asia: A New Axis of Knowledge, Accra, Ghana, September 25.

Adebajo, Adekeye. 2010. *The Curse of Berlin: Africa After the Cold War*. London: Hurst.

Adel, Abderrezak, Thierry Pairault and Fatiha Talahite. eds. 2017. *La Chine en Algérie: Approches socio-économiques*. Paris: MA Éditrions–ESKA.

Adem, Seifudein. ed. 2013. *China's Diplomacy in Eastern and Southern Africa*. Aldershot: Ashgate.

Abraham, Kinfe. ed. 2005. *China Comes to Africa: The political economy and diplomatic history of China's relation with Africa*. Addis Ababa: Ethiopian International Institute for Peace and Development.

Alden, Chris. 2001. "Solving South Africa's Chinese Puzzle: Democratic Foreign Policy Making and the 'Two Chinas' Question." In *South Africa's Foreign Policy: Dilemmas of a New Democracy*, edited by Jim Broderick et al., 119–138. Basingstoke: Palgrave.

Alden, Chris. 2005. "China in Africa." *Survival* 47(3): 147–164.

Alden, Chris and Martyn Davies. 2006. "A Profile of the Operations of Chinese Multinational Corporations in Africa." *South African Journal of International Affairs* 13(1): 83–96.

Alden, Chris. 2007. *China in Africa*. London: Zed Books.

Alden, Chris and Ana Cristina Alves. 2008. "History and Identity in the Construction of China's Africa Policy." *Review of African Political Economy* 35(115): 43–58.

Alden, Chris, Daniel Large and Ricardo Soares de Oliveira. eds. 2008. *China Returns to Africa: A Continent and a Rising Power Embrace*. London and New York: Hurst Publishers and Columbia University Press.

Alden, Christopher and Park, YoonJung. 2013. "Upstairs and downstairs dimensions of China and the Chinese in South Africa." In *State of the Nation: South Africa 2012–2013*, edited by Udesh Pillay, Gerard Hagg, Francis Nyamnjoh, Jonathan D. Jansen. Pretoria: HSRC.

Alden, Chris and Sérgio Chichava, eds. 2014. *China and Mozambique: From Comrades to Capitalists*. Auckland Park: Fanele.

Alden, Chris and Yu-Shan Wu. 2016. "South African Foreign Policy and China: Converging Visions, Competing Interests, Contested Identities." *Journal of Commonwealth and Comparative Politics* 54(2): 203–231.

Ali, Ali Abdalla. 2006. *The Sudanese-Chinese Relations Before and After Oil*. Khartoum: National Centre for Scientific Research.

Ampiah, Kweku and Sanusha Naidu. eds. 2008. *Crouching Tiger, Hidden Dragon? Africa and China*. Scottsville: University of KwaZulu-Natal Press.

April, Funeka Yazini and Garth Shelton. eds. 2015. *Perspectives on South Africa-China Relations*. Pretoria: Africa Institute of South Africa.

Ash, Robert, David Shambaugh, and Seichiro Takagi. 2007. "International China Watching in the Twenty-first Century: Coping with a Changing Profession." In *China Watching: Perspectives from Europe, Japan and the United States*, edited by Robert Ash, David Shambaugh and Seiichiro Takagi, 243–248. Abingdon: Routledge.

Ashcroft, Bill. 2014. "Knowing Time: Temporal Epistemology and the African Novel." In *Africa-Centred Knowledges: Crossing Fields and Worlds*, edited by Brenda Cooper and Robert Morrell, 64–77. Oxford: James Currey/Boydell and Brewer.

Asongu, Simplice A. 2015. "Knowledge Economy Gaps, Policy Syndromes, and Catch-Up Strategies: Fresh South Korean Lessons to Africa." *Journal of the Knowledge Economy* 8(1): 211–253.

Ayers, Alison J. 2013. "Beyond Myths, Lies and Stereotypes: The Political Economy of a 'New Scramble for Africa.'" *New Political Economy* 18(2): 227–257.

A'Zami, Darius. 2015. "'China in Africa': From Under-researched to Under-theorised?" *Millennium: Journal of International Studies* 43(2): 724–734.

Bayart, Jean-Francois. 2000. "Africa in the World: A History of Extraversion." *African Affairs* 99(395): 217–267.

Bell, Daniel A. 2016. *The China Model: Political Meritocracy and the Limits of Democracy*. Princeton, NJ: Princeton University Press.

Benazeraf, David. 2014. "The Construction by Chinese Players of Roads and Housing in Nairobi: The Transfer of Town planning practices between China and Kenya." *China Perspectives* 1: 51–59.

Beri, Ruchita. ed. 2014. *India and Africa: Enhancing Mutual Engagement*. New Delhi: Institute for Defence Studies and Analyses.

Bodomo, Adams. 2012. *Africans in China: A Sociocultural Study and its Implications on Africa-China Relations*. New York: Cambria Press.

Bodomo, Adams. ed. 2015. Special Issue of the *Journal of Pan African Studies* 7(10), May.

Bräutigam, Deborah. 1998. *Chinese Aid and African Development: Exporting Green Revolution*. London: Macmillan Press Ltd.

Bräutigam, Deborah. 2009. *The Dragon's Gift: The Real Story of China in Africa*. Oxford: Oxford University Press.

Bräutigam, Deborah and Tang Xiaoyang. 2012. "Economic Statecraft in China's New Overseas Special Economic Zones: Soft Power, Business or Resource Security?" *International Affairs* 88(4): 799–816.

Bräutigam, Deborah and Haisen Zhang. 2013. "Green Dreams: Myth and Reality in China's Agricultural Investment in Africa." *Third World Quarterly* 34(9): 1676–1696.

Bräutigam, Deborah. 2015. *Will Africa Feed China?* Oxford: Oxford University Press.

Briggs, Ryan C. and Scott Weathers. 2016. "Gender and Location in African Politics Scholarship: The Other White Man's Burden?" *African Affairs* 115(460): 466–489.

Brown, William and Sophie Harman. eds. 2013. *African Agency in International Politics*. London: Routledge.

Burgis, Tom. 2015. *The Looting Machine*. London: William Collins.

Cabestan, Jean-Pierre. 2016. "Burkina Faso: Between Taiwan's active public diplomacy and China's business attractiveness." *South African Journal of International Affairs* 23(4): 495–519.

Cao, Qing. 2013. "From Revolution to Business: China's Changing Discourse on Africa." In *The Morality of China in Africa: The Middle Kingdom and the Dark Continent*, edited by Stephen Chan, 60–69. London: Zed.

Castillo, Roberto. 2016. "'Homing' Guangzhou: Emplacement, Belonging and Precarity among Africans in China." *International Journal of Cultural Studies* 19 (May): 287–306.

Chan, Stephen. 2013a. "The Middle Kingdom and the Dark Continent: An Essay on China, Africa and Many Fault Lines." In *The Morality of China in Africa: The Middle Kingdom and the Dark Continent*, edited by Stephen Chan, 3–46. London: Zed.

Chan, Stephen. 2013b. "Afterward: The future of China and Africa." In *The Morality of China in Africa: The Middle Kingdom and the Dark Continent*, edited by Stephen Chan, 140–145. London: Zed.

Chaturvedi, Sachin, Thomas Fues and Elizabeth Sidiropoulos. eds. 2012. *Development Cooperation and Emerging Powers: New Partners or Old Patterns*. London: Zed.

Cheeseman, Nic, Carl Death and Linsay Whitfield. 2016. "Notes on Researching Africa." *African Affairs* https://doi.org/10.1093/afraf/adx005.

Chen, Wenjie, David Dollar, Heiwai Tang. 2016. "Why is China Investing in Africa? Evidence from the Firm Level." *The World Bank Economic Review*, doi:10.1093/wber/lhw049.

Cheru, Fantu and Cyril Obi. eds. 2010. *The Rise of China and India in Africa*. London: Zed.

China-DAC Study Group. 2011. *Economic Transformation and Poverty Reduction: How It Happened in China, Helping it Happen in Africa Volume One: Main Findings and Policy Implications.* Beijing: IPRCC/OECD.

Cissé, Daouda. 2013. "South-South Migration and Sino-African Small Traders: A Comparative Study of Chinese in Senegal and Africans in China." *African Review of Economics and Finance* 5(1): 17–28.

Clowes, Edith W. and Shelly Jarrett Bromberg. 2016. "Introduction: Area Studies after Several 'Turns'." In *Area Studies in the Global Age: Community, Place, Identity*, edited by Edith W. Clowes and Shelly Jarrett Bromberg, 1–12. DeKalb: Northern Illinois University Press.

Comaroff, Jean and John L. Comaroff. 2012. "Theory from the South: Or, how Euro-America is Evolving Toward Africa." *Anthropological Forum* 22(2): 113–131.

Cooley, John. 1965. *East Wind Over Africa: Red China's African Offensive.* New York: Walker and Company.

Cooper, Brenda and Robert Morrell. 2014. "Introduction: The Possibility of Africa-centred Knowledges." In *Africa-Centred Knowledges: Crossing Fields and Worlds*, edited by Brenda Cooper and Robert Morrell, 1–20. Oxford: James Currey/Boydell and Brewer.

Corkin, Lucy. 2013. *Uncovering African Agency: Angola's Management of China's Credit Lines.* Farnham: Ashgate.

Cornelissen, Scarlett. 2000. "The Political Economy of Chinese and Japanese Linkages with Africa: A Comparative Perspective." *Pacific Review* 13(4): 615–633.

Cornelissen, Scarlett, Fantu Cheru and Timothy M. Shaw. 2012. "Introduction: African and International Relations in the 21st Century: Still Challenging Theory?" In *Africa and International Relations in the 21st Century*, edited by Scarlett Cornelissen, Fantu Cheru and Timothy M. Shaw, 1–17. Basingstoke: Palgrave Macmillan.

Cox, Robert. 1981. "Social Forces States and World Orders: Beyond International Relations Theory." *Millennium: Journal of international Studies* 10(2): 126–155.

Damachi, Ukandi G., Guy Routh and Abdel-Rahman E. Ali Taha. eds. 1976. *Development Paths in Africa and China.* London: The Macmillan Press Ltd.

Davies, Martyn. 1998. "South Africa and Taiwan: Managing the Post-diplomatic Relationship." Working Paper Series 21, East Asia Project, University of the Witwatersrand, Braamfontein.

Davies, Martyn and Lucy Corkin. 2007. "China's Entry into Africa's Construction Sector: The Case of Angola. " In *China in Africa: Mercantilist Predator or Partner in Development?* edited by Garth Le Pere and Peter Draper. Braamfontein: SAIIA/Institute for Global Dialogue.

Dent, Christopher, ed. 2011. *China and Africa Development Relations.* Abingdon: Routledge.

Desai, Gaurav. 2013. *Commerce with the Universe: Africa, India, and the Afrasian Imagination.* New York: Columbia University Press.

Dittgen, Romain. 2015. "Of Other Spaces? Hybrid Forms of Chinese Engagement in Sub-Saharan Africa." *Journal of Current Chinese Affairs* 44(1): 43–73.

Dobler, Gregor. 2017. "China and Namibia, 1990 to 2015: How a New Actor Changes the Dynamics of Political Economy." *Review of African Political Economy* 44(153): 449–465.

Draper, Peter and Garth le Pere. eds. 2006. *Enter the Dragon: Towards a Free Trade Agreement between China and the Southern African Customs Union.* Midrand: Institute for Global Dialogue and South African Institute for International Affairs.

Dubey, Ajay Kumar and Aparajita Biswas. eds. 2016. *India and Africa's Partnership: A Vision for a New Future.* New Delhi: Springer.

Duffield, Mark. 2014. "From Immersion to Simulation: Remote Methodologies and the Decline of Area Studies." *Review of African Political Economy* 41(1): 75–94.

Ehizuelen, Michael M. O. and Hodan Osman Abdi. 2017. "Sustaining China-Africa Relations: Slotting Africa into China's One Belt, One Road Initiative Makes Economic

Sense." *Asian Journal of Comparative Politics*September 18: https://doi.org/10.1177/2057891117727901.

Engel, Ulf and Gorm Rye Olsen. eds. 2005. *Africa and the North: Between Globalization and Marginalization.* London: Routledge.

Farole, Thomas and Lotta Moberg. 2014. "It Worked in China, So Why Not in Africa? The Political Economy Challenge of Special Economic Zones." UNW WIDER Working Paper 2014/152, November.

Ferchen, Matt. 2016. "China, Economic Development, and Global Security: Bridging the Gaps." Carnegie-Tsinghua Center for Global Policy. https://carnegietsinghua.org/2016/12/09/china-economic-development-and-global-security-bridging-gaps-pub-66397.

Fourie, Elsje. 2015. "China's Example for Meles' Ethiopia: When Development 'Models' Land." *The Journal of Modern African Studies* 53(3): 289–316.

Flores-Macias, Gustavo A. and Sarah E. Kreps. 2013. "The Foreign Policy Consequences of Trade: China's Commercial Relations with Africa and Latin America, 1992–2006." *The Journal of Politics* 75(2): 357–371.

French, Howard. 2014. *China's Second Continent.* New York: Alfred A. Knopf.

Friedman, Edward. 2009. "How Economic Superpower China Could Transform Africa." *Journal of Chinese Political Science* 14: 1–20.

Foucault, Michel. 1980. *Power/knowledge: Selected Interviews and Other Writings, 1972–1977.* New York: Pantheon Books.

Gadzala, Aleksandra. ed. 2015. *Africa and China: How Africans and Their Governments are Shaping Relations with China.* Lanham: Rowman and Littlefield.

Gallagher, Julia. 2011. *Britain and Africa Under Blair: In Pursuit of the Good State.* Manchester: Manchester University Press.

Gaye, Adama. 2006. *Chine-Afrique: Le Dragon et L'Autruche.* Paris: Editions L'Harmattan.

Giese, Karsten and Alena Thiel. 2015. "The Psychological Contract in Chinese-African Informal Labor Relations." *The International Journal of Human Resource Management* 26(14): 1807–1826.

Giese, Karsten and Alena Thiel. 2015. "Chinese Factor in the Space, Place and Agency of Female Head Porters in Urban Ghana." *Social & Cultural Geography* 16(4): 444–464.

Giese, Karsten. 2015. "Adaptation and Learning among Chinese Actors in Africa." *Journal of Current Chinese Affairs* 44(1): 3–8.

Gill, Bates and James Reilly. 2007. "The Tenuous Hold of China Inc. in Africa." *The Washington Quarterly* 30(3): 37–52.

Gill, Bates, Chin-hao Huang and J. Stephen Morrison. 2007. *China's Expanding Role in Africa: Implications for the United States.* Washington, DC: CSIS.

Greenpeace. 2015. "Africa's Fisheries' Paradise at a Crossroads: Investigating Chinese Companies' Illegal Fishing Practices in West Africa." Beijing and Dakar: Greenpeace.

Gu, Jing. 2009. "China's Private Enterprises in Africa and the Implications for African Development." *European Journal of Development Research* 21(4): 570–587.

Halvorsen, Tor and Jorun Nossum. eds. 2016. *North-South Knowledge Networks: Towards Equitable Collaboration between Academics, Donors and Universities.* Cape Town: African Minds.

Hamrell, Sven and Carl Gosta Widstrand. 1964. *The Soviet Bloc, China and Africa.* Uppsala: The Scandinavian Institute of African Studies.

Haugen, Heidi Østbø and Jorgen Carling. 2005. "On the Edge of the Chinese Diaspora: the Surge of Baihuo Business in an African City." *Ethnic and Racial Studies* 28(4): 639–662.

Haugen, Heidi Østbø. 2011. "Chinese Exports to Africa: Competition, Complementarity and Cooperation between Micro-Level Actors." *Forum for Development Studies* 38(2): 157–176.

Haugen, Heidi Østbø. 2012. "Nigerians in China: A Second State of Immobility." *International Migration* 50(2): 65–80.

Haugen, Heidi. 2013. "China's Recruitment of African University Students: Policy Efficacy and Unintended Outcomes." *Globalisation, Societies and Education* 11(3): 315–334.

Hart, Gillian. 2002. *Disabling Globalization: Places of Power in Post-Apartheid South Africa.* Berkeley, CA: University of California Press.

Hartig, Falk. 2015. *Chinese Public Diplomacy: The Rise of the Confucius Institute.* London: Routledge.

Hevi, Emmanuel John. 1963. *An African Student in China.* London: Pall Mall Press.

Hevi, Emmanuel John. 1967. *The Dragon's Embrace: The Chinese Communists and Africa.* London: Pall Mall Press.

Hilson, Gavin, Abigail Hilson, and Eunice Adu-Darko. 2014. "Chinese Participation in Ghana's Informal Gold Mining Economy: Drivers, Implications and Clarifications." *Journal of Rural Studies* 34: 292–303.

Hirono, Miwa and Shogo Suzuki. 2014. "Why Do We Need 'Myth-Busting' in the Study of Sino-African Relations?" *Journal of Contemporary China* 23(87): 443–461.

Ho, Elaine L.E. 2017. "The Geo-Social and Global Geographies of Power: Urban Aspirations of 'Worlding' African Students in China." *Geopolitics* 22(1): 15–33.

Hountondji, Paulin J. 2009. "Knowledge of Africa, Knowledge by Africans: Two Perspectives on African Studies." *RCCS Annual Review* [Online], 1. September 1, 2009. www.ces.uc.pt/publicacoes/annualreview/media/2009%20issue%20n.%201/AR1_6.PHountondji_RCCS80.pdf.

Hsu, Jennifer, Timothy Hildebrandt and Reza Hasmath. 2016. "'Going Out' or Staying In? The Expansion of Chinese NGOs in Africa." *Development Policy Review* 34(3): 423–439.

Hu, Weixing. 2016. "Xi Jinping's 'Big Power Diplomacy' and China's Central National Security Commission." *Journal of Contemporary China* 25(98): 163–177.

Huynh, T. Tu. 2015. "'It's Not Copyrighted,' Looking West for Authenticity: Historical Chinatowns and China Town Malls in South Africa." *China Media Research* 11(1): 99–111.

Huynh, T. Tu. 2015. "A 'Wild West' of Trade? African Women and Men and the Gendering of Globalisation from Below in Guangzhou." *Identities: Global Studies in Culture and Power* 23(5): 501–518.

Hutchinson, Alan. 1975. *China's Africa Revolution.* London: Hutchinson.

IPPR. 2006. *The New Sinosphere: China in Africa.* London: IPPR.

Jackson, Terence. 2012. "Postcolonialism and Organizational Knowledge in the Wake of China's Presence in Africa: Interrogating South-South Relations." *Organization* 19(2): 181–204.

Jerven, Morten. 2016. "Research Note: Africa by Numbers: Reviewing the Database Approach to Studying African Economies." *African Affairs* 115(459): 342–358.

Kaplinsky, Raphael with Anne Posthuma. 1994. *Easternisation: The Spread of Japanese Management Techniques to Developing Countries.* Ilford: Frank Cass.

Kaplinsky, Raphael. 2008. "What Does the Rise of China do for Industrialization in SSA?" *Review of African Political Economy* 35(1): 7–22.

Kato, Hiroshi. 2017. "Japan and Africa: A Historical Review of Interaction and Future Prospects." *Asia-Pacific Review* 24(1): 95–115.

Kim, Soyeun and Gray, Kevin. 2016. "Overseas Development Aid as Spatial Fix? Examining South Korea's Africa Policy." *Third World Quarterly,* 37(4): 649–664.

King, Kenneth. 2013. *China's Aid and Soft Power in Africa: The Case of Education and Training.* Oxford: James Currey/Boydell and Brewer.

King, Charles. 2015. "The Decline of International Studies." *Foreign Affairs,* June 16.

Kragelund, Peter. 2015. "Towards Convergence and Cooperation in the Global Development Finance Regime: Closing Africa's Policy Space?" *Cambridge Review of International Affairs* 28(2): 246–262.

Lan, Shanshan. 2017. *Mapping the New African Diaspora in China: Race and the Cultural Politics of Belonging*. London: Routledge.

Large, Daniel. 2008. "Beyond 'Dragon in the Bush': The Study of China-Africa Relations." *African Affairs* 107(426): 45–61.

Large, Daniel and Luke Patey. eds. 2011. *Sudan Looks East: China, India and the Politics of Asian Alternatives*. Oxford: James Currey.

Larkin, Bruce D. 1971. *China and Africa 1949–1970: The Foreign Policy of the People's Republic of China*. Berkeley, CA: University of California Press.

Lee, Ching Kuan. 2014. "The Spectre of Global China." *New Left Review* 89: 29–65.

Le Pere, Garth and Garth Shelton. 2007. *China, Africa and South Africa*. Midrand: Institute for Global Dialogue.

Le Pere, Garth. ed. 2007. *China in Africa: Mercantilist Predator or Partner in Development?* Midrand: Institute for Global Dialogue, and the South African Institute for International Affairs.

Li, Anshan. 2007. "China and Africa: Policy and Challenges." *China Security* 3(3): 69–93.

Li, Xiaoyun. 2011. "Zhongguo yao gaibian ziji de feizhou guan" ["China needs to change its own views of Africa"], *Fenghuang zhoukan*, June 15.

Li, Xiaoyun, Dan Banik, Lixia Tang and Jin Wu. 2014. "Difference or Indifference: China's Development Assistance Unpacked." *IDS Bulletin* 45(4): 22–35.

Li, Xing and Abdulkadir Osman Farah. eds. 2013. *China-Africa Relations in an Era of Great Transformations*. Farnham: Ashgate.

Lin, Justin Yifu and Wang Yan. 2016. *Going Beyond Aid: Development Cooperation for Structural Transformation*. Cambridge: Cambridge University Press.

Lin, Yunsheng and Shu Zhan. 1994. "From Transformation to Economic Prosperity: Lessons from the Chinese Experience." Working Paper Series 3, East Asia Project, University of the Witwatersrand, Braamfontein.

Liu, Yuncheng and Mou Hong. 1996. "China: Marching to a Socialist Market Economy in the Asia-Pacific Region." Working Paper Series 11, East Asia Project, University of the Witwatersrand, Braamfontein.

Liu, Haifang. 2008. "China-Africa Relations through the Prism of Culture: The Dynamics of China's Cultural Diplomacy with Africa." *China aktuell* 37: 9–44.

Lopes, Carlos. 2016. "Reinserting African Agency into Sino-Africa Relations." *Strategic Review for Southern Africa* 38(1): 50–68.

Lumumba-Kasongo, Tukumbi. 2010. *Japan–Africa Relations*. New York: Palgrave Macmillan.

Marks, Stephen. ed. 2009. *Strengthening the Civil Society Perspective: China's African Impact*. Cape Town: Fahamu.

Martin, William G. and Brendan Innis McQuade. 2014. "Militarising – and Marginalising? – African Studies USA." *Review of African Political Economy* 41(141): 441–457.

Mason, Robert. 2017. "China's Impact on the Landscape of African International Relations: Implications for Dependency Theory." *Third World Quarterly* 38(1): 84–96.

Matambo, Emmanuel and Khondlo Mtshali. 2016. "The Relevance of Emmanuel Hevi: China in Contemporary Sino-African Relations." *Africology: The Journal of Pan African Studies* 9(4): 219–237.

Mathews, Gordon. 2015. "Taking Copies from China Past Customs: Routines, Risks, and the Possibility of Catastrophe." *Journal of Borderlands Studies* 30(3): 423–435.

Mawdsley, Emma. 2007. "China and Africa: Emerging Challenges to the Geographies of Power." *Geography Compass* 1(3): 405–421.

Mawdsley, Emma. 2008. "Fu Manchu versus Dr Livingstone in the Dark Continent? Representing China, Africa and the West in British Broadsheet Newspapers." *Political Geography* 27(5): 519–520.

Mawdsley, Emma and Gerard McCann. eds. 2011. *India in Africa: Changing Geographies of Power*. Oxford: Fahamu.

Mazrui, Ali A. and Seifudein Adem. 2013. *Afrasia: A Tale of Two Continents*. Lanham: University Press of America.

Metz, Thaddeus. 2015. "Values in China as Compared to Africa." In *The Rise and Decline and Rise of China: Searching for an Organising Philosophy*, edited by Hester du Plessis, 75–116. Johannesburg: Real African Publishers.

Modi, Renu. 2017. "India-Africa Forum Summits and Capacity Building: Achievements and Challenges." *Journal of Asian and African Studies* 16: 139–166.

Mohan, Giles. 2013. "Beyond the Enclave: Towards a Critical Political Economy of China and Africa." *Development and Change* 44(6): 1255–1272.

Mohan, Giles, Ben Lampert, May Tan-Mullins and Oaphne Chang. 2013. *Chinese Migrants and Africa's Development: New Imperialists or Agents of Change?* London: Zed.

Mohan, Giles and Ben Lampert. 2012. "Negotiating China: Reinserting African Agency into China-Africa Relations." *African Affairs* 112(446): 92–110.

Monson, Jamie. 2009. *Africa's Freedom Railway*. Bloomington: Indiana University Press.

Monson, Jamie and Stephanie Rupp. 2013. "Introduction: Africa and China: New Engagements, New Research." *African Studies Review* 56(1): 21–44.

Makungu, M., Nuah, M. 2013. "Is the Democratic Republic of Congo being Globalised by China? The Case of Small Commerce at Kinshasa Central Market." *Quarterly Journal of Chinese Studies* 2(1): 89–101.

Muekalia, Domingos Jardo. 2004. "Africa and China's Strategic Partnership." *African Security Review* 13(1): 5–11.

Mazrui, Ali A. and Seifudein Adem. 2013. *Afrasia: A Tale of Two Continents*. Lanham, MD: University Press of America.

Naim, Moises. 2007. "Rogue Aid." *Foreign Policy*, March 1.

Odoom, Isaac and Andrews, Nathan. 2016. "What/who is Still Missing in International Relations Scholarship? Situating Africa as an Agent in IR Theorising." *Third World Quarterly* 38(1): 42–60.

Ogunsanwo, Alaba. 1974. *China's Policy in Africa, 1958–71*. Cambridge: Cambridge University Press.

Ottaway, Marina and David Ottaway. 1986. *Afrocommunism*. New York: Holmes and Meier.

Pang, Zhongying and Hongying Wang. 2013. "Debating International Institutions and Global Governance: The Missing Chinese IPE Contribution." *Review of International Political Economy* 20(6): 1189–1214.

Park, Yoon. 2009. "Chinese Migration in Africa," *South African Institute of International Affairs* Occasional Paper no. 24 (January): 1–18.

Patey, Luke. 2017. "Learning in Africa: China's Overseas Oil Investments in Sudan and South Sudan." *Journal of Contemporary China* 26(107): 756–768.

Pedersen, Morten Axel, and Morten Nielsen. 2013. "Trans-Temporal Hinges: Reflections on an Ethnographic Study of Chinese Infrastructure Projects in Mozambique and Mongolia." *Social Analysis* 57(1): 122–142.

Pieke, Frank N. 2012. "Immigrant China." *Modern China* 38(1): 40–77.

Li, Lianxing. 2016. "The Image of Africa in China: The Emerging Role of Chinese Social Media." *African Studies Quarterly* 16(3–4): 149–160.

Power, Marcus, Giles Mohan and May Tan-Mullins. 2012. *China's Resource Diplomacy in Africa: powering Development?* Houndmills: Palgrave Macmillan.

Power, Marcus and Ana Alves. 2012. *China and Angola: A Marriage of Convenience?* Oxford: Pambazuka Press.

Puppin, Giovanna. 2017. "Making Space for Emotions: Exploring China–Africa 'Mediated Relationships' Through CCTV-9's Documentary African Chronicles (Feizhou jishi 非洲 纪事)." *Journal of African Cultural Studies* 29(1): 131–147.

Qi Gubo, Tang Lixia, Zhao Lixia, Jin Leshan, Guo Zhanfeng, and Wu Jin. 2012. *Agricultural Development in China and Africa: A Comparative Analysis*. New York: Routledge.

Rounds, Zander and Hongxiang Huang. 2017. "We Are Not So Different: A Comparative Study of Employment Relations at Chinese and American firms in Kenya." Working Paper No. 2017/10. China Africa Research Initiative, School of Advanced International Studies, Johns Hopkins University, Washington, DC. Retrieved from www.sais-cari.org/publications.

Rotberg, Robert, ed. 2008. *China into Africa: Trade, Aid and Influence*. Washington, DC: Brookings Institution Press.

SAIIA. 2009. *The China Africa Toolkit: A Resource for African Policy Makers*. Johannesburg: SAIIA.

Sautman, Barry and Yan Hairong. 2008. "The Forest for the Trees: Trade, Investment and the China-in-Africa Discourse." *Pacific Affairs* 81(1): 9–29.

Selden, Mark, ed. 1997. "Asia, Asian Studies, and the National Security State: A Symposium." *Bulletin of Concerned Asian Scholars* 29(1). http://criticalasianstudies.org/assets/files/bcas/v29n01.pdf.

Shelton, Garth and Chris Alden. 1994. "From Transformation to Economic Prosperity: Lessons from the Chinese Experience." Working Paper Series 9, East Asia Project, University of the Witwatersrand, Braamfontein.

Shen, Gordon C. and Victoria Y. Fan. 2014. "China's Provincial Diplomacy to Africa: Applications to Health Cooperation." *Contemporary Politics* 20(2): 182–208.

Shinn, David and Joshua Eisenman. 2012. *China and Africa: A Century of Engagement*. Philadelphia: University of Pennsylvania Press.

Sidiropoulos, Elizabeth, ed. 2006. "China in Africa" special edition of *South African Journal for International Affairs* 13(1).

Smith, Karen. 2009. "Has Africa Got Anything to Say? African Contributions to the Theoretical Development of International Relations." *The Round Table* 98(402): 269–284.

Smith, Karen. 2012. "Africa as an Agent of International Relations Knowledge." In *Africa and International Relations in the 21st Century*, edited by Scarlett Cornelissen, Fantu Cheru and Timothy M. Shaw, 21–35. Basingstoke: Palgrave Macmillan.

Snow, Philip. 1988. *The Star Raft: China's Encounter with Africa*. London: Weidenfeld and Nicolson.

Soares de Oliveira, Ricardo. 2015. *Magnificent and Beggar Land: Angola Since the Civil War*. London: Hurst.

Sun, Irene Yuan, Kartik Jayaram and Omid Kassiri. 2017. *Dance of the Lions and Dragons: How are Africa and China Engaging, and How Will the Partnership Evolve?* McKinsey & Company, June 2017.

Strauss, Julia C. and Martha Saavedra. eds. 2009. *China and Africa: Emerging Patterns in Globalization and Development*. Cambridge: Cambridge University Press.

Strauss, Julia C. 2009. "The Past in the Present: Historical and Rhetorical Lineages in China's Relations with Africa." *China Quarterly* 199(2009): 777–795.

Strauss, Julia C. 2013. "China and Africa Rebooted: Globalization(s), Simplification(s), and Cross-cutting Dynamics in 'South-South' Relations." *African Studies Review* 56(1): 155–170.

Swedlund, Haley J. 2017. "Is China Eroding the Bargaining Power of Traditional Donors in Africa?" *International Affairs* 93(2): 389–408.

Szanton, David. 2004. "Introduction: The Origin, Nature, and Challenges of Area Studies in the United States." In *The Politics of Knowledge: Area Studies and the Disciplines*, edited by David Szanton, 1–33. Berkeley: University of California Press.

Taylor, Ian. 1997. "The 'Captive States' of Southern Africa and China: the PRC and Botswana, Lesotho and Swaziland." *Journal of Commonwealth and Comparative Politics* 35(2): 75–95.

Taylor, Ian. 1997. "China and SWAPO: The Role of the People's Republic in Namibia's Liberation and Post-independence Relations." *South African Journal of International Affairs* 5 (1): 110–122.

Taylor, Ian. 1997. "Mainland China-Angola Relations: Moving from Debacle to Détente." *Issues and Studies: Journal of Chinese Studies and International Affairs* 33(9): 64–81.

Taylor, Ian. 2006. *China and Africa: Engagement and Compromise*. Abingdon: Routledge.

Taylor, Ian. 2009. *China's New Role in Africa*. Boulder, CO: Lynne Rienner.

Taylor, Ian. 2014. *Africa Rising? BRICS – Diversifying Dependency*. Oxford: James Currey.

Tjonneland, Elling et al. 2006. "China in Africa: Implications for Norwegian Foreign and Development Policies." Bergen, NJ: Chr. *Michelsen Institute Report*.

Walder, Andrew G. 2004. "The Transformation of Contemporary China Studies, 1977–2002." In *The Politics of Knowledge: Area Studies and the Disciplines*, edited by David Szanton, 314–340. Berkeley: University of California Press.

Wekesa, Bob. 2017. "New Directions in the Study of Africa–China Media and Communications Engagements." *Journal of African Cultural Studies* 29(1): 11–24.

Wu, Yu-Shan. 2012. 'The Rise of China's State-led Media Dynasty in Africa', *Occasional Paper 117*, Braamfontein: SAIIA. 1–31.

Xiao, Allen Hai. 2015. "In the Shadow of the States: The Informalities of Chinese Petty Entrepreneurship in Nigeria." *Journal of Current Chinese Affairs* 44(1): 75–105.

Yu, George T. 1975. *China's African Policy: A Study of Tanzania*. New York: Praeger.

Zhao, Suisheng. 2014. "A Neo-Colonialist Predator or Development Partner? China's Engagement and Rebalance in Africa." *Journal of Contemporary China* 23(90): 1033–1052.

Zhou, Min, Tao Xu and Shabnam Shenasi. 2016. "Entrepreneurship and Interracial Dynamics: A Case Study of Self-employed Africans and Chinese in Guangzhou, China." *Ethnic and Racial Studies* 39(9): 1566–1586.

From China in Africa to global African studies

2

FROM FIELD WORK TO ACADEMIC FIELD

Personal reflections on China–Africa research

George T. Yu

Contemporary China–Africa relations began about 60 years ago, during the late 1950s and early 1960s. To date, China's relations with Africa constitutes one of the most enduring and successful China initiated foreign policy operations. It is a story that denotes the extraordinary evolutionary development of Chinese foreign policy against multiple internal and external challenges, on the one hand, and how China responded and sought to overcome the trials through the adoption of concurrent separate foreign policies and varied foreign policy instruments, on the other. Foremost, it is a tale of China's global rise, as seen through developing relations with Africa. Globally, China–Africa relations also represents a new pattern in international relations.

China's relations with Africa were born amidst the Cold War and the Sino-Soviet conflict. Politically and militarily isolated, and surrounded by major powers, China sought to establish and safeguard its independence and security, claim its legitimacy and sovereignty, and establish and enhance its global presence through multiple foreign policy initiatives. One such step was China's outreach to Africa, to win friends and build alliances to meet varied challenges. Employing a diverse range of foreign policy instruments, linking aid in its infinite variety to national interests, from ideological/political interactions, economic and technical developmental aid, military assistance and other activities, China sought to confront the varied disputes. Africa became a major battleground for China against its adversaries, welcomed by select African countries that utilised China's international role and resources for their own development and national interests.

At its beginnings, China–Africa academic studies in the West was a solitary field, consisting mainly of a small handful of scholars primarily interested in China and international relations; today, it is a global 'growth industry', led by international academics and attracting innovative scholars in a variety of disciplines from a growing number of countries. In the 60 plus years since its birth, China-Africa studies has been transformed from a largely single-issue, foreign policy/international relations, peripheral subject to a multi-disciplinary mainstream academic field of study.

Hindsight is a rare opportunity to both re-examine and question one's motives and behaviour of the past. This chapter is an attempt to recall and describe the challenging decisions I took concerning my research and the road, academic and personal, I made in pursuing this. It is divided into six primary sections, covering the beginnings of the research project, operationalizing and conducting the research, the Tanzanian 'laboratory', persistent research questions and conclusion. While the consequence and results of my decisions are transparent, the research choices were determined both by academic components and the situational-environmental conditions in the field, on the ground.

The beginnings: Africa as a gateway to China

My research into China–Africa relations began at the University of North Carolina, Chapel Hill, where I began my academic career, in 1961. Following revision of my dissertation, subsequently published by the University of California Press (Yu 1966), I began searching for a new research subject. In the beginning, I considered continuing research in modern, post-1949 Chinese politics. Given the state of U.S.–China relations, however, fieldwork was impossible. In the 1960s and 1970s, research for American academics on contemporary China was mainly limited to library investigation; the closest one could get to China were Hong Kong and Taiwan, centres for studying China until the opening of China in the early 1980s.

What I sought was a topic and environment where I could conduct field research directly observing Chinese behaviour and attempt to assess and verify developments first hand, on the ground. China's relations with Africa soon caught my attention, especially media reports of China's battles with Taiwan in Africa over political issues of diplomatic recognition and sovereignty, namely, what constituted China.

In the 1960s and 1970s, the primary geographic fields for studying China's foreign policy did not include Africa; the field or sub-field of China–Africa had yet to be recognised. My own sense was that the addition and inclusion of Africa was within the conceptual paradigm of either the Cold War or the Sino-Soviet conflict, which included Taiwan's struggle with China. Thus, while the field, Africa, was new, a micro study of China–Africa relations fit well into macro research of China's foreign policy, or into the existing academic disciplinary studies, international relations/foreign policy field of Political Science.

Through the early 1970s, Taiwan conducted an aggressive and successful campaign to win diplomatic recognition of the new African states. Indeed, Taiwan was able to sustain the relationship, through international support and utilization of foreign aid, despite China's incremental increase in activities on the continent, gaining recognition from the African states between 1949 and 1971. In 1970, for example, 22 out of the 47 independent African states maintained diplomatic relations with Taiwan, with only 15 recognizing China (five African states recognised neither). The tide of African recognition in favour of China only shifted following China's admission, with African support, to the United Nations in 1971.

My first project on China–Africa relations, based on library research, was published in 1963 (Yu, 1963). Encouraged by the reception of my initial publication, I began to consider how to proceed with the research; it quickly became apparent that the subject was broader in context and scope than following 'Peking' and 'Taipei'. I considered a number of questions. First, the 'competition' between China and Taiwan was not merely between the two parties across the Taiwan Straits, nor regional, but global in nature. Africa (and other regions) had become a battleground between the two parties, each seeking to gain and/or retain diplomatic recognition and international legitimacy.[1] For China, going global by reaching out to Africa was a new initiative; a new pattern of international relations was being formed, challenging Taiwan and the West, while expanding and enhancing China's global presence. Second, the 'competition' between China and Taiwan was a segment of the larger Cold War, between the West and the Communist world, in this instance, between the United States and China, which sought the latter's international isolation. The 'competition' between China and Taiwan in Africa was an extension of the Cold War by proxies, each representing the larger camps. It was the beginning of a global Chinese foreign policy.

Subsequently, China's global foreign policy, namely, the use of Africa as a battleground to fight with its adversaries, was not limited to challenging the United States and Taiwan. In the 1970s, following the emergence of open Sino-Soviet discord in the 1960s, Africa also became a combat zone challenging the Soviet Union, as a revolutionary state supporting wars of liberation and accusing it of being an aggressive, expansionist imperialist power, seeking, among others, African natural resources. The combative relationship with the Soviet Union added another conflict front to China's role in Africa.

Third, would China continue and expand its relations with Africa; how would China meet the multiple varied challenges on each front, and how would Africa respond? Was China's relation with Africa a subject and environment where I could conduct field research directly observing Chinese behaviour and Africa's response, and assess and verify developments? How would I operationalise the research, including conducting field research in Africa?

These research considerations, namely the multiple fronts of conflict, China's foreign policy initiatives and Africa's responses, and China's practices of utilizing select foreign policy instruments all led to the basic question of the whys and hows of Chinese foreign policy: namely, what factors contributed to China's adoption of different foreign policies to meet diverse conflict challenges, and what instruments of foreign policy had China utilised to achieve policy goals? In the language of globalisation, the proposed research would examine the extensity, intensity and impact of China's relations with Africa and its global reach.

Reflecting upon my choice and selection of the new research project in the early 1960s, I cannot say that I knew what lay ahead, either for my academic career or the success or failure of employing China's African policy as a measurement for better understanding China's foreign policy and behaviour, and China's emerging global role. I knew I was entering uncharted territory, which required an innovative and flexible mind, a knowledge and understanding of both Africa and China and, in the course of the research, a willingness to accept uncertainties, be prepared for setbacks and adapt to varied environments. Was I up for the challenges? Was my choice of research a valid approach to study China? And what future would the research have?

Operationalizing research on China–Africa

I had settled on a new research topic, China's relations with Africa; I had also changed my focus in Political Science from comparative politics to foreign policy/international relations, and I was also adding a new area studies field, Africa. The next step was operationalizing the research project, including field research in Africa to separate facts of China's role from fiction. I had begun basic research, including studying contemporary Africa, at Chapel Hill; subsequently, the main body of research was continued at the University of Illinois at Urbana-Champaign, which offered me a tenured appointment in 1965. Two years following arrival at Illinois, I applied for a year's leave to conduct field research in Africa during 1967–1968. I was awarded a Social Science Research Council (SSRC) grant, while the University of Illinois granted me research leave and additional funding. In subsequent years, to continue and complete the research, I received additional funding from both.

The initial location choice of my field research was East Africa, including Kenya, Tanzania and Uganda; later, I would expand the geographical area to include Botswana, Zambia and Zimbabwe. The selection of East Africa to study China's relations with Africa was founded upon three primary relevant factors. First, all three African nations had only recently won independence from British colonial rule: Tanzania in 1961, Uganda in 1962, and Kenya in 1963. Each sought to create a new international identity, build a new nation, and sought new nation building models. In short, East Africa was in political flux, each new country seeking to create and control their respective destinies. Second, beginning in the early 1960s, China began an intensive campaign to 'win friends and influence people' in East Africa, conducting diplomatic campaigns, securing recognition from Uganda in 1962, followed by Kenya in 1963 and Tanzania in 1964. This provided a fresh 'laboratory' to observe and study first-hand China's foreign policy and practices. Finally, since I planned to conduct field research for nearly a year and a half, there was never any doubt that my family would travel with me. We determined that East Africa possessed a basic social infrastructure system, including communications, housing, public health, schools, and security, which provided assurance that our basic needs and security would be met. In addition, being former British colonies, English was widely spoken.

The selection of East Africa as the primary research site was no problem; the difficulty was securing East African governments' permission to conduct research. In the 1960s, research by foreign scholars required approval by East African governments; this provided government control of research topics and the number of foreign scholars permitted. I applied for permission six months in advance.

The 1960s was a high tide of the Cold War era. While the new East African nations maintained strong linkages with their former colonial rulers, they sought also to establish separate relationships with all nations, including China, to avoid being identified with any one political bloc. But some nations did not wish to publicise their new relationships with China, since 'Communist' China had yet to be accepted and integrated into the global political system. Indeed, in the 1960s, China was perceived by many as the world's adversary, seen by both the West and the Soviet Union–led Communist world as an uncontainable threat.

Therefore, it was not totally unexpected, though disappointing, that my application for a research visa to Kenya (where I planned to begin my research) was rejected. The year 1967 was an especially trying period in Sino-Kenyan relations, after the outbreak of China's Cultural Revolution. In October, following an earlier dispute in Beijing between a staff of the Kenyan Embassy and members of the 'Red Guard' and the distribution in Kenya of Chinese Cultural Revolution literature (which the Kenya government regarded as interfering in Kenya's internal affairs), Kenya withdrew its diplomatic staff from China, declared a member of the Nairobi Chinese Embassy 'persona non grata', and imposed limits on the distribution of all Chinese literature.

But all was not lost. I secured an appointment as Visiting Lecturer in the Rockefeller Foundation Program to Kenya, teaching American Politics at Nairobi College during the spring quarter of 1968; I was otherwise left to conduct my research. Following Kenya, my family and I would visit Tanzania and Uganda on extended visitor's visas, applied on site. We encountered no difficulties from East African governments with the arrangements during our nine-month sojourn.

Field research

In the summer of 1967, I left the United States together with my family to begin exploration of my new research project; the first leg of our journey was London. There I was associated with the School of African and Oriental Studies (SOAS), and interacted with the African Studies faculty. SOAS proved a great beginning for my research, with an abundance of human and material resources relating to Africa. As Africa was a new research area, I gathered much new knowledge in preparation for the upcoming fieldwork. I also contacted British journalists specializing in Africa; Colin Legum, former Commonwealth Editor for *The Observer* (London), was especially generous, readily introducing me to a select group of East African leaders, which proved extremely helpful upon arrival in Africa. The information gained and the contacts made were of great research benefit later in the field.

Following a productive stay, we departed London in December 1967 for Nairobi, Kenya, the first of three East African nations (the others being Tanzania and Uganda) where I planned to conduct field research. We remained in East Africa through August 1968.

In East Africa, my life soon fell into a routinised pattern. In all three countries, Kenya, Tanzania and Uganda, I was associated with the three national universities/colleges. I met faculty and students, who in turn introduced me to government officials and local inhabitants. At first, I was looked upon as an anomaly and with suspicion; few had met a Chinese-American academic; some suspected that I might be a spy (I am not certain for which side!). In general, I was well received; I enjoyed many good exchanges with African faculty members and others.

In Kenya, and later in Uganda, I began my field research, seeking to uncover and follow China's presence and impact, while attempting to construct a conceptual framework to examine and explain China's foreign policy and behaviour, including use of select foreign policy instruments. This proved to be a trying experience, especially in Kenya. As mentioned, China–Kenya relations were at a low point in 1967, including the imposition of a limit on China's activities in Kenya. Kenyan society was also in turmoil in 1968, due to the government's new immigration and labour policies relating to non-citizens and the forced expulsion of individuals of South Asian heritage. Overnight, office staff, shopkeepers and school students were suddenly removed from their usual stations. The massive and sudden economic and social upheaval, together with strained relations with China, did not provide an ideal research environment. Indeed, Kenyan interests in and preoccupation with domestic political and social conditions trumped interests in and attention towards foreign affairs. My initial introduction to Africa had made known the trials and tribulations of field research. It was also a 'Dragon in the Bush' moment: I was aware of China's presence in Kenya but had great difficulty discerning its exact manifestations (Yu 1968). My research was on hold.

Tanzania: the laboratory

My next field research site was Tanzania. The political and social uncertainties and research difficulties in Kenya took a complete turn for the better in Tanzania. Though neighbours, and both newly independent from British colonial rule, Tanzania's relations with China assumed a vastly different direction compared to Kenya. First, China–Tanzania relations were thriving, symbolised by the 1965 Sino-Tanzanian Treaty of Friendship and the arrival of the initial group of Chinese personnel and materials to begin construction of the Tanzanian-Zambian Railway (TAZARA) in the summer of 1968. This was the 'laboratory' I was seeking, China's foreign policy and behaviour at its maximum and Tanzania's response at its fullest. Second, through the generosity of Colin Legum I was introduced to Joan Wicken, personal assistant to President Nyerere of Tanzania, who graciously welcomed me to State House, the presidential office in Dar es Salaam. During our stay in Dar es Salaam in 1968, and in subsequent visits, she never failed to receive me, providing me with a greater understanding of Tanzania and its relations with China. I also requested her to comment on drafts of my research. I will be forever grateful for her wise counsel and generous assistance. When I published the results of my research in the 1970s, I asked Miss Wicken if I could publically acknowledge her assistance. She declined. Instead, she requested that I include the following: 'Appreciation is also due to the government officials and scholars of Tanzania.' My work on China–Tanzania relations and China's relations with Africa had been enhanced through my relationship with Miss Wicken.

Finally, it was during research in Dar es Salaam, amidst generally negative views of Tanzania and Tanzanian foreign policy and a general absence of formal Tanzanian research on its budding relations with China, that I decided upon a conceptual model of cooperative interaction, as opposed to the then more conventional conflict approach of China's foreign policy and behaviour to study China–Tanzania relations. Indeed, for both China and Tanzania, two struggling developing states seeking international legitimacy, recognition and support, cooperative interaction was a likely behavioural pattern: it enhanced both parties' domestic and international needs and role, real and symbolic.

In the context of the then prevailing Cold War literature, there was a predominance of conflict-oriented research on China; yet China's foreign policy and behaviour was surely more complex. With an enhanced appreciation of Tanzanian foreign policy and domestic needs, together with on the ground observations of China's policy towards Tanzania, I concluded that China's cooperative foreign policy did not contradict its conflict foreign policy mode. Similar to other powers, China conducted concurrently different modes of foreign policy, depending upon countries, issues and timing. While, in the instance of Africa, China engaged in battles on several fronts, relating to different issues, with varied adversaries, on other fronts, China could and did conduct cooperative interaction with other nations, to achieve different foreign policy needs and goals. More recently, China's 'flexibility' in the conduct of foreign policy was further demonstrated when, for example, Foreign Minister Wang Yi stated in early 2016 that relations with the United States included both elements of competition and cooperation.

In the context of global politics, cooperative and conflictual interactions were but different forms of behaviour on the same international interaction continuum. There was a need to examine and understand both modes of foreign policy. This was especially true of China's foreign policy and behaviour in the 1960s, which had been seen primarily in conflict modes, in the context of the Cold War and the Sino-Soviet conflict. An opportunity had presented itself to study China in a different light, in terms of why and how it engaged in cooperation interaction.

Reenergised in my research, I did my utmost to focus on the intensity dimension of the relationship, collecting, observing and examining all sectors of China–Tanzania interaction. In 1967, Tanzania reported that China extended aid to 23 projects, totalling US$2.6 million.

At the beginning of our arrival in Dar es Salaam, a temporary difficulty arose with the American Embassy in Tanzania. This was not unexpected, given the tenuous state of U.S.–Tanzania relations. The forced removal of an American satellite tracking station on Zanzibar in 1964, the expulsion of American diplomats from Tanzania in 1965 and Tanzania's opposition to 'imperialist aggression' in Vietnam must have made the Embassy extra cautious, seeking to prevent and avoid new issues that could further complicate relations between the two countries. The problem arose with the Embassy over the fact that I was conducting research on a sensitive subject without an approved research visa. The Embassy even wrote a formal letter stating that should I meet with 'difficulties' in Tanzania they would be unable to provide assistance. In 1968, the Embassy was led by a *charge d'affaires*, Thomas R. Pickering. Subsequently, with no trouble on my part, the Embassy's attitude towards my presence changed, inviting my wife and I to cocktails and even meetings with Embassy staff.

It was especially exciting to be in Dar es Salaam in 1968, being witness to the deepening relationship between China and Tanzania. Aside from observing the beginning of China's assistance with building the TAZARA, seen with the arrival of Chinese ships in the Dar es Salaam harbour bringing human and material resources for the massive railway project, I focused on the operational side of China's foreign policy, building political relations and the use of various types of foreign aid to enhance Tanzania's national development, including agriculture, economic/manufacturing, educational, infrastructure development, medical and public health, military and other forms of assistance. Tanzania's response to China's outreach was also noted; during the summer of 1968, President Nyerere made his second state visit to China. The one disappointment during our stay in East Africa was our inability to meet with the Chinese. During the early years of their presence, Chinese diplomatic representatives and technical personnel did not, as a rule, interact with individuals outside their immediate community. (During subsequent research visits, Chinese embassies in East Africa generously entertained us.) Otherwise my research in Tanzania was exceptionally successful, providing the basis for my research and publications (such as Yu 1970, 1975).

In 1968, Dar es Salaam was a vibrant centre of international visitors. Tanzania had taken on the role of a new, progressive African state, regarded as an emerging 'socialist' brother by the Nordic countries, and welcomed by the 'progressive' international elements for its anti-Vietnam war stand. We became friends with Tanzanian and expatriate members of the university faculty, met members of the diplomatic corps and NGO representatives, doctors, engineers and other aid workers. We also became acquainted with members of the Canadian military mission; one interesting aspect of Canada's role was training Tanzanians in the use of Chinese military equipment, as Canadian legislation governing military assistance prohibited it from providing military supplies classed as combat equipment. The wide range of foreign aid programmes and representatives provided contextual balance to China's role in Tanzania and Tanzania's foreign policy.

The presence of an estimated one million plus Chinese inhabitants in Africa has often been noted in recent research. In 2015, it was estimated that 30,000 plus China lived in Tanzania. In 1968, we were introduced to the local Chinese community when we first attended an expatriate church in Dar es Salaam. There we met and became friends with a Chinese family who had migrated to Tanzania before independence; the husband, originally from Shandong, China, owned and managed a textile plant in Dar es Salaam. Through them, we were introduced to a small professional group of local Chinese, all working for different agencies of the United Nations. In subsequent years, I also met a 'local' Chinese, a carpenter by trade, who was hired by the British in India, arriving in Tanzania after World War I. Dar es Salaam also had a Chinese-owned restaurant and 'book store'. The total Chinese population in Dar es Salaam was less than 20 in 1968. Until recent years, outside of South Africa, there were few Chinese living in Tanzania and Africa.

My initial field research in Africa was completed in August 1968. From the 1970s through the 1990s, I returned repeatedly to Africa. During each subsequent research visit, I would notice the increased presence and influence of China, including being openly welcomed by local Chinese embassies. Following China's opening, I was invited jointly by Peking University and the Chinese Academy of Social Sciences to lecture on Africa in 1981. During the warm summer months of July and August, I lectured to Africa and other Chinese area specialists on African studies in America. In the Fall of 1981, I led an American Africanists delegation visiting China; during the visit, agreement was reached that the delegation would seek support in America to organise academic exchanges between American and Chinese African studies centres.

For a decade beginning in 1982, I directed a Ford Foundation funded pro-gramme to train and retrain Chinese Africanists in academic and government research institutions. Peter Geithner of the Foundation was a strong supporter of the programme. A U.S. China National Committee on Africa Studies whose membership included leading American Africanists was established to administer the programme; in addition to organizing joint conferences in America and China, an academic exchange programme between American and Chinese African studies centres was established. Chinese Africanists from universities and research institu-tions were invited to study and conduct research (up to one academic year) at various American university African studies centres; American Africanists were invited to lecture at African studies centres in China; and the Committee also funded Ph.D. studies for students from China in disciplinary studies with an Africa specialization.

The Committee's programmes made major contributions to the regeneration of Africa studies in China. Graduates of the exchange programme were appointed leaders of major university and government African studies centres, several were appointed ambassadors and senior foreign service officers in China's embassies in Africa, and others returned to their academic and government research positions, reinvigorating Africa studies and providing training to a generation of academic and government personnel. Without doubt, the Ford Foundation funded Africa studies programme, together with its other assisted area studies projects, contributed to China's reintroduction and resurgence in area studies, both in terms of new knowledge introduced and the training and re-training of human resources.

At long last, I had established connectivity with China's Africanists and institu-tions, including visits to China's Africa studies centres, meeting with China's Africa scholars, and was accepted and able to conduct research in China. My academic journey in search of studying Chinese foreign policy and behaviour, with a focus on interaction with Africa, had come full circle: through Africa, I had been introduced to China.

Persistent questions in China–Africa research

For students of China-Africa relations, a persistent question has been how best to explain and understand the relationship. Reflecting upon my research experience, while never hesitating from explaining the relationship in macro terms (e.g. Yu 1966, 2010), I appreciate both the contributions and limitations of such studies. They constitute generalizations, with select examples, some based on high levels of abstraction, of complex relationships within diverse and dynamic economic and social environments. Macro analysis has the advantage of identifying and analyzing large-scale patterns and trends, such as Africa's economic development and the growth and extent of China's reach into Africa. But it is also important to remember that the African continent comprises 54 independent countries; not all share common needs and hold common goals, including neighbouring countries that share select cultural and/or economic and social characteristics.

The contributions of microanalysis, for example country studies, were brought home to me in my field research in East Africa. Initially, I had planned to conduct research on China's relations with Kenya, Uganda and Tanzania. But the political and economic conditions at the time, especially in Kenya, were not conducive to field research. But across Kenya's southern border into Tanzania, completely different conditions and environment were present. China–Africa relations, represented by China–Tanzania interaction, were a model of China–Africa cooperative interaction, as opposed to the conflict relation model between China and Kenya.

In short, Africa comprised what became 54 disparate sovereign states, each with its own identity and domestic and foreign political and economic needs and goals. Even Kenya and Tanzania, two neighbours, with shared political and cultural backgrounds, differed significantly in their respective domestic and foreign policies. The selection to conduct research of one country's interaction pattern with China versus another would produce vastly different conclusions on China's capacity and capability to conduct cooperative interaction.

Another vexing research problem has been the vital role of foreign aid, the primary instrument of China's foreign policy, in China–Tanzania relations. There can be no doubt China's response to Tanzania's needs and requests, from assistance in building the TAZARA to providing rural medical teams, from offering agriculture projects to delivering military aid, the extensity of foreign policy instruments employed had been both diverse and massive. Recognition of, providing for and sustaining Tanzania's developmental needs and requests over time through a continued flow of resources had been a major Chinese achievement and challenge, at significant self-sacrifice. Though not all aid projects achieved success (e.g., agriculture and the textile industry), and others required persisted assistance (e.g., the TAZARA), sustained aid contributed to the success of the political connection. In principle, China's aid to Tanzania (and Africa) had few limits.

China's human and material assistance also raised the question of the practical limits of aid. As mentioned, the failure of aid projects was acknowledged; it is estimated that at least 50 per cent of the aid projects did not achieve their original goal. One common problem had been the shortage of trained local staff to continue the work of a project, following transfer of the mission to the Tanzanians; others included the inability of the local economy to absorb and integrate the aid projects. In 2016, an article in *China Today* acknowledged the difficulties and limits besetting China's flagship aid project, the TAZARA, including the shortage of trained local professional staff, the failure of the staff to fully utilise and follow 'the rules for the railway's operation and personnel management', under capitalization of the railway company, and the company's management structure; despite continued Chinese technical and management assistance, the railway in 2012–2013 was only operating at 17 per cent capacity from its peak volume in 1977–1978 (Lu, 2016). However, in the words of the commentator, '… specific issues over how to operate the TAZARA should never overshadow the railway's status as a monument to Sino-African friendship'. Put differently, foreign aid served foreign policy. The limit of China's foreign aid to Tanzania was not determined by its success or failure, singularly or collectively, but the extent to which aid achieved and supported cooperative interaction. In sum, the primary function of China's foreign aid was political; as one Chinese official explained to the author, China's aid was 80 per cent political and 20 per cent economic (see Yu 1975, 147–148; and Interview, 20 May 2017).

My research is not intended to advocate 54 separate studies of China–Africa relations. One case study does not represent the whole; nor does one research approach describe the total. However, it does support the thesis that to explain and understand the whys and hows of relationship variations, whether by region, nation, political, economic and social sectors and other variables, it is imperative that we accept and recognise that, similar to the human condition, China and Africa and their relationships are both complex and diverse, requiring multiple approaches *and* particularised attention.

Conclusion

Looking back to the 1960s, beginning with a small cohort of scholars one would never have expected that over the course of 60 plus years, China–Africa studies would achieve recognised status within academia and become a promising field of studies, whether as an interdisciplinary field of studies, such as China studies and African studies, or as a sub-field in an existing disciplinary field, such as Anthropology. My own research has sought to be located within the discipline of Political Science. Whether my and other studies of China Africa interaction have contributed to the building or formation of a new and separate body of literature and field of studies, only the totality of past, present and future scholarship in the field will ascertain. My sense is that as the body of innovative scholarship continues to develop and expand, the field holds the promise of contributing to the creation of a new and separate field of study.

Reflecting on my academic journey and research, I am satisfied that I was among the few scholars in the United States to begin and sustain research on China's relations with Africa, recognizing at an early date the multiple roles and importance of Africa in Chinese foreign policy, including China's rise and the globalisation of Chinese foreign policy. By 2017, China–African studies had become a multiple disciplinary study area and China–Africa relations had become a 'growth' industry. China's promise to deliver investment and increase aid at the 2015 Johannesburg summit of the Forum on China–Africa Cooperation was expected to insure continued and deeper growth of China–Africa relations. China's new 'Belt and Road Initiative' was expected to insure enhanced relations with the African continent. With the potential of further benefitting all participants in the increasingly multi-dimensional complex relationship, Africa's economic and social development and China's domestic and foreign policy needs and goals, the relationship held great promise. It also assured greater scrutiny, including attracting new members to the China–Africa studies community.

Note

1 Africa as a battleground between China and Taiwan continues to the present. On 3 March 2016, Gambia announced that it was resuming diplomatic relations with China, after it had earlier broken its ties with China and recognised Taiwan.

References

Lu, Rucai. 2016. 'Du Jian: Witness to the Birth of TAZARA', *China Today* 65(1): 39–42.

Yu, George T. 1963. 'Peking versus Taipei in the World Arena: Chinese Competition in Africa', *Asian Survey* 3(9): 439–453.

Yu, George T. 1966. *Party Politics in Republican China: The Kuomintang, 1912–1924*. Berkeley: University of California Press.

Yu, George T. 1966. 'China's Failure in Africa', *Asian Survey* 6(8): 461–469.

Yu, George T. 1968. 'Dragon in the Bush: Peking's Presence in Africa', *Asian Survey* 8(12): 1018–1026.

Yu, George T. 2010. 'China's Africa Policy', in *China, the Developing World, and the New Global Dynamic*, edited by Lowell Dittmer and George T. Yu, 129–156. Boulder and London: Lynne Rienner.

Yu, George T. 1970. *China and Tanzania: A Study in Cooperative Interaction*. Berkeley: University of California Center for Chinese Studies.

Yu, George T. 1975. *China's African Policy: A Study of Tanzania*. New York: Praeger.

3

AFRICAN STUDIES IN CHINA IN THE TWENTY-FIRST CENTURY

A historiographical survey

Li Anshan

Academic studies are always a reflection of reality. The fast development of China–Africa relations has seen Africanists outside China show great interest in China–Africa academic engagement. Within China, the study of Africa might be divided it into four phases: contacting Africa (before 1900), sensing Africa (1900–1949), supporting Africa (1949–1965), understanding Africa (1966–1976) and studying Africa (1977–2000) (Li 2005). Although China's trade with Africa increased from $10.5 billion in 2000 to 220 billion in 2014, African studies in China did not have the same fortune as trade. However, the dramatic development of relations has provided Chinese Africanists with new opportunities and challenges. This chapter elaborates on what Chinese Africanists have been studying in the period 2000–2015. What subjects have been studied? What are the achievements and weaknesses of this scholarship? The chapter is divided into four parts: focus and new interests, achievements, young scholars, references and afterthoughts.

During the past 15 years, the focus has been mainly on China–Africa relations and African nations. FOCAC has greatly promoted bilateral economic relations. The increasing numbers of Chinese companies in the continent need to know more about Africa and its people. Quite a number of studies have been carried out on China–Africa relations and the current situation of African countries. Between 2000–2005, 232 books on or about Africa were published (Chen Hong and Zhao Ping 2006). If we add books published in 2006–2015, the total number should be much more covering a wide range of fields. Given space constraints, this chapter is necessarily only a review of African studies in China.[1]

China--Africa relations

China–Africa relations is a hot topic within China and abroad. A few books were published either on general study, cooperation plus international development cooperation, or bilateral migration. As early as 2000, a four-volume 'Series of Investment Guide for Development of African Agriculture' was published to cele-brate the opening of FOCAC (Lu Ting-en, Wen Yunchao, He Xiurong, Wang Xiuqing and Li Ping 2000: all published in Yao Guimei and Chen Zonde, eds 2000). Li studied the linkage between African Economic Zones and Chinese Enterprises (Li Zhibiao 2000). A few investment guides were also published in various fields such as mining, oil and gas, and emerging markets. A journalist in Africa for eight years, Li Xinfeng traveled extensively. Exploring Zheng He's voyage to Africa, his publications stirred up excitement both in East Africa and China. In other works, he provided a fresh image of Africa with reports on important events (Li Xinfeng 2005, 2006). In 2012, another book tried linking Zheng He and Africa through data, maritime silk-road and various records (Li Xinfeng et al. 2012). Cooperation between China–Kenya archaeologists, headed by Qin Dashu of Peking University, undertook to explore the Kenyan coast and made discoveries (Qin Dashu and Yuan Jian 2013).

Bilateral China–Africa migration is another focus. In 2000, the first history of overseas Chinese in Africa was published covering early China–Africa relations and the origin of Chinese communities in Africa, the survival and adaptation of Chinese in Africa, their transformation and integration. The first part of this book was translated into English (Li Anshan 2000, 2012). A sister volume was published with records, reminiscences, articles in early journals and Chinese newspapers in Africa (Li Anshan 2006). Now an increasing number of works are written on this subject, some are by young scholars. Chinese scholars are also involved in the study of African communities in China (Li Zhigang 2009, 2012; Bodomo and Ma 2010, 2012; Ma Enyu 2012; Xu Tao 2013; Li Anshan 2015a).

China–Africa relations are characterised by summit diplomacy, equality, co-development and institutionalisation of cooperation. After the 2006 Beijing Summit, a Chinese–-English bilingual appraisal was published (Li Weijian 2008). In 2009, a collection by Chinese scholars was published in English to celebrate 50 years of China–Africa cooperation (Liu Hongwu and Yang Jiemian 2009). Another book analysed the theory, strategy and policies of China–Africa development coopera-tion (Liu Hongwu and Luo Jianbo 2011). A monograph on FOCAC deals with Africa's position in the international arena, China's strategy towards Africa, the funding of FOCAC, and the pattern of China–Africa cooperation (Zhang Zhong-xiang, 2012). Qi Jianhua's book on the new China–Africa partnership deals with bilateral cooperation in various aspects, including economic/financial, political/ legal, security/military, cultural/social, African regional integration, and the con-tributors include scholars of African French-speaking countries (Qi Jianhua 2014). A survey of African studies in various disciplines in China (1949–2010) was included in a collection celebrating the Institute of West Asia and Africa (IWAA 2011).

What is the implication of China–Africa economic diplomacy to the global value chain? Tang's work probed the issue from the perspective of trade, infrastructure, mining, agriculture, economic zone, manufacturing, and social transformation (Tang Xiaoyang 2014, 2014a, 2014b, 2014c). Zhang Zhe (2014) examined the

development of China–Africa economic and trade relations. China–Africa cooperation in low-carbon development strategy is studied in terms of the international rule,
international cooperation, and African low-carbon development strategy (Zhang
Yonghong, Liang Yijian, Wang Tao & Yang Guangsheng 2014). Another important
work deals with the strategy of China–Africa economic and trade cooperation in the
new situation (Shi Yongjie, 2015). However, a different view argued that China
lacked an African strategy and that 'there is everything Chinese in Africa except a
strategy' (Li Anshan 2011). In addition, China's Achilles' heel lies in the shortage of
strategic means and specific measures to realise its aim (He Liehui 2012).

Several important works have covered various aspects of China–Africa relations.
Zhang's work deals with the economic cooperation between Africa and big economies, including developed economies and new economies such as India, Russia,
Brazil, and China. It also made a comparison of economic cooperation between
Africa and different powers (Zhang Hongming 2012a, 2012b). Covering a wide
range of fields, Yang's work studied the comprehensive strategy of China–Africa
economic cooperation in terms of historical heritage, trade, investment, project
contract, assistance, science, and technology (Yang Lihua et al. 2013). There are
comparative studies of poverty and poverty reduction between China and Africa
(Li Xiaoyun 2010a, 2010b).

As for international development aid, several works were published including
studies of Chinese and Western aid to Africa (Zhang Yongpeng 2012) and Chinese
medical cooperation with Africa focusing on Chinese medical teams and a campaign
against Malaria (Li Anshan 2011). One study used the concept of 'development-
guided assistance' to describe China's model (Zhang Haibin 2013). Other studies
partly deal with China–Africa development cooperation (Zhou Hong 2013; Liu
Hongwu and Huang Meibo et al. 2013). China–Africa relations are studied
from various perspectives such as African integration (Luo Jianbo 2006), NGOs
(Liu Hongwu and Shen Peili 2009), or infrastructure (Hu Yongju and Qiu Xin
2014). 'Entering Africa to Seek Development' became a theme for conferences
held by the Chinese Association of African Studies (CAAS), which has published
continuously.

Country studies

A special committee was set up by the Chinese Academy of Social Sciences (CASS)
in 2002 in charge of a Guide to the World States. The outline of the content is
uniform, focusing on seven aspects: land and people, history, politics, economy,
military, education (with cultural aspects), and foreign relations. In 2006, the
Institute of West Asia and Africa Studies (IWAAS) of CASS, Chinese Society of
African Historical Studies (CSAHS) and Peking University's Center for African
Studies decided to carry out a project on the bibliography of African studies in
China between 1997 and 2005. Regarding graduate theses on individual countries,
it found that there were 152 titles on 29 countries. South Africa is at the top, with
36 theses (Chen Hong and Zhao Ping 2006).

As the statistics indicate, the study on African countries is concentrated on big countries, notably South Africa and Egypt. Among over 4,000 articles published in more than 800 journals, five countries attract more attention and articles on those countries count for more than one quarter of the total. South Africa is at the top with about half of the total articles, comprising 620 on South Africa out of 1,256 (Chen Hong and Zhao Ping 2006).

The study on Portuguese-speaking African countries has been long neglected owing to language barriers. This situation is changing. The fourth volume of the series published by the Center for African Studies at Peking University was a collection of articles on the development of these countries (Li Baoping et al. 2006). There are studies of individual countries as well, such as the history of Ghana (Chen Zhongdan 2000), Nigeria (Liu Hongwu, et al. 2014) and Egypt (Wang Haili 2014), development of Tanzania (Li Xiangyun 2014), South Africa's politics and urbanisation (Qin Hui 2013).

Current situation

It is necessary to provide a survey of the current situation in different fields. Recently, there have been quite a few studies on subjects like African transportation, African tourism (Luo Gaoyuan 2010), African agriculture (Jiang Zhongjin 2012), industry and mining in Africa (Zhu Huayou, et al., 2014), African education (Liu Yan 2014; Lou Shizhou 2014; Wan Xiulan and Li Wei 2014), law (Hong Yonghong 2014), international organisations (Li Bojun 2014), security regime in Africa (Mo Xiang 2014), resources and environment, and AIDS (Cai Gaojiang, 2014).

The most important work is the *Oxford Handbook of Africa and Economics*, edited by two prominent economists Célestin Monga and Justin Yifu Lin. Raising the issue of linkage between economics and Africa, it conveys several firm convictions, including that Africa as a region is still under-researched, and scholarship has thus far neglected African contribution to economic knowledge. Realising Africa is on the verge of take-off, the book attempts to serve as useful knowledge in guiding Africa's new phase of development and provide clear guidance to policymakers (Monga and Lin 2015).

According to the above-mentioned 2006 statistics, most of the articles are on current issues. Among 1,256 articles, 424 are on economy, about one-third of the total, while 208 articles are on politics and law and 127 on foreign affairs.

TABLE 3.1 Statistics of Graduate Theses on African Countries (1981–2005)

Egypt 35	Kenya 2	Nigeria 18	Mali 3	South Africa 36
Sudan 5	Somali 1	Cameroon 1	Cogo (B) 1	Lesotho 2
Libya 2	Tanzania 5	Benin 1	Congo (K) 4	Madagascar 3
Algeria 4	Uganda 2	Togo 1	Mozambique 7	Mauritius 2
Morocco 1	Burundi 1	Ghana 2	Botswana 2	Zimbabwe 2
Ethiopia 4	Niger 1	Cote d' Ivoire 1	Zambia 3	Total 152

It is noticeable that articles about economics stand out as No.1 in the list for all five countries, reflecting China's focus today. There are more works on politics/ law or foreign affairs for Ethiopia and Nigeria. History occupies the second place in Egypt since Egyptology is included in the subject. More works on the culture and society of South Africa indicate that more Chinese are familiar with the country. It is interesting that the writing on South Africa is on the top, with 620 items. The graduate theses (1981–2005) have some implications. Among 238 M.A. and PhD. Theses, 73 titles are about Africa in general, 17 on politics, 13 on economy, 26 foreign affairs, 12 history and five culture. There are four on East Africa, seven on West Africa and two on Southern Africa. Now more studies are focused on security, environment and climate change.

African integration is another focus (Luo Jianbo 2010). CSAHS held its annual conference on 'China–Africa Cooperation and African Integration' in 2013. The collection of papers is divided into Pan-Africanism and African Unity, African integration and China–Africa Cooperation (Zhai, Wang and Pan 2013). There are studies on regional integration as well. Xiao Hongyu emphasised the linkage between African regional integration and economic development. Taking West Africa as a case, she studied the interaction between integration and modernisation (Xiao Hongyu 2014). African economic integration is an important phenomenon, covered in Zhang Jing's case study of 30 years' development of SADC (Zhang Jing 2014).

Monographs and achievements

African history

Although contemporary Africa now attracts more attention from Chinese scholars, the study of African history is still important in China. Several scholars of the old

TABLE 3.2 Classification of Articles on Specific African Countries (1997–2005)

Subject/Country	Egypt	Ethiopia	Kenya	Nigeria	South Africa	Total
Politics & Law	42	2	15	31	118	208
Economy	107	31	41	44	201	424
Foreign Affairs	61	14	3	6	43	127
Ethnicity		2			20	22
Religion	15				5	20
Military	5	1			33	39
History & Archaeology	68	3		1	7	79
Culture	44	2	15	19	99	179
Society	10	6	25	10	67	118
Important Figures	9	2	2		27	40
Total	361	63	101	111	620	1256

generation published their works. Ai Zhouchang finished a book on modernisation in South Africa (Ai Zhouchang et al. 2000). Lu Ting-en compiled his articles into a volume of four sections covering African history in the colonial period, history of African parties and politics, African economic history and the history of China–Africa relations (Lu Ting-en 2005). Zheng Jiaxing taught South African history at Peking University from the beginning of the 1980s. As a summary of his teaching, his book studies the history from the establishment of Cape Town until the formation of the New South Africa government (Zheng Jiaxing, 2010). His volume of colonialism in Africa forms part of the 'Series of History of Colonialism' (Zheng Jiaxing 2000). Gao Jinyuan, a senior researcher in CAAS, published two works, one a collection of his study of Africa examining colonialism and the liberation movement, area/country study, contemporary politics; and the other on Britain–Africa relations from the slave trade to the present (Gao Jinyuan 2007, 2008). Xu Yongzhang compiled his early articles on the history of China–Africa relations, and published a comprehensive history on African countries (Xu Yongzhang 2004, 2014).

Shu's work deals with the structural adjustment in Africa, an important chapter in African development. After an analysis of the interference of international financial system, and the response of African countries, he concluded that the World Bank's structural adjustment was a failure (Shu Yunguo 2014). The history of African economic development gives a survey from the nineteenth century to the 1990s, with additional chapters on South Africa, African economic relations with China and other countries (Shu Yunguo and Liu Weicai 2013). A history of Pan-Africanism is an important work (Shu Yunguo 2012, 2014).

Li's book on rural protest in Ghana during the colonial period was the first monograph in English by a Chinese Africanist. Based on government documents and fieldwork, he explored protests of the Ghanaian people against colonial government, commoners against chiefs, religious leaders against secular authority and lesser local leaders against paramount chiefs (Li Anshan 2002). Another work introduced ancient kingdoms in different parts of the African continent (Li Anshan 2012). The 'World Modernization Series' volume on Africa covers the process from different perspectives of history, politics, economy, nation-building and integration, with case studies of Ethiopia, South Africa, Nigeria, Ghana, Tanzania, Zambia, Angola and French-speaking countries (Li Anshan 2013d; Li Anshan et al., 2013).[2]

An excellent study on African intellectuals of modern time with a focus on the eighteenth–nineteenth century, Zhang's book deals with the ideological background of the slave trade through important figures of the eighteenth century, and then the three cultural trends of Westernisation, Africanisation and Integration of the nineteenth century (Zhang Hongming 2008). Chen Xiaohong contributed a study on De Gaulle and African decolonisation analysed change of the international situation, struggle of the colonies, demand of French monopolised capitalism and the change of social configuration. Sun Hongqi's study tried to analyse the role of colonialism in Africa (Sun Hongqi 2008).

Politics, international relations and law

At the beginning of the twenty-first century, the Chinese government called for a grand diplomatic effort requiring effort, experience and ideas from all walks of life. 'It is recognizable that there should be more cooperation between practical work and academic research. The government needs information, analysis and assessment, while the academia needs funding, stimulus and feedback' (Li Anshan 2005). Since then, the situation has been developing dramatically. Scholars were asked to give lectures to top leaders or for opinions on the draft of state leaders' speeches in FOCAC.[3] Africanists took up projects from various ministries in order to provide them with thinking and ideas about how to carry out development cooperation with Africa. The Ministry of Education promoted the formation of think-tanks in universities. All of these trends show the adjustment of the government to a changing situation and increasing interaction with academia.

Zhang Hongming's work discussed the internal and external factors of African politics. As for the internal, he illustrated the relation between politics and the state, tribalism, traditional culture and religion. The external factors covered Western political culture, Eastern political culture and Islamic political culture and their linkage with political development (Zhang Hongming 1999). Xia Jisheng of Peking University explored the structure and function of the parliamentary systems of South Africa and Egypt (Xia Jisheng 2005). Li Baoping's book on African culture and politics dealt with traditional culture, political transformation and the cases of Tanzania and South Africa (Li Baoping 2011).

Studying the origin and evolution of nationalism in Africa, Li approached the subject from its various expressions (national intellectual, religion, peasantry, nation-building, democratisation, international politics) and its different forms (such as Pan-Africanism, African nationalism, state nationalism and local nationalism). Using 'local nationalism' to replace 'tribalism', he argued that local nationalism had its origin in pre-colonial society and was strengthened by indirect rule. After independence, the poor distribution of power, economic difficulties and external interference strengthened ethnic conflicts (Li Anshan 2004). Amidst increasing interest on democratisation in Africa, He Wenping's work on the subject enriches our understanding of the process. She argues that different countries take different forms and ways in pursuing democracy, using case studies of South Africa, Nigeria, Kenya and Uganda, stating: "There is the common desire for democracy, but there is not the common way for realizing it. The democracy that people has been empowered must be built by people themselves." (He Wenping 2005a; authors own translation).

To understand early communist leaders' view on Africa, a book was compiled of the sayings of Marx, Engels, Lenin and Stalin on the Middle East and Africa (Cui Jianmin 2010). There are studies of the early generation of African leaders (Lu Ting-en 2005) and contemporary leaders such as President Museveni (Mu Tao and Yu Bin, 2013). Nyerere's important works were translated (Nyerere 2015). Another focus is on African diplomacy and foreign relations. The first diplomatic history of South Africa deals with foreign policy during the apartheid and international reactions, South African's neighbouring policy, adjustment of De Klerk's 'new diplomacy' and foreign policy of new South Africa (Mu Tao 2003). Relations between modern Egypt and the U.S., Russia, Israel, Saudi Arabia and China were studied (Chen Tiandu et al. 2010). There are studies on Nigeria's foreign relations (Yang Guangsheng 2014) and the new South Africa, including the political economy of South Africa's land issue (Sun Hongqi 2011). The Darfur issue was probed in terms of its origin, relation with north and south Sudan, and with oil, geopolitics, China and the 2008 Beijing Olympics (Jiang Hengkun 2014). Political systems have been examined, such as Ethiopia's federalism and political transformation (Zhang Xiangdong 2012), Islamic socialism in Libya (Han Zhibin et al. 2014). The AU's role is also studied in terms of African economy, conflict management, common foreign policy, collective development and its contribution to the world politics (Luo Jianbo 2010).

Despite one view questioning the existence of African legal systems, Hong has studied African law for more than ten years (Hong Yonghong 2005, 2014). Another study covered the various legal systems practiced in the continent, such as ancient Egyptian law, Islamisation of African law, African customary law, common law, civil law, and mixed jurisdiction (He Qinhua and Hong Yonghong 2006). Hong Yongong contributed an important work on the International Criminal Tribunal for Rwanda (ICTR) which helped the Chinese understand the organ and the case. Hong carried out studies on African law with his colleagues and continuously published works, including translations (Hong Yonghong and Xia Xinhua 2010; Mancuso and Hong Yonghong 2009; Zhu Weidong 2011). *West Asia and Africa* ran a special column on the study of African law for more than ten years (Institute of West Asia and Africa 2011).

Chinese scholars have also contributed to the field of African geography. Jiang's work offers a comprehensive survey of African agriculture, assessment of agricultural natural resources, analysis of social economic conditions in agriculture, the history of agricultural development, regional distribution and economic type of agriculture. It deals with various topics of agricultural natural resources, food crop, husbandry, forestry, fishery, agricultural food processing, consumption of agricultural products and nutrition security. This work probes the relations between people, culture of agricultural economy and environment (Jiang Zhongjin 2012). As a focus of research of Nanjing University's Center for African Studies, the 'Series of Security Study on China–Africa Resource Development and Energy Cooperation' covers various subjects, such as China–Africa energy cooperation and security (Jiang Zhongjin and Liu Litao 2014), African agriculture and development (Jiang Zhongjin 2014), port economy and urban development (Zhen Feng 2014), land resource and food security (Huang Xianjin et al. 2014), and modern African human geography (Jiang Zhongjin 2014). Cultural geography also became a subject of research (Chong Xiuquan 2014).

African art is a rich subject. Various translations were published yet few serious studies have been done. Seven volumes of African arts were published in 2000 yet they are more for eyes than for thoughts. Quite a few African art works, especially Egyptian art and architecture, are edited or translated. There are several cultural studies, either in general (Ai and Mu 2001; Ai and Shu 2008; Zheng Jiaxing 2011) or specific countries (Yang and Zheng 2001; Jiang Dong 2005), and related subjects (Liu Hongwu and Li Shudi 2010). The most important work is a history of South African literature by Li Yongcai, a scholar involved in the study of African literature (Li Yongcai 2009) since a long time, and there are articles of study in African art, sculpture, film, literature, dance, or music.

English publications and young scholars

An increasing number of Chinese scholars take an active part in international academia. Some of them become editors of books related to China–Africa relations, some published articles in journals, as chapters, or in networks.

Engagement with international scholarship is another achievement. There are a few books in English written or edited by Chinese scholars in international academia (Li Anshan 2002, 2011, 2012, 2013; Li Anshan et al., 2012; Li and April 2013; Berhe and Liu 2013; Monga and Lin 2015; Shelton et al. 2015). Chinese participation in international conferences is increasing and their views have appeared in international journals and magazines (Zhang Hongming et al., 2001; Zeng Qiang, 2002, 2010; Li Anshan 2005, 2007, 2008, 2009b, 2010, 2013a, 2015a, 2015b; Li Baoping 2008; Liu Haifang 2006, 2008, 2015a; Yang Lihua 2006. More and more Chinese scholars have been involved in international cooperation and their works are included in English books or conference paper collections (He Wenping 2005, 2007a, 2008a, 2009a, 2010a, 2010b, 2012, 2012a, 2012b; Li Baoping 2007; Hong Yonghong 2007, 2010; Zhang Yongpeng 2007; Li Zhibiao 2007; An Chunying 2007; Li Anshan 2007b, 2008a,b,2009, 2010a, 2011a,b,e, 2012a, 2013, 2013b,c, 2015; Zeng Qiang 2010; Zhi and Bai 2010; Zhang Xinghui 2011; Liu Hongwu 2012; Tang Xiaoyang 2012; Zhang Chun 2012; Liang Xijian 2012; Pang Zhongying 2013; Wang Xuejun 2013; Lin J.Y. 2015; Xu Liang and Akyeampong 2015).[4]

After a conference on 'China–Africa Relations: Past, Present and Future' held at South Africa in November 2005, a collection of papers was edited by the prominent Ghanaian Africanist Kwesi Prah; several Chinese scholars contributed their ideas (Prah, 2007). 'The China–African Civil Society Dialogue' conference was held in Nairobi in April 2008 by Heinrich Böll Foundation and 10 Chinese scholars were invited (see Harneit-Sievers, et al., 2010). A seminar was held in Nairobi by the Inter Region Economic Network as a concrete result of the China–Africa Joint Research and Exchange Program. The meeting was attended by a delegation from China whose speeches were included in an edited volume (Shikwati 2012). In October 2012, the China–Africa Think Tanks Forum (CATTF) met in Ethiopia, co-hosted by Institute for Peace and Security Studies (IPSS) of Addis Ababa University and the Institute of African Studies (IAS), Zhejiang Normal University, and a collection was subsequently published including that of Chinese scholars (Berhe and Liu 2013). Some of them are actively involved in English networks, such as He, Liu Haifang, Luo Jianbo and Zhang Xiaomin. Some Chinese students are studying or collecting data in Africa and are starting to show their academic capability (Cheng Ying 2014, 2016; Zhang Qiaowen 2015, 2015a; Xu Liang and Akyeampong 2015, 2015a).

More young students are engaged in African studies and are benefiting from better opportunities to work in Africa. Luo Jianbo has done work on African integration and China–Africa relations. A few PhD students of different disciplines finished their dissertation or based on fieldwork, such Chen Fenglan (2011) and Chen Xiaoying (2012) of sociology, Ding Yu of archeology (2013, 2015) and Yang Tingzhi (2015) and Shen Xiaolei of political science (2015). Chinese anthropologists or social scientists went to Africa through different channels or did fieldwork there, such as Si Lin and Xu Wei (Shi Lin 2012; Xu Wei 2014).

Academic monographs by young scholars

The new generation of Chinese Africanists has more opportunities for international contact, favourable academic environment and better time to undertake African studies. Bi Jiankang studied the linkage between Egyptian modernisation and political stability between 1805 and the 1990s, analysed different political regimes, i.e., military regime and president regime, and related issues such as political participation, political parties, Islam, political violence, or urbanisation. He also analysed the impact of economy, unemployment and external factors on political stability. The first Chinese MA graduate in Hausa awarded in Ahmadu Bello University, Nigeria, Sun published her PhD dissertation on British educational policy in northern Nigeria during the colonial period. She studied the interaction between power and language. Analysing language policy, the examination system, and the development of Hausa and educational policy, she explained how the British colonial government used Hausa language as a tool of colonial administration (Sun Xiaomeng 2004, 2014). Luo published two related books and a new one on China's responsibility. One analysed the achievement, problem and perspective of African integration and also tries to explore the linkage between China–Africa relation and African integration process, and the other African Union's relation with its member states in terms of development, economic cooperation, conflict management, foreign policy and its significance to the world (Luo Jianbo 2006; 2010).

Indigenous knowledge of Africa is a new subject and Zhang Yonghong made a detailed study of its role in various fields and its relation with development. Li Weijan probed the history of the spread of Islam in West Africa and traced the historical origin of Islam in West Africa in ancient times, the Jihad movement in the nineteenth century, Sufism, Islam during the colonial period and the contemporary time. Zhu excelled himself in the study of African law. Besides translation of related works, he also published two books (Zhu Weidong 2011). Jiang has studied Sudan for a long time and his work on Darfur examined the causes, process, condition for peaceful solution and impact of the crisis (Jiang Hengkun 2014). Wang Tao studied the Lord's Resistance Army in Uganda in terms of its origin, development, influence and international connections. Based on knowledge of Arab and English languages, Huang Hui has studied Berberism in Algeria.

Several important works on African economics have been written by young scholars. An Chunying studied poverty and anti-poverty measures in Africa, concluding that pro-poor growth is the solution for reducing poverty. Yang Baorong researched the linkage between liabilities and development in Africa from an international relations perspective, and studied the theory, origin and development of debt in Africa, debt issue in international affairs and the effect and impact of the debt-relief programme, as well as adjustment of policy and development with debt. Comparative advantage is a different perspective regarding African economic development. Liang Yijian argued that Africa can develop only through its own path, not copying others.

Annual report, memoirs and references

In China, associations and institutions of African studies have their academic activities annually and usually publish their works in a form of collection of papers.

Memoirs and references

With the opening up of China, discipline gradually loosened and officials started to write reminiscences or life memoires, especially when retired. Some diplomats with experience of work in Africa contributed articles to a volume with a subtitle 'A Glorious Passage of China–African Friendly Relations' (Lu Miaogeng et al., 2006). Several series of works by diplomats serve as supplementary data, such as 'Witness the History: Republican Ambassadors' Narrations', a series of ambassadors' life experiences. Wang Shu recounted his life as reporter in Africa during the late 1950s and the early 1960s, including his experience during the Congo incident (Wang Shu 2007). Guo Jing-an and Wu Jun's work is included in the 'Diplomats Look at the World' series. The former Ambassador in Ghana, Guo, described experience in African countries, such as the severing of diplomatic relations with Liberia because of Taiwan, his mission as special envoy in Somali, and as an ambassador in Ghana (Guo Jing-an and Wu Jun 2006). The 'Chinese Diplomats Series' attracts students of IR. *Chinese Diplomats in Africa* includes 19 articles by diplomats who worked in Africa. The collection covers different topics, sacred 'mission impossible' (Botswana), their suffering (Zambia), witnesses of important events in Ghana, Tanzania–Zambia railway, Cameroon and South Africa, reminiscences of their life, etc. (Li Tongcheng and Jin Buoxiong 2005). Former vice-premier and foreign minister, Huang Hua, one of the early diplomats and Chinese ambassador to Ghana and Egypt, published his memoir (Huang Hua 2008). Several Ambassadors described their life in African countries vividly. A few Chinese ambassadors and diplomats in Africa also told their stories and reminiscences (Cheng Tao and Lu Miaogen 2013).

Former vice-premier Qian Qichen's memoir is by no means less important, since he started his diplomatic career in Africa. Through Qian's memoir, we know things that do not appear in other writings; for example, former President Jiang Zemin once wrote four letters to President Mandela, in order to promote friendship and establish diplomatic relations between China and South Africa (Qian Qichen 2003, 245–287). A report of Jiang Zeming's visits abroad vividly describes the president's visits to African countries, especially his two important visits and talks with several African leaders in 1996 and 2002 (Zhong Zhicheng 2006). Former vice-minister of Commerce, Wei Jianguo, devoted most of his career to Africa; his book records different events and life experience (Wei Jianguo 2011).

Different dictionaries and encyclopedia have been published during this period. Two important dictionaries of diplomacy were most useful for their Africa related items. The *Dictionary on China's Diplomacy* contains various diplomatic contacts between China and Africa in history (Tang Jiaxuan 2000). The *Dictionary on World's Diplomacy* comprises important events, treaties and figures in African diplomacy (Qian Qichen 2005). The *Encyclopedia of Overseas Chinese*, containing over 15 million Chinese words, was finished in 2002. The monumental work includes 12 volumes on different subjects. Each volume contains some items on Chinese overseas in Africa (Zhou Nanjing 1999–2002). Gu Mingyuan's *Dictionary of World's Educational Events* covers schools and educational events in Africa.

Conclusion

There are several new features regarding African studies in China during the past 15 years. First, with the increase of monographs, more academics now concentrate on current politics, economy, culture and society. Economics is the most studied subject, and South Africa the most studied state. While a proliferating number of publications have emphasised the importance of research quality, Chinese scholars have a long way to go. Second, various studies have expanded interdisciplinary approaches, which implies the significance of methodology, and long-time field-work for which local language capability is required. Third, more Chinese scholars are engaged in international academic exchanges and their views are gradually catching attention from outside, yet this is merely concentrated on China–Africa relations. Young scholars are growing up with better opportunities to study Africa and some within this new generation have displayed their academic capability. African study in China is promising, but requires more effort and hard work.

Notes

1 This is a revised version of an article published in the *Brazilian Journal of African Studies* 1 (2), 2016: 48–88. As Chair of Chinese Society of African Historical Studies, I would like to thank the members who responded to my email accordingly with the information of their own publications. Owing to the shortage of space, non-Chinese works with a Chinese translation have the publication year in a bracket and articles in Chinese are not included with few exceptions.
Professor Qi Shirong, the Vice Chairman of the Association of Chinese Historians, praised it highly in his keynote speech at the conference 'World History Study in China in the 20th Century', Peking University, April 2000.

2 *World Modernization Process: Volume of Africa* was published in early March 2013 by Jiangsu People's Press. After President Xi Jinping raised the concept of the 'African dream' during

his visit to Africa, March 24–30, 2013, Jiangsu People's Press decided to republish it under a different title, *The African Dream: In Search of the Road to Modernization*.

3 In May 2004, Li Anshan was invited to give two lectures about African history to former President Jiang Zemin.

4 For example, Li Anshan was invited by the Director-General of UNESCO, Ms. Irina Bokova on 8 November 2013 and became a member of International Scientific Committee of UNESCO General History of Africa. He was elected Vice Chair of the Committee at its first session, held at Salvador, Brazil, 20–24 November 2013.

Bibliography

Ai Zhouchang *et al.* 2000. *Nanfei Xiandaihua Yanjiu* (A Study on Modernization in South Africa). Shanghai: East Normal University Press.

Ai Zhouchang and Mu Tao. 2001. *Zoujin Hei Feizhou* (Enter into Black Africa). Shanghai: Shanghai Literature and Art Publishing House.

Ai Zhouchang and Shu Yunguo. 2008. *Feizhou Heiren Wenmin* (African Black Civilization). Fuzhou: Fujian Educational Press.

An Chunying. 2007. 'Mining Industry Cooperation between China and Africa: Challenges and Prospects'. In *Afro-Chinese Relations: Past, Present and Future*, edited by Kwesi Kwaa Prah, 309–330.

Berhe Mulugeta Gebrehiwot and Liu Hongwu. eds. 2013. *China-Africa Relations: Governance, Peace and Security*. Addis Ababa and Jinhua: Institute for Peace and Security Studies (Addis Ababa University) and Institute of African Studies (Zhejiang Normal University).

Bodomo, A. and Grace Ma. 2010. 'From "Guangzhou" to "Yiwu": Emerging Facets of the "African Diaspora" in China'. *International Journal of African Renaissance Studies* 5(2): 283–289.

Bodomo, Adams and Grace Ma. 2012. 'We Are What We Eat: Food in the Process of Community Formation and Identity Shaping among "African Traders" in "Guangzhou" and "Yiwu"'. *African Diaspora* 5(1): 1–26.

Cai Gaojiang. 2014. *Feizhou Aizibing Wenti Yanjiu* (Research on AIDS in Africa). Hangzhou: Zhejiang People's Press.

Chen Fenglan. 2011. Wenhua Congtu yu Kuaguo Qianyi Qunti de Shiyin Celue: Yi Nanfei Zhongguo Xinyimin Wei Li (Cultural Clash and the Adaptive Stratagems of Transnational Group of Migration). *Huaqiao Huaren Historical Studies*, No. 3.

Chen Hong and Zhao Ping. eds. 2006. *Feizhou Wenti Yanjiu Zhongwen Wenxian Mulu (1997–2006)* (Bibliography of Chinese Writings on African Issues, 1997–2006), Institute of West Asia and Africa of CASS, Chinese Society of African Historical Studies and Center for African Studies, Peking University.

Chen Tiandu *et al.* 2010. *Dangdai Aiji yu Daguo Guanxi* (Modern Egypt's Relations with Great Powers). World Affairs Press.

Chen Xiaoying. 2012. 'Nanfei Zhongguo Xinyimin Mianlin de Kunjing Jiqi Yuanyin Tanxi.' (Chinese New Immigrants' Dilemma and Causes in South Africa) *Huaqiao Huaren Historical Studies*, No. 2.

Chen Zhongdan. 2000. *Jiana: Xunzhao Xiandaihua de Genji* (Ghana: Looking for a base for Modernization). Chengdu: Sichuan People's Publishing House.

Cheng Tao and Lu Miaogen. 2013. *Chinese Ambassadors Tell African Stories*. Beijing: World Affairs Press.

Cheng Ying. 2014. 'Bàrígà Boys' Urban Experience: Making Manifest (Im)mobility Through "Mobile" Performances'. *SOAS Journal of Postgraduate Research* 7 (Fall: 48–62).

Cheng Ying. 2016. '"Naija Halloween or wetin?": Naija Superheroes and a Time-traveling Performance'. *Journal of African Cultural Studies* 28(3): 275–282.

Chong Xiuquan. 2014. *Dangdai Feizhou Jishi Sheying Wenhua Dili* (The Cultural Geography of Contemporary African Documentary Photography). Hangzhou: Zhejiang People's Press.

Cui Jianmin. ed. 2010. *Makesi, Engesi, Lienin, Sidalin Lun Xiya Feizhou* (Marx, Engels, Lenin, Stalin on West Asia and Africa). China Social Sciences Press.

Ding Yu (with QinDashu and Xie Rouxing). 2013. '2010 Niandu Beijing Daxui Keniya Kaogu ji Zhuyao Souhuo' (Peking University archaeological discovery in Kenya in 2010). In *Annual Review of African Studies in China (2012)*, edited by Li Anshan and Liu Haifang. Social Sciences Academic Press.

Ding Yu (with Qin Dashu). 2015. 'Keniya Binghaishen Manbuluyi Yizhi de Kaogufajue yu Zhuyao Shouhuo' (Archaeological discovery in Manburui of Kenya Coast). In *Annual Review of African Studies in China (2014)*, edited by Li Anshan and Pan Huaqiong. Social Sciences Academic Press.

Gao Jing-an and Wu Jun. 2006. *The Years as Diplomats in Africa*. Chengdu: Sichuan People's Publishing House.

Gao Jinyuan. 2007. *Gao Jinyuan Ji* (Selection of Gao Jinyuan's Works). China: Social Sciences Press.

Gao Jinyuan. 2008. *Yingguo-Feizhou Guanxi Shilue* (A Brief History of Britain-Africa Relations). China: Social Sciences Press.

Han Zhibin *et al.* 2014. *Libiya Yisilan Shehuizhuyi Yanjie* (The Study of Libyan Islamic Socialism). Hangzhou: Zhejiang People's Press.

He Liehui. 2012. *Zhongguo de Feizhou Zhanlue* (China's Africa Strategy). China's Science and Culture Press.

He Qinhua and Hong Yonghong. 2006. *Fezhou Falü Fada Shi* (A History of Development of African Law). Law Press China.

He Wenping. 2002. "China and Africa: Cooperation in 50 Years", *Asia and Africa Today (in Russian)*, Russian Academy of Sciences, No. 12.

He Wenping. 2005. "All Weather Friends: A Vivid Portrayal of Contemporary Political Relations Between China and Africa." In *China Comes to Africa: The Political Economy and Diplomatic History of China's Relation with Africa*, edited by Kinfe Abraham. Addis Ababa: Ethiopian International Institute for Peace and Development.

He Wenping. 2005a. *Feizhou Guojia Minzhuhua Jincheng Yanjiu* (The Study of Democratization Process in African Countries). Beijing: Current Affairs Publisher.

He Wenping. 2006. 'China-Africa Relations Moving into an Era of Rapid Development'. *Inside AISA* 3&4: 3–6.

He Wenping. 2007. 'The Balancing Act of China's Africa Policy'. *China Security* 3(3): 23–40.

He Wenping. 2007a. '"All Weather Friend": The Evolution of China's African Policy'. In *Afro-Chinese Relations*, edited by Prah, 24–47.

He Wenping. 2008. 'Promoting Political Development through Democratic Change in Africa'. *Contemporary Chinese Thought* 40(1).

He Wenping. 2008a. 'China's Perspectives on Contemporary China-Africa Relations'. In *China Returns to Africa: A Superpower and a Continent Embrace*, edited by Chris Alden, Daniel Large and Ricardo Soares de Oliveira. London: Hurst.

He Wenping. 2009. 'China's African Policy: Driving Forces, Features and Global Impact'. *Africa Review* 1(1), Jan.–June.

He Wenping. 2009a. 'A Chinese Perception of Africa', In *China, Africa, and the African Diaspora: Perspectives*, edited by Sharon T. Freeman. Washington, DC: AASBEA Publishers.

HeWenping. 2010. 'Darfur Issue and China's Role', In *Chinese and African Perspectives*, edited by Harneit-Sievers *et al.*, 176–193.

He Wenping. 2010a. 'The Darfur Issue: A New Test for China's Africa Policy'. In *The Rise of China and India in Africa*, edited by Fantu Cheru and Cyril Obi. Zed Books.

He Wenping. 2010b. 'China's Aid to Africa: Policy Evolution, Characteristics and its Role'. In *Challenging the Aid Paradigm: Western Currents and Asian Alternatives*, edited by J. Stillhoff Sørensen. Houndmills: Palgrave Macmillan.

He Wenping. 2010c. 'Overturning the Wall: Building Soft Power in Africa', *China Security* 6(1).

He Wenping. 2012. 'Infrastructure and Development Cooperation: Take China in Africa as an Example', In *Emerging Asian Approaches to Development Cooperation*, edited by Lim Wonhyuk. Korea Development Institute.

He Wenping. (with Sven Grimm). 2012a. 'Emerging Partners and their Impact on African Development.' In *Africa Toward 2030: Challenges for Development Policy*, edited by Erik Lundsgaarde. Palgrave Macmillan.

He Wenping. 2012b. 'From "Aid Effectiveness" to "Development Effectiveness": What China's Experiences Can Contribute to the Discourse Evolution?' *Global Review* 9.

He Wenping. 2012c. 'China-Africa Economic Relations: Current Situation and Future Challenges'. In *China-Africa Partnership*, edited by Shikwati, 7–12.

Harneit-Sievers, Axel *et al*. eds. 2010. *Chinese and African Perspectives on China in Africa*. Pambazuka Press.

Hong Yonghong. 2005. *Feizhou Xingfa Pinglun* (Review on Criminal Law in Africa), China Procuratorial Press.

Hong Yonghong. 2007. 'The African Charter and China's Legislation: A Comparative Study of Ideas of Human Rights', In *Afro-Chinese Relations*, edited by Prah, 88–100.

Hong Yonghong. 2010. 'Trade, Investment and Legal Cooperation between China and Africa'. In *Chinese and African Perspectives*, edited by Harneit-Sievers et al., 82–90.

Hong Yonghong. 2014. *Dangdai Feizhou Falü* (Contemporary African Law). Zhejiang: People's Press.

Hong Yonghong and Xia Xinhua. 2010. *Feizhou Falü yu Shehui Fazhan Bianqian* (Africa Law and Social Development). Xiangtang University Press.

Hu Yongju and Qiu Xin. 2014. *Feizhou Jiaotong Jichu Sheshi Jianshe ji Zhongguo Canyu Celue* (The Present Situations and Developing Trend of African Transportation Infrastructure and China's Participation Strategies). Hangzhou: Zhejiang People's Press.

Huang Hua. 2007. *My Reminiscences*. Beijing.

Huang Xianjin, *et al*. eds. 2014. *Feizhou Tudi Ziyuan yu Liangshi Anquan* (African Land Resource and Food Security). Nanjing University Press.

IWAA CASS. 2011. *Zhongguo de Zhongdong Feizhou Yanjiu (1949–2010)* (Middle East & African Studies in China [1949–2010]), Social Sciences Academic Press.

Jiang Hengkun. 2014. *Daerfue Weiji: Yuanyin, Jincheng ji Yingxiang* (The Darfur Crisis: Causes, Processes and Impacts). Hangzhou: Zhejiang People's Press.

Jiang Dong. 2005. *Niriliya Wenhua* (Nigerian Culture). Culture and Art Publishing House.

Jiang Zhongjin. ed. 2012. *Feizhou Nongye Tuzhi* (Graphical Records of African Agriculture). Nanjing University Press.

Jiang Zhongjin. ed. 2014. *Xiandai Feizhou Renwen Dili* (Modern African Human Geography). Nanjing University Press.

Jiang Zhongjin and Liu Litao. eds. 2014. *Zhong-Fei Hezuo Nengyuan Anquan Zhanlue Yanjiu* (A Study on Strategy of China-Africa Energy Cooperation and Security). Nanjing University Press.

Li Anshan. 2000. *Feizhou Huaqiao Huaren Shi* (A History of Chinese Overseas in Africa). Beijing: Overseas Chinese Publishing House.

Li Anshan. 2002. *British Rule and Rural Protest in Southern Ghana*. New York: Peter Lang.

Li Anshan. 2004. *Feizhou Minzu Zhuyi Yanjiu* (Study on African Nationalism), Beijing: China's International Broadcast Publisher.

Li Anshan. 2005. 'African Studies in China in the Twentieth Century: A Historiographical Survey', *African Studies Review*, 48(1): 59–87.

Li Anshan. 2006. *Feizhou Huaqiao Huaren Shehui Shi Ziliao Xuanji (1800–2005)* (Social History of Chinese Overseas in Africa: Selected Documents 1800–2005). Hong Kong Press for Social Science Ltd.

Li Anshan. 2007. 'China and Africa: Policies and Challenges', *China Security*, 3(3): 69–93.

Li Anshan. 2007a. 'Transformation of China's Policy towards Africa', CTR Working Paper, Hong Kong University of Science and Technology.

Li Anshan. 2007b. 'African Studies in China in the Twentieth Century'. In *The Study of Africa, Global and Transnational Engagements*, edited by Paul Tiyambe Zeleza. Dakar: CODESRIA.

Li Anshan. 2008. 'Gli studi africanistici in Cina agli inizi del XXI secolo', *Afriche e Orienti*, No. 2 (part of Cristiana Fiamingo. ed., *La Cina in Africa*).

Li Anshan. 2008a. 'China-Sudan Relations: The Past and Present', *Symposium on Chinese-Sudanese Relations*. London: Center for Foreign Policy Analysis, 4–12.

Li Anshan. 2008b. 'China's New Policy towards Africa'. In *China into Africa: Trade, Aid, and Influence*, edited by Robert Rotberg, 21–49. Washington, DC: Brookings Institution Press.

Li Anshan. 2009. 'China's Immigrants in Africa and China's Africa Policy: Implications for China-African Cooperation'. In *China, Africa, and the African Diaspora: Perspectives*, edited by Sharon T. Freeman, 94–105. Washington, DC: AASBEA Publishers.

Li Anshan. 2009a. 'Zhong-Fei Guanxi Yanjiu Sanshinian' (The Study of China-Africa Relations in the Past Thirty Years), *West Asia and Africa* 4: 5–15.

Li Anshan. 2009b. 'What's to be Done After the Fourth FOCAC', *China Monitor*, Nov: 7–9.

Li Anshan. 2010. 'Control and Combat: Chinese Indentured Labor in South Africa, 1904–1910'. *Encounter* 3(Fall): 41–61.

Li Anshan. 2010a. 'African Studies in China: A Historiographical Survey'. In *Chinese and African Perspectives on China in Africa*, edited by Axel Harneit-Sievers *et al.* 2–24. Pambazuka Press.

Li Anshan. 2011. *Chinese Medical Cooperation with Africa: With a Special Emphasis on Chinese Medical Team and Anti-Malaria Campaign*. Uppsala: Nordiska Afrikainstitutet.

Li Anshan. 2011a. 'La coopération médicale Sino-Africaine: une autre forme d'aide humanitaire', In *Dans l'œil des Autre: Perception de l'action humanitaire et de MSF*, edited by Caroline Abu-Sada. Suisse: Editions Antipodes.

Li Anshan. 2011b. 'From "How Could" to "How Should": The Possibility of a Pilot U.S.-China Project in Africa'. In *China's Emerging Global Health and Foreign Aid Engagement in Africa*, edited by Charles W. Freeman III and Xiaoqing Lu Boynton, 37–46. Center for Strategic and International Studies.

Li Anshan. 2011c. 'Cultural Heritage and China's Africa Policy'. In *China and the European Union in Africa*, edited by Jing Men and Benjamin Barton, 41–59. Ashgate.

Li Anshan. 2011d. *Feizhou Gudai Wangguo* (Ancient Kingdoms in Africa). Peking University Press.

Li Anshan. 2011e. 'Zhongguo Zoujing Feizhou de Xianshi yu Zhengxiang' (China's Entry into Africa: Reality and Truth), *Social Observation* 8: 27–29.

Li Anshan. 2012. *A History of Overseas Chinese in Africa till 1911*. New York: Diasporic Africa Press.

Li Anshan. 2012a. 'China and Africa: Cultural Similarity and Mutual Learning'. In *China-Africa Partnership: The Quest for a Win-win Relationship*, edited by James Shikwati, 93–97. Nairobi: Inter Region Economic Network.

Li Anshan. 2012b. 'Neither Devil Nor Angel: The Role of the Media in Sino-African Relations', Opinion, http://allafrica.com/stories/201205180551.html.

Li Anshan. 2012c. *Zhong-Fei Guanxi Yanjiezhong de Fangfalun Chuyi – JiantanZiliao Liyong Wenti* (On the Methodology of the Study of China-Africa Relations-How to Exploit the Data), 3.

Li Anshan. 2013. 'BRICS: Dynamics, Resilience and Role of China', In *BRICS-Africa: Partnership and Interaction*, 122–134. Moscow: Institute for African Studies, Russian Academy of Sciences.

Li Anshan. 2013a. 'Book Review: The Dragon's Gift: The Real Story of China in Africa'. *Pacific Affairs* 86(1): 138–140.

Li Anshan. 2013b. 'China's African Policy and the Chinese Immigrants in Africa'. In *Routledge Handbook of the Chinese Diaspora*, edited by Tan Chee-Beng, 59–70. London: Routledge.

Li Anshan. 2013c. 'Chinese Medical Cooperation in Africa from the pre-FOCAC Era to the Present', In *Forum*, edited by Li Anshan and Funeka Yazini April, 64–80.

Li Anshan. ed. 2013d. *Shijie Xiandaihua Licheng Feizhoujuan*, (World Modernization Process: Volume of Africa). Jiangshu People's Press.

Li Anshan. 2015. 'A Long-Time Neglected Subject: China-Africa People-to-People Contact'. In *FOCAC 2015*, edited by Shelton, April, Li, 446–475.

Li Anshan. 2015a. 'African Diaspora in China: Reality, Research and Reflection'. *The Journal of Pan African Studies* 7(10): 10–43.

Li Anshan. 2015b. 'Contact between China and Africa before Vasco da Gama: Archeology, Document and Historiography'. *World History Studies* 2(1).

Li Anshan. 2015c. '10 Questions about Migration between China and Africa', China Policy Institute, http://blogs.nottingham.ac.uk/chinapolicyinstitute/2015/03/04/10-questions-about-migration-between-china-and-africa/.

Li Anshan, An Chunying and Li Zhongren. eds. 2009. *Zhong-Fei Guanxi yu Tangdai Shijie* (China-Africa Relations and the Contemporary World). Taiyuan: Chinese Society of African Historical Studies.

Li Anshan and Funeka Yazini April. eds. 2013. *Forum on China-Africa Cooperation: The Politics of Human Resource Development*. Pretoria: Africa Institute of South Africa.

Li Anshan *et al.* 2012. *FOCAC Twelve Years Later Achievements, Challenges and the Way Forward*, Uppsala: The Nordic Africa Institute.

Li Anshan *et al.* 2013. *Feizhou Meng: Tansuo Xiandaihua zi Lu* (African Dream: Search for Modernization). Jiangshu People's Press.

Li Baoping. 2007. 'Sino-Tanzanian Relations and Political Development', In *Afro-Chinese Relations*, edited by Prah, 126–141.

Li Baoping. 2008. 'Sulla questione della cooperazione tra Africa e Cina nel settore dell'istruzione', *Afriche e Orienti*, No. 2 (as part of Cristiana Fiamingo, ed., *La Cina in Africa*).

Li Baoping, 2011. *Chuantong yu Xiandai: Feizhou Wenhua yu Zhengzhi Bianqian* (Tradition and Modern: African Culture and Political Transformation). Peking University Press.

Li Baoping *et al.* eds. 2006. *Yafei Puyu Guojia Fazhang Yanjiu* (A Study on the Development of Portuguese-speaking Countries in Asia and Africa). World Affairs Press.

Li, Bojun. 2014. *Dangdai Feizhou Guoji Zhuzhi* (Contemporary African International Organizations). Hangzhou: Zhejiang People's Press.

Li Tongcheng and Jin Buoxiong. 2005. *Chinese Diplomats in Africa*. Shanghai: Shanghai People's Publishing House.

Li, Weijian. ed. 2008. *Beijing Summit & the Third Ministerial Conference of the Forum on China-Africa Cooperation: Appraisal and Prospects*. Shanghai Institutes for International Studies.

Li, Xiangyun. 2014. *Dangdai Tansangniya Guojia Fazhan Jincheng* (Tanzania State-Building and Development). Hangzhou: Zhejiang People's Press.

Li, Xiaoyun. ed. 2010a. *Zhongguo he Feizhou de Fazhan yu Huanpin* (Comparative Perspectives in Development and Poverty Reduction in China and Africa). China Financial and Economic Publishing House.

Li, Xiaoyun. ed. 2010b. *Zhongguo he Feizhou Fazhan Pinkun yu Jianpin* (Development, Poverty and Poverty Alleviation in China and Africa). China Financial and Economic Publishing House.

Li, Xinfeng. 2005. *Feizhou Ta Xun Zheng Ho Lu* (Following Zheng He's Footprints through Africa). Chenguang Press.

Li, Xinfeng. 2006. *Feifan Zhouyou-Wo Zai Feizhou Dang Jizhe* (An Unusual Journey Across the African Continent). Chenguang Press.

Li, Xinfeng. 2012. 'Feizhou Huaqiaohuaren Shuliang Yanjiu'. (A Quantitative Study of Chinese Overseas) *Chinese Overseas* 1–2: 7–12.

Li, Xinfeng *et al.* 2012. *Zheng He yu Feizhou* (Zheng He and Africa). China's Social Sciences Press.

Li, Yongcai. 2009. *Nanfei Wenxue Shi* (A History of South African Literature).

Li, Zhigang *et al.* 2009. 'An African Enclave in China: The Making of a New Transnational Urban Space'. *Eurasian Geography and Economics* 50(6): 699–719.

Li, Zhigang *et al.* 2012. China's 'Chocolate City': An Ethnic Enclave in a Changing Landscape. *African Diaspora* 5: 51–72.

Li Zhibiao. ed. 2000. *Feizhou Jingjiquan yu Zhonguo Qiye* (African Economic Zone and Chinese Enterprises). Beijing Press.

Li Zhibiao. ed. 2007. 'Contemporary Economic and Trade Relations between China and Africa'. In *Afro-Chinese Relations: Past, Present and Future*, edited by Kwesi Kwaa Prah, 280–293, The Centre for Advanced Studies of African Society.

Liang Yijian, 2012. 'Sustainable Development and Sino-African Low-carbon Cooperation: China's Role'. In *China-Africa Partnership*, edited by Shikwati, 40–45.

Lin, Justin Yifu, 2015. '"China's Rise and Structural Transformation in Africa": Ideas and Opportunities', In *The Oxford Handbook of Africa and Economics 2*, edited by Monga and Lin, 815–829.

Liu Haifang, 2006. 'China and Africa: Transcending "Threat or boon"', *China Monitor* March. Stellenbosch: The Centre for Chinese Studies.

Liu Haifang, 2008. 'China-Africa Relations through the Prism of Culture: The Dynamics of China's African Cultural Diplomacy'. *Journal of Current Chinese Affairs*.

Liu Haifang, 2015. 'Rising China, Foreign Aid and the World'. *China International Development Research Network Policy Recommendation* 7.

Liu Haifang, 2015a. 'FOCAC VI: African Initiatives Toward a Sustainable Chinese Relationship'. *China Monitor*. Stellenbosch: The Centre for Chinese Studies.

Liu Hongwu, 2012. 'New Impetus of African Development and New Path to Sustainable Development of China-Africa Relations'. In *China-Africa Partnership*, edited by Shikwati, 177–181.

Liu Hongwu and Bao Mingying. 2008, 2014. *Dongfei Siwaxili Wenhua Yanjiu* (Studies on Swahili Civilization in the East Africa). Hangzhou: Zhejiang People's Press.

Liu Hongwu and Luo Jianbo. 2011. *Zhong-Fei Fazhan Hezuo Lilun Zhanlue yu Zhengce Yanjiu* (Sino-African Development Cooperation: Studies on the Theories, Strategies and Policies). China's Social Sciences Press.

Liu Hongwu *et al.* 2014. *Niriliya Jianguo Bainian Shi (1914–2014)* (A Century History of Nigeria since its Foundation, 1914–2014). Hangzhou: Zhejiang People's Press.

Liu Hongwu and Li Shudi. eds. 2010. *Feizhou Yishu Yanjiu* (African Art Research). Yunnan People's Press.

Liu Hongwu and Shen Peili. eds. 2009. *Feizhou Feizhngfu Zhuzhi yu Zhong-Fei Guanxi* (The African NGOs and Sino-African Relations). World Affairs Press.

Liu Hongwu and Yang Jiemian. eds. 2009. *Fifty Years of Sino-African Cooperation: Background, Progress and Significance-Chinese Perspectives on Sino-African Relations*. Yunnan University Press.

Liu Hongwu and Huang Meibo *et al.* 2013. *Zhongguo Duiwai Yuanzhu yu Guoji Zeren de Zhanlue Yanjiu* (A Study on the Strategy of Chinese Foreign Aid and International Responsibility). China's Social Sciences Press.

Liu Yan. 2014. *Houzhimin Shidai Feizhou Jiaoyu Gaige Moshi Yanjiu* (A Study of African Education Reform Modes in the Postcolonial Era). Hangzhou: Zhejiang People's Press.

Lou Shizhou. 2014. *Saineijiae Gaodeng Jiaoyu Yanjiu* (Studies on High Education in Senegal). Hangzhou: Zhejiang People's Press.

Lu Miaogeng, Huang Shejiao and Lin Ye, eds. 2006. *United Hearts as Gold.* Beijing: World Affairs Press.

Lu Ting-en. 2005. *Feizhou Wenti Lunji* (Treatises on Africa). World Affairs Publishers.

Luo Gaoyuan. 2010. *Dandai Feizhou Lüyou* (Contemporary Tourism in Africa). World Affairs Press.

Luo Jianbo. 2006. *Feizhou Yitihua yu Zhong-Fei Guanxi* (African Integration and Sino-African Relations). Social Sciences Academic Press.

Luo Jianbo. 2010. *Tongxiang Fuxing zi Lu: Feimeng yu Feizhou Yitihua Yanjiu* (The Road to Renaissance: Studies on the African Union and the African Integration). Chinese Social Sciences Press.

Ma Enyu. 2012. 'Yiwu Mode and Sino-African Relations'. *Journal of Cambridge Studies* 7(3): 93–108.

Mancuso, S. and Hong Yonghong. 2009. *Zhongguo Dui Fei Touzhi Falü Huanjing Yanjiu* (Research on Legal Environment for Chinese Investments in Africa). Xiangtan University Press.

Mo Xiang. 2014. *Dangdai Feizhou Anquan Jizhi* (The Research on Security of Contemporary Africa). Hangzhou: Zhejiang People's Press.

Monga, Célestin and Justin Yifu Lin. eds. 2015. *The Oxford Handbook of Africa and Economics*, Vols. 1–2. Oxford University Press.

Mu Tao, 2003. *Nanfei Duiwai Guanxi Yanjiu* (A Study on South African External Relations). East China Normal University Press.

Mu Tao and Yu Bin. 2013. *Musaiweini Zhongtong yu Wuganda* (Y.K. Museveni: President of the Permanent Snow on the Equator and the Pearl of Africa-Uganda). Shanghai Dictionary Press.

Nyerere, Julius Kambarage, 2015. *Nileier Wenxian* (Selected Works of Julius Kambarage Nyerere). East China Normal University.

Pang Zhongying, 2013. 'The Non-interference Dilemma: Adapting China's Approach to the New Context of African and International Realities', In *China-Africa Relations: Governance*, edited by Berhe Mulugeta Gebrehiwot and Liu Hongwu, 46–54.

Prah, Kwesi Kwaa. ed. 2007. *Afro-Chinese Relations: Past, Present and the Future*, CASAS Publisher.

Qi Jianhua. 2014. *Fazhan Zhongguo yu Feizhou Xinxing Quanmian Hezuo Guanxi* (Developing China-Africa Partnership of New and Comprehensive Cooperation), World Affairs Press.

Qian Qichen. 2003. *Weijiao Shiji* (Ten Stories of a Diplomat). World Affairs Press.

Qian Qichen. ed. 2005. *Shijie Waijiao Da Cidian* (Dictionary on World's Diplomacy), 2 Vols. World Affairs Press.

Qin Dashu and Yuan Jian. 2013. *Gu Sichou Zhilu* (Ancient Silk Road). Singapore.

Qin Hui. 2013. *Nanfei de Qishi* (South Africa's Revelation). Jiangsu Literature and Art Publishing House.

Shelton, Garth, Funeka Yazini April and Li Anshan. eds. 2015. *FOCAC 2015: A New Beginning of China-Africa Relations.* Pretoria: Africa Institute of South Africa.

Shen Xiaolei, 2015. 'Shixi Zhongguo Xinyimin Rongru Jinbabuwei de Kunjing' (Dilemma of Chinese New Immigrants' Integration in Zimbabwe). *International Politics Quarterly* 5.

Shi Lin, 2012. 'The Ethnographic Study of the Contemporary Africa from the Perspective of China', In *China-Africa Partnership*, edited by Shikwati, 104–109.

Shi Yongjie, 2015. *Tuchu Baowei de Qiangguo Zhilu: Xin Xingshixia de Zhong-Fei Jingmao Hezuo Zhanlue Yanjie* (Strengthening the Nation Through Breaking the Siege: A Study on China-Africa Economic and Trade Cooperation in the New Situation). China Commerce and Trade Press.

Shikwati, James. ed. 2012. *China-Africa Partnership: The quest for a win-win relationship*. Nairobi: Inter Region Economic Network.

Shu Yunguo. 2012. *Feizhoushi Yanjiu Rumen* (An Introduction to African History). Peking University Press.

Shu, Yunguo. 2014. *Fanfei Yundong Shi 1900–2002* (History of Pan-Africanism 1900–2002). The Commercial Press.

Shu, Yunguo and Liu Weicai. eds. 2013. *20th Shiji Feizhou Jingji Shi* (The Economic History of Africa in 20th Century). Zhejiang People's Press.

Sun, Hongqi. 2008. *Zhiminzhuyi yu Feizhou Zhuanlun* (On Colonialism and Africa).

Sun, Hongqi. 2011. *Tudi Wenti yu Nanfei Zhengzhi Jingji* (Land Problem and the Political Economy of South Africa).

Sun Xiaomeng. 2004. *Rubutaccen Wasan Kwaikwayo a Rukunin Adabin Hausa: Muhimmancinsa da habakarsa*, (A Study on Written Hausa Drama). MA thesis, Ahmadu Bello University.

Sun Xiaomeng. 2014. *Yuyan yu Quanli: Zhimin Shiqi Haosayu zai Bei Niriliya de Yunyong*, (Language and Power: The Application of Hausa in Northern Nigeria during the British Administration), Beijing: Social Sciences Academic Press.

Tang Jiaxuan, ed. 2000. *Zhongguo Waijiao Cidian* (Dictionary on China's Diplomacy), Beijing: World Affairs Press.

Tang Xiaoyang. 2012. 'African Regional Integration and Sino-Africa Cooperation: Opportunities and Challenges'. In *China-Africa Partnership*, edited by Shikwati, 13–19.

Tang Xiaoyang, 2014. *China-Africa Economic Diplomacy and Its Implication to Global Value Chain*. World Affairs Press.

Tang Xiaoyang, (with Deborah Brautigam). 2014. 'Going Global in Groups: China's Special Economic Zones Overseas'. *World Development* 63: 78–91.

Tang Xiaoyang. 2014a. 'Models of Chinese Engagement in Africa's Extractive Sectors and Their Implications'. *Environment: Science and Policy for Sustainable Development* 56 (2): 27–30.

Tang Xiaoyang. 2014b. *The Impact of Asian Investment on Africa's Textile Industries*, Carnegie-Tsinghua Center for Global Policy.

Tang Xiaoyang. 2014c. 'Investissements chinois dans l'industrie textile tanzanienne et zambienne'. *Afrique Contemporaine* 250: 119–136.

Tang Xiaoyang, (with Jean-Jacques Gabas). 2014. 'Coopération agricole chinoise en Afrique subsaharienne'. *Dépasser les idées reçues*. N° 26. CIRAD, Montpellier, Perspective, 4. http s://hal.archives-ouvertes.fr/hal-01538131/document.

Wan Xiulan, Li Wei *et al.* 2014. *Bociwana Gaodeng Jiaoyu Yanjiu* (Studies on Higher Education in Botswana). Hangzhou: Zhejiang People's Press.

Wang Haili, 2014. *Aiji Tongshi* (The History of Egypt). Shanghai Academy of Social Sciences Press.

Wang Shu. 2007. *Stories of Five Continents*, Shanghai: Shanghai Dictionary Publications.

Wang Xuejun. 2013. 'The Corporate Social Responsibility of Chinese Oil Companies in Nigeria: Implications for the Governance of Oil Resources'. In *China-Africa Relations: Governance*, edited by Mulugeta Gebrehiwot Berhe and Liu Hongwu, 128–145.

Wei Jianguo. 2011. *Africa: A Lifetime of Memories - My Experiences and Understanding of Africa*. Beijing: China Commerce and Trade Press.

Xia Jisheng. 2005. *Feizhou Liang Guo Yihui* (Two African Parliaments). China's Financial and Economic Press.

Xiao Hongyu. 2014. *Feizhou Titihua yu Xiandaihu de Hudong* (Interactive Nature of Modernization and Integration in Africa: The Case of Regional Integration in West Africa). Social Sciences Academic Press.

Xu Liang and Emmanuel Akyeampong. 2015. 'The Three Phases/Faces of China in Independent Africa: Re-conceptualizing China-Africa Engagement'. In *The Oxford Handbook of Africa and Economics Volume 2: Policies and Practices*, edited by Celestin Monga and Justin Yifu Lin, 762–779. Oxford: Oxford University Press.

Xu Liang. 2015a. 'Historical Lessons, Common Challenges and Mutual Learning: Assessing China-Africa Cooperation in Environmental Protection.' In *FOCAC 2015*, edited by Shelton and Li, 425–445.

Xu Tao. 2013. *Zaihua Feizhou Shangren de Shehui Shiying Yanjie* (The Social Adaptions of African Merchants in China). Hangzhou: Zhejiang People's Press.

Xu Wei. 2011. 'Bociwana Yeyiren Tianye Diaocha Jiqi Guojia Jiangou zi Shikao'. (Fieldwork on the Wayeyi of Botswana and Thought on State-building). *African Studies* 1.

Xu Wei. 2014. *Bociwana Zuqun Shenghuo yu Shehui Bianqian* (Ethnicity, Everyday Life and Social Change in Botswana). Hangzhou: Zhejiang People's Press.

Xu Yongzhang. 2004. *Zhongguo yu Ya Fei Guojia Guanxi Shi Kao Lun* (Research on the History of Relations between China and Asia-African Countries). Hong Kong Press for Social Sciences Ltd.

Xu Yongzhang. 2014. *Feizhou Wu-shi-si Guo Jianshi* (A Brief History of Fifty-four African Countries). Hangzhou: Zhejiang People's Press.

Yang Guangsheng. 2014. *Niriliya Duiwai Guanxi Yanjiu* (A Study on the Foreign Relations of Nigeria). Hangzhou: Zhejiang People's Press.

Yang Lihua. 2006. 'Africa: A View from China.' *South African Journal of International Affairs* 13(1): 23–32.

Yang Lihua *et al.* 2013. *Zhongguo yu Feizhou Jingmao Hezuo Fazhan Zhongti Zhanlue Yanjiu* (A Comprehensive Strategic Study on Development of China-Africa Economic Cooperation). China's Social Sciences Press.

Yang Tingzhi. 2015. *Zanbiya Qiuzhang Zhidu de Lishi Bianqian* (On the Historical Evolution of Zambian Chieftaincy). China Social Sciences Press.

Yang Xuelun and Zheng Xizhen. 2001. *Tu-ni-si Wenhua* (The Culture of Tunisia). Culture and Arts Press.

Yao Guimei and Chen Zongde. eds. 2000. *Geguo Feizhou Nongye Gaikuang* (The Series of Development of African Agriculture), I and II. China Financial & Economic Publishing House.

Zeng Qiang, 2002. 'Some Reflections on Expanding Sino-African Trade and Economic Cooperative Relations in the New Century (The Viewpoint of a Chinese Scholar)'. *TINABANTU: Journal of African National Affairs* 1(1), May.

Zeng Qiang, 2010. 'China's Strategic Relations with Africa'. In *Chinese and African Perspectives*, edited by Harneit-Sievers et al, 56–69.

Zhai Fengjie, Wang Yuhua and Pan Liang. 2013. *Feizhou Yitihua Beijing Xia de Zhong-Fei Hezuo* (China-Africa Coopertion in the Integration of Africa). World Affairs Press.

Zhang Chun. 2012. 'China's Engagement in African Post-Conflict Reconstruction: Achievements and Future Developments'. In *China-Africa Partnership*, edited by Shikwati, 55–62.

Zhang Jing. 2014. *Feizhou Quyu Jingji Yitihua Tansuo: Nanbu Fezhou Fazhan Gongtongti 30 years* (Regional Economic Integration in Africa: Thirty years of Southern African Development Community). Hangzhou: Zhejiang People's Press.

Zhang Haibin. 2013. *Fazhan Yingdaoxing Yuanzhu: Zhongguo Dui Fei Yuanzhu Moshi Yanjiu* (Development-Guided Assistance: A Study on the Model of Chinese Aid to Africa).

Zhang Hongming. 1999. *Duowei Shiye zhong de Feizhou Zhengzhi Fazhan* (African Political Development in Multiple Perspective). Social Sciences Academic Press.

Zhang Hongming. 2008. *Jindai Feizhou Sixiang Jingwei: 18–19 Shiji Feizhou Zhishi Fenzi Sixiang Yanjiu* (A Study on the Thoughts of African Scholars: 18th and 19th Centuries). Social Sciences Academic Press.

Zhang Hongming. ed. 2012. *Zhongguo he Shijie Zhuyao Jingjiti yu Feizhou Jingmao Hezuo Yanjiu* (China and World Major Economies' Economic and Trade Cooperation with Africa). World Affairs Press.

Zhang Hongming. ed. 2012b. *Feizhou Huangpishu No.14 2011–2012* (Yellow Book of Africa No.14 2011–2012). Social Sciences Academic Press.

Zhang Hongming, Liu Lide and Xu Jiming. 2001. 'Focus: Sino-African Relations'. *Africa Insight* 31(2): 33–42.

Zhang Qiaowen. 2015. 'China Africa Development Fund: Beyond a Foreign Policy Instrument'. *CCS Commentary*, January 13.

Zhang Qiaowen. 2015a. 'Responsible Investing in Africa: Building China's Competitiveness.' CCS Commentary, February 18.

Zhang Xiangdong. 2012. *Aisaiebiya Lianbangzhi: 1950–2010* (A Study on Ethiopia's Federalism: 1950–2010). China Economic Publishing House.

Zhang Xinghui. 2011. 'China's Aid to Africa: A Challenge to the EU?' In *China and the European Union in Africa*, edited by Jing Men and Benjamin Barton, 209–224.

Zhang Yonghong, Liang Yijian, Wang Tao and Yang Guangsheng. 2014. *Zhong-Fei Ditan Fazhan Hezuo de Zhanlue Beijing Yanjiu* (A Study on the Background of China-Africa Low-Carbon Development Cooperative Strategy). World Affairs Press.

Zhang Yongpeng. 2007. 'Reality and Strategic Construction: Globalisation and Sino-African Relations'. In *Afro-Chinese Relations*, edited by Prah, 268–279.

Zhang Yongpeng. 2012. *Guoji Fazhan Hezuo yu Feizhou: Zhongguo yu Xifangyuanzhu Feizhou Bijiao Yanjiu* (International Development Cooperation and Africa: A comparative Study on Chinese and Western Aid to Africa). Social Sciences Academic Press.

Zhang Zhe. 2014. *Zhong-Fei Jingmao Guanxi Fazhan Yanjiu* (The Development of China-Africa Economic and Trade Relations). Hangzhou: Zhejiang People's Press.

Zhen Feng, 2014. *Feizhou Gangkou Jingji yu Chengshi Fazhan* (African Port Economy and Urban Development). Nanjing University Press.

Zheng Jiaxing. 2000. *Zhiminzhuyi Shi* (History of Colonialism: Africa). Peking University Press.

Zheng Jiaxing. 2010. *Nanfei Shi* (A History of South Africa). Peking University Press.

Zheng Jiaxing. 2011. *Yifang Suitu Yangyu Yifang Wenming: Feizhou Wenming ZiLu* (The Path of African Civilization). People's Press.

Zhi Yingbiao and Bai Jie. 2010. 'The Global Environmental Institute: Regulating the Ecological Impact of Chinese Overseas Enterprises'. In *Chinese and African Perspectives*, edited by Harneit-Sievers, 247–254.

Zhong, Zhicheng. 2006. *For a More Beautiful World: Record of Jiang Zemin's Visit Abroad*. Beijing: World Affairs Publishers.

Zhou, Hong. ed. 2013. *Zhongguo Yuanwai 60 Nian* (China's Foreign Aid 60 Years in Retrospect). Social Sciences Academic Press.

Zhou, Nanjing, ed. 1999–2002. *Encyclopedia of Chinese Overseas*, 12 volumes. Beijing: Chinese Overseas Publishing House.

Zhu, Huayou *et al.* 2014. *Dangdai Feizhou Gongkuangye* (Contemporary Industry and Mining in Africa). Hangzhou: Zhejiang People's Press.

Zhu, Weidong. 2011. *Nanfei Guoji Maoyi Falu ZHidu Zhuanti Yanjiu* (A Specific Study on Legal System of International Trade in South Africa). Xiangtan University.

4

THEMES AND THOUGHTS IN AFRICANISTS' DISCOURSE ABOUT CHINA AND AFRICA

Kweku Ampiah

In September 2011 I was riding in a taxi that was caught up in irritating traffic on Oxford Street in Osu. Among the pedestrians on one side of the road I spotted two men in their late twenties strolling along with much excitement; they were animated, talking, laughing and seemingly teasing each other, oblivious to their surroundings. The men were dressed no differently from the rest of the crowd. I was intrigued by the energy between them, one East Asian, the other African and, in all likelihood, a Ghanaian. Even more strikingly, as I discovered when they got closer to the taxi I was in, they were speaking in Mandarin. The world, some would divine, is getting smaller.

Later that evening at a Chinese restaurant, a conversation with the proprietor revealed how dissatisfied he was with his Ghanaian workers, and conversely, the workers told me, how unrewarding it was working at the restaurant. There was some uneasiness, the result of a mutual lack of trust between the employer and his local employees.

In early 2013 I encountered a sprightly young Chinese man in his early thirties at a hotel in Niamey, Niger. He had done a degree in French, "because I couldn't get my head around English," he explained, as if to apologise for his poor English. The company he worked for also operated in most of the countries in West Africa, but he had been given the responsibility of Francophone countries, specifically to promote the sale of green tea in the region. He told me his primary target was women. Apparently, weight-conscious West African women with big wallets were consuming green tea for weight loss, and the Chinese company was determined to grow the market.

At the Inaugural Association of Asian Studies in Africa conference in Accra, in September 2015, I encountered another variant of the Chinese presence in Africa. A panel I chaired involved a Chinese PhD student who had been conducting research in Ghana. In his presentation he whipped himself into a frenzy when he spoke about the case of illegal Chinese gold prospectors in Ghana, incensed by accusations levelled against the Chinese by Ghanaian and international media. Indeed, in 2014 the Government of Ghana deported over 4,000 Chinese illegal immigrants, most of whom had been prospecting for gold illegally and causing many environmental hazards, including water pollution, in the villages where they worked (Afua Hirsch, "Ghana Deports Thousands in Crackdown on Illegal Chinese Goldminers." *The Guardian*, July 15, 2013).

These accounts reflect how intertwined and yet how expansive and diverse the evolving relations between China and Africa have become since the early 2000s. For example, while there were only a handful of Chinese restaurants in Accra in 2004, today people speak of the ten best Chinese restaurants in a city that now has such restaurants sprouting like bamboo shoots. There are Chinese nationals selling cooked meals across the length and breadth of Accra, including in the slum areas of the city. Thus, there is a need for Africanists and scholars with an interest in Sino-African relations to start appreciating the diversity of Chinese actors in Africa and their accompanying narratives and ambitions, juxtaposed with the aspirations of the African countries on the one hand, and the anxieties of the region's traditional partners about China's growing importance in Africa on the other.

This chapter evaluates the discourse regarding China's role in Africa's recent socio-economic development, and maps out how Africanist scholars have discussed and interrogated the unfolding relations between the relevant stakeholders. It notes how the drivers of the "China in Africa" discourse approached the debate largely from a Eurocentric perspective, igniting a counter-discourse especially involving African researchers. Since the latter part of the 2000s, assessments of the evolving role of China in African political economy have thus included more research and contributions by Africans (including those in the diaspora), which are largely supportive of China's growing presence in Africa. While acknowledging the economic content of the relationship as manifestly important to the discourse, the following analysis also evokes the important role of culture in the relationship between China and Africa. Thus the discussion is designed to exhort Africanist scholars to transcend the perennial focus on trade and development-based scholarship that is too often analyzed in the discipline of IR, and to engage with and interrogate the underlying socio-cultural imperatives of the China–Africa matrix, the circulation of knowledge engendered through that medium and its power implications, as Shanshan Lan has done in her 2017 publication. The first part of the following discussion explores how Africanists and other researchers have evaluated the evolving relationship so far, and the second part attempts to invoke the cultural content of the relationship that has been largely missing from the discourse (Sautman 1994; Monson and Rupp 2013; Zhou et al. 2016).

Terms of engagement in the Africanist debate about China–Africa relations

The general discourse regarding China–Africa relations is dominated largely by analysis of its economic content (Taylor 2008), although the discussion is dictated by the political connotations of the relationship, and invariably by Africa's place and roles in the global political economy. Consequently the discussion tends to focus on the following themes: China's interest in, and exploitation of, Africa's natural resources; the dumping of Chinese manufactured products on Africa; infrastructure and public works projects (constructed through Chinese government and private sector funding and investment); and Chinese government loans to African governments. The political implications of these engagements, which tend to be correlated to Beijing's principle of "non-interference" and its influence on African countries (Taylor 2008; Halper 2010; Mbaye 2011; Thompson 2012), are also a constant feature of the discourse. The analysis is often infused with narratives about scandals originating from concerns about "deals" between the donor and the recipient, respectively.

Research about the social implications of the growing presence of Chinese migrants in Africa is limited but growing (Park 2009; Mohan and Tan-Mullins 2009; Mohan et al. 2013). In the meantime, such factors are being chronicled in local popular discourse throughout the length and breadth of Africa. In one instance, for example, Rei Helder, the Angolan semba and kizomba musician, sings about how rich Chinese men are courting Angolan women at the expense of local men.[1] On the other hand, the music video also metaphorically suggests that even though Beijing is the more powerful part of the partnership between China and Angola, Luanda has nevertheless managed to seduce China. The video shows an East Asian couple dancing semba, Angola's popular music genre, an indication of how China's presence in Africa is impacting on the social lives of both Africans and Chinese. Martha Saavedra's work (2009) on how Africa is represented in a Hong Kong soap opera also suggests there are unique insights that can be gained from such media to help enrich our understanding of the dynamics of China's growing influence in Africa.

Raw materials, loads of money, and corrupt deals

The discourse's obsession with the narrative of China as a threat is in part attributable to empirical factors such as the following. According to Chinese Ministry of Commerce indicators, the net annual flow of Chinese direct investment into Africa increased eightfold between 2005 and 2014 to $3.2 billion, and the total stock of Chinese investments grew twentyfold to $32 billion over the same period, making China one of the biggest investors in Africa (*The Economist*, "Chinese Investment in Africa: Not As Easy As It Looks," November 21, 2015). The analysis recognises the extent of Beijing's involvement in development projects across Africa at least since early 2000. The response of the West to this, both in the media and in academic discussions, is largely characterised by a sense of unease, not least because, as *The Economist* (ibid.) notes, "Western policymakers worry about whether they can compete with a flood of Chinese cash" in Africa. The fact that the Chinese contribution to the region's economic development has been mixed, tainted in places by "failed project[s]," "… evidence of the mixed success of China's scramble for

Africa," has tempered those anxieties to some extent. There is a sense of relief amongst China's competitors in Africa that after all "Chinese firms that leap into Africa are struggling with the same problems Africa has long given Western investors." In other words, the Chinese do not have a "unique ability to make their projects in Africa work where Western firms cannot" (*The Economist*, ibid.).

The earlier discourse, mainly led and dominated by Anglo-American observers, reflected an exaggerated sense of panic (Kristof 1993; Roy 1994; Shambaugh 1996) and a tone that was, and remains, deeply critical of the form and content of the relationship between China and the African countries (Taylor 2006, 2006; French 2007). In a tone reflecting the West's hysteria regarding the rise of China, Howard French's 2007 essay in *African Affairs* encapsulates the popular narrative about the perceived "neo-colonialist" "invasion" of Africa by Beijing. Between 2001 and 2005, according to French, China

> surpassed the United Kingdom to become the continent's third leading commercial partner, after the United States and France. Just as suddenly, the Chinese can be seen almost everywhere in Africa. With about 100,000 nationals living and working on the continent, the newcomers have become more numerous than Britons in places such as Nigeria, London's most prized West African possession.
>
> *(French 2007, 127)*

In general the drivers of the discourse reiterated how China "has challenged Western pre-eminence in a region that had long served as Europe's 'chasse garde' and, for the United States, as an increasingly important source of its energy needs." As a global manufacturer and trader of "essential commodities" China has thus become the envy of the Western world, not least because it questions the relevance of the West in the region, as indicated by "the West's putative marginalisation on the continent" (Alden et al. 2008, 23).

One 2008 essay noted how China's rise in Africa was often equated with the advance of a fire-spitting "dragon in an unvariegated African bush stripped of historical and political content" (Large 2008, 46). The flipside of this imagery in the popular discourse depicts China as a pied piper followed by the unsuspecting, infantile Africa, implying that Africa should be saved from its naiveté. Convinced that researchers attracted to the topic were on the cutting edge of understanding China's new role in Africa, Daniel Large rallied them to develop a culture of serious research beyond their then-current "dragon in the bush" preoccupations, and to engage with a complex subject that was about to enter the mainstream discourse around the political economy of Africa. Effectively, he challenged Africanists and researchers with an interest in China–Africa relations to get off the safari tour and engage more seriously in forensic research about the different facets of Beijing's involvement in the region, including the African agency in the engagement.

The challenge was taken up by some researchers, while another pioneer involved in the discourse also had good reason to reappraise his focus, analysis and approach. Consequently, Ian Taylor's *China's New Role in Africa* (2008) addresses specific issues such as "Oil Diplomacy," the impact of "Cheap Chinese Goods" on Africa's manufacturing sector, and "The Arms Trade," in contrast to the rather broad country themes in his 2006 publication. However, the analyses and insights in this book evoke the same desperate concerns about China's influence on Africa, even as it attempts to inject nuance into the discussion. For example, it posits that "The PRC's policies arguably jar with Africa's increasing attempts [by which it meant the West's initiatives] to promote human rights and good governance, as crystallised in NEPAD" (p. 98). This refers, of course, to Beijing's cosy relationship with autocratic regimes, such as the governments of Sudan and Zimbabwe. China is therefore complicit in Africa's failing attempts to embrace and implement the principles of human rights in so far as it supports regimes that stifle these rights. On the other hand, the assessment suggests that allegations about China's negative impacts on Africa "Upon close inspection ... appear far much less salient and accurate" (p. 179). This is reinforced with the argument that allegations against China "... are often balanced out in any case by positive impacts." An interesting thought! Yet it is difficult to understand how this is accounted for empirically, or how the balance is divined. It is worth noting that through correlation analysis, for example, it should be possible to ascertain how, and the extent to which, China's accommodation and support of the interests of autocratic African governments "... undermin[e] opportunities to promote good governance" (p. 23) in the region, beyond mere speculation.

Another publication from 2008 that also grasped the nettle in the discourse's attempt to focus on specific issues is *China Returns to Africa: A Rising Power and a Continent Embrace*, edited by Chris Alden et al. In this volume, Elizabeth Hsu's study of Chinese medicine in Tanzania provides interesting insights into the diversity of the Chinese private sector in Africa, as does Deborah Bräutigam's chapter, "'Flying Geese' or 'Hidden Dragon': Chinese Business and Africa's Industrial Development." The latter discusses China's potential as a stimulus for Africa's industrial growth, implicitly likening China's role in Africa to Japan's in East Asia as a driver of the region's industrialisation, based on the Flying Geese paradigm developed by the Japanese economist Kaname Akamatsu in the 1930s. Bräutigam stands out as one of the leading researchers of Africa–China dynamics who explores the positives in the relationship (2009). Indeed she does so without losing sight of the fact that the Chinese presence in the region, including in the form of joint ventures in Tanzania, Angola, and Uganda, is also facing extreme competition from Asian exports, which are potentially undermining fledgling manufacturing ventures in these African countries. That notwithstanding, Irene Yuan Su (2017) is very optimistic about Chinese investments in Africa as a catalyst for the region's industrialisation.

The sustainability of these African manufacturing ventures, as Bräutigam notes, will depend on "the degree to which [African] governments establish a more conducive environment for business," including the provision of the basic infrastructure, physical and non-physical, required for growth. Industrial policy, the "concern with the structure of domestic industry and with promoting the structure that enhances the [country's] international competitiveness" (Johnson 1987, 19), may also be imperative in the process, essentially to nurture and support fledgling manufacturing firms, in the short- and medium-term, to become more competitive.

An important deviation from the above economic and development-oriented themes is Stephanie Rupp's piece in another 2008 edited publication because it touches on the cultural content of the Africa–China relationship. According to Rupp (2008, 77): "Social relations between ordinary Africans and Chinese are marked by a tension between mutual admiration and mutual loathing." Her explanation for Chinese admiration of Africans, however, hardly adds up. How does one process Chinese admiration for the majesty of African animals as admiration for Africans, when the backwardness of Africa as understood in the Chinese conception of modernity is invariably linked to the animals that apparently roam free in Africa? It should be noted that the popular narrative regarding Zheng He's voyage to East Africa, which according to Sverdrup-Thygeson (2017) is being adroitly used by Beijing as a tool in China's engagement with Africa, is incomplete without the poignant image of Africa's tallest mammal, the giraffe, *on a rope and led triumphantly by a diminutive Ming Dynasty attendant.* Symbolically this suggests a civilizing mission. Given the asymmetrical nature of the contemporary relationship between China and the African countries, it is difficult not to conclude that despite the rhetoric about "mutual cooperation" China sees itself as leading Africa out of darkness. Referring again to Rupp's point above, how does a society that is desperately copying all things European (Liu 2008, 16), which in the modernist tradition is conceptually the extreme opposite of everything that is African, not inherit and internalise the arrogance that epitomises Europe's attitude toward Africa? I will return to the issue of the cultural content of the relationship below.

The 2008 volume edited by Ampiah and Naidu and published by the University of Kwazulu Natal emerged from concerns among some Africanists that the African perspective on China's engagement with Africa was marginalised in a discourse that was largely dominated by American, British, and European researchers and media outlets (Ampiah and Naidu 2008). Thus the publication was developed with a vision to inject African voices into the narrative, and, as with the other publications that highlighted country case studies, it aimed to debunk representations of Africa as a single construct vis-à-vis China, as well as to assess African agency in the relationship.

By 2009 the discourse in the West had begun to recognise the need to deconstruct the narrative of China as a monolithic entity in Africa and to tease out the variegated interests, many of them conflicting, that guide China's relations with Africa. This was to some extent articulated in *The China Quarterly*'s special issue on China and Africa (see Strauss and Saavedra 2009). While highlighting concerns about China's predatory advances toward Africa, this collection includes articles that hint at the proliferation of separate Chinese actors in Africa, all of whom carry with them their bespoke ambitions and problems, beyond the control of the central Chinese government. Ching Kwan Lee complements this with an analysis that questions the concept of a "China Incorporated" that imposes its interests on the

African countries (Lee 2009, 665), a point that is reinforced in Ching Kwan Lee's work (2017). Indeed, the Chinese migrants involved in the illegal gold rush in Ghana and in polluting rivers and farmland in Dunkwa and other regions, while toting machine guns to defend themselves and their loot against the locals who oppose them, are as divorced from Beijing's policies toward Ghana as they are from the communities they have "invaded." In essence, it is futile to continue evoking China's official five principles of peaceful coexistence in a discussion of every Chinese presence and initiative outside China, whether in Dunkwa, Niamey or Carlyle.

The counter-discourse revealed an interesting Africanist dimension to China's expanding interests in sub-Saharan Africa, which is that contrary to the alarmist portrayal of a correlation between Chinese aid and resource interests, there is no obvious link, for example, between China's expansive assistance toward Ethiopia's economic development and resource exploitation. This is not least because "Ethiopia does not produce raw materials critical to China," nor does Addis Ababa have large-scale bilateral trade with Beijing. That is one of the conclusions of *China's Diplomacy in Eastern and Southern Africa*, edited by Seifudein Adem (2013). Adem assesses China's interest in Ethiopia in the context of the latter's diplomatic capital as the center of Africa's international relations, and partly attributes it to the charisma of Ethiopia's president, Meles Zenawi, an alchemist in models of development, and an ardent supporter of the East Asian development model (Ohno and Ohno 2013). This builds on the developmentalist content of the evolving arrangements between China and the host African countries, which Bräutigam has endeavoured to inject into the discourse. Complementing this is Adams Bodomo's defiant position that "China's re-engagement with Africa … constitutes a paradigm shift that has completely altered the investment stratosphere …" (2017, 35–36).

The alarmist trope persists, as evident in Howard French's *China's Second Continent: How a Million Migrants Are Building a New Empire in Africa* (2014). The publication is a catalogue of corrupt activities perpetrated by the Chinese and their African agents, mostly elites. French quotes an informant as saying that the president of Mozambique, Armando Guebuza, took over the country's logging and land deals because, as he apparently explained to his family, "he has two more years in office, so this is the time [for him] to make money" (French 2014, 215). Much earlier, Stephen Ellis and ter Haar audaciously argued that with Africa, "the essence of rumours is that they are credible" (2004, 28). Thus, the defining truth in African political economy, as is so often the case, is divined through rumour and gossip because rumour is good enough evidence in African studies, apparently. In a more nuanced assessment, however, Bräutigam (2015) commented on the myths about Chinese investments in Africa, and the tendency to dramatically overestimate the size of Chinese loans and grant aid to Africa.

The cultural imperative in the China–Africa matrix

Despite advances in research the Africanist discourse regarding China's relations with the African countries remains full of gaps, two of which I will speak to in the following discussion. The first concerns the fact that there is minimal consideration in the discourse of Beijing's policymaking process in regards to the African countries, either in respect of its bilateral relations or its multilateral initiatives conducted through the Forum on China Africa Cooperation (FOCAC). Nor does current research contain insights into the potential bureaucratic interests and conflicts in Beijing regarding Africa, which poses the question of how consensus is arrived at in Beijing regarding its approaches toward the African countries. More specifically, for example, what, if any, were the conflicts of interests between the Ministry of Finance and the Ministry of Foreign Affairs or the Ministry of Commerce regarding China's relations with Angola or the Democratic Republic of Congo? Furthermore, while the Ministry of Foreign Affairs might well endorse the ambitions of Hanban (Office of Chinese Language Council International), to grow the number of Confucius Institutes in Africa, how is this negotiated in budgetary terms? In other words, how does the Ministry of Finance respond to such ambitions, or are all decisions simply taken by the Political Bureau of the CPC Central Committee by way of a simple dialogue over green tea? The answer is that we do not know. The contribution of Chinese scholarship to the discourse provides little insight into these issues because Chinese researchers, either in China or elsewhere, do not seem to have any understanding about what takes place within policymaking circles or in the corridors of bureaucratic power in Beijing,[2] much less what goes on within the Politburo about Africa. Aside from that, the lack of critical commentary in the research by Chinese scholars about China's role in Africa is striking; as Stephen Chan has noted, "Chinese academic research in Africa can be extremely superficial and parochial" (2013, 7). Indeed, we cannot fail to acknowledge that Chinese scholars are by and large agents and advocates of the state's policies, as is evident in the 2014 edited volume, *China–Africa Relations: Review and Analysis* (Zhang, 2014). Informative though the essays in the volume may be, they are all devoid of any critical assessment of the evolving relations between China and Africa. It is well known that as a developmental state the PRC is not a fan of political satire, so Chinese scholars may be sensibly avoiding provoking the state with critical comments about China's role in Africa. On the other hand, Chinese researchers might not critically evaluate China's role in Africa for reasons of nationalism, which recalls a presentation by a Chinese researcher on a panel I chaired in Accra in September 2015, which I will discuss below.

Taking advantage of my position as Chair of a panel at the Inaugural "Africa–Asia: A New Axis of Knowledge" Conference in Accra, in a conversation with the Chinese PhD researcher I mentioned above, I asked whether he was an employee of the PRC or an independent researcher. It was a rhetorical question that I was forced to ask because all too often Chinese researchers and academics behave as though they are representing the PRC. As mentioned earlier, this student had spoken emotively about the criticisms that had been levelled at the illegal Chinese gold prospectors in Ghana. During the course of our conversation we both agreed that as an independent researcher he should at least try to be less emotional in his academic presentations. However, he claimed that his reaction was "cultural," by which I presume he meant that as a Chinese citizen, he is intuitively obliged to defend

China in every situation. In short, he evoked culture as an attempt to legitimise feelings of nationalism.

That brings me to the other gap in the discourse I wish to address, which concerns the lack of research about the cultural content of the relationship between China and African countries, and which invariably brings into perspective the role and function of Confucius Institutes (and Confucius Classrooms) in Africa. The Chinese government believes in the efficacy of culture as a medium of public diplomacy (Liu, 2008), as is evident in the prominent role of the Confucius Institute (CI) in China's engagement with the global public. As of 2015, there were 500 CIs worldwide, representing half of the 2020 target of 1,000. Of the CIs already established, 42 are in 30 sub-Saharan African countries, which is indicative of China's resolve to mold African minds in its favour.[3] Thus the CI serves as a tool in China's bold engagement with the African countries (Procopio 2015). Its functions include teaching Chinese language, and propagating cultural norms such as etiquette, aesthetics, games, and food culture, including the tea ceremony and use of chopsticks. The functions of the CI, however, prompt questions about cultural asymmetry and mutual respect between Africans and the Chinese in relation to which the Zambian political analyst Sishuwa Sishuwa (2015) wrote a cathartic assessment in the *New African* magazine.

Indeed while China promotes its image through the CI, there is no comparable arrangement for African countries to attempt to mold the minds of and educate Chinese citizens to respect African norms and values, despite Beijing's official soundbite about how "the hearts of Chinese and African people are connected, and they will live together in harmony and promote inter-cultural dialogue" (*China Daily*, 2015). As it stands, the Confucian ideals of harmony have done little to mitigate the racist tendencies of the Chinese against Africans (Chan 2013, 30), whether in China itself (Min Zhou et al. 2016; Lan, 2017) or in the African countries where they live or work.

Here is a thought. When Ariana Miyamoto, a multi-racial woman of Japanese and African-American parentage, was crowned Miss Japan in 2015 amid some controversy in the Japanese media, a colleague in Chinese Studies was surprised that Japan had chosen "a black person" as Miss Japan, against the background of Japan's nationalist history and the popular assumption that the Japanese view themselves as a homogenous ethnic group. I could not help but respond that, despite the controversy generated by the selection of Miyamoto as Miss Japan (which incidentally was much exaggerated by the Western media), Japan is the only country in East Asia that is socio-culturally liberal enough to appreciate the beauty of a mixed-race person to the extent of crowning her Miss Japan. Upon reflection, it is difficult to imagine how the Chinese public would have reacted if this had happened in China.

The above points necessitate further discussion of the paucity of interest in, and analysis of, culture in China–Africa relations. In this regard, Stephen Chan and Jerry C.Y Liu's engagement with Confucian ideals in their separate assessments of the growing partnership between China and the African countries is a useful point of reference. They also bring into focus possibilities for a new form of scholarship in

African studies and East Asian studies, and the potential cross fertilization of the two area studies in respect of the expanding discourse on China–Africa relations, and perhaps a joint honors undergraduate programme in East Asian Studies and African Studies.

Both encourage Africanists and other researchers to consider the role of culture in their analysis of China's relations with African countries (see Chan 2013; Liu 2013). According to Liu, Chinese foreign policy is more inclined toward a humanistic way of reasoning, compared to the realist tradition that seems to guide Western approaches to foreign policy. This is based on the conception that "humanistic reason prioritises not the calculative scientific or logical calculation, scientific or logical articulation of interest for an individual or a specific group, but a general and sympathetic understanding of human desires, thinking and feeling as a whole" (Liu 2013, 56–57). Thus, as Liu further argues, Chinese humanist thought "emphasises moral-ethical cultural values, and Confucian ideals permeate traditional Chinese statecraft and governance and … inform contemporary Sino-African relations," implying a compassionate Chinese foreign policy toward the African countries. This is suggestive of a Chinese moral compass, but in almost the same breath Liu dampens this sense of optimism as he "accept[s] that the Confucian way of doing things in today's Africa is different from the way of 1405," the year of Zheng He's first expedition to the Horn of Africa. He is perhaps right about the contemporary Confucian way of doing things in today's Africa, if there is such a contemporary Confucian way to speak of. Kwame Anthony Appiah might also say that the authority of such wisdom as attributed to Confucianism may "lie largely in its purported antiquity, not in the quality of the reasoning – or the evidence – that sustains it" (Appiah 1992, 91). Liu's point also recalls Michel Foucault's conception of the system of government of the ancient Judeo-Christian nations which, as he argued, was grounded in the ideals of "pastoral care," evocative of the shepherd's tender care for his flock, which in principle amounts to the same Confucian ideals of compassion. The point to stress here is that the modern state, Christian or Confucian, operates within a framework that upholds, if not glorifies, the national interests of the state, and with that primary objective as a guiding principle there is no other viable tool of engagement but power, of which soft power is the messenger.

Nevertheless, Chan and Liu's essays raise interesting questions about the socio-cultural tensions between Chinese and Africans, and the racist bravado of the Chinese in their encounters with Africans, as mentioned earlier, whether in Nanjing in the 1980s, or more recently in Guangzhou (Min Zhou et al. 2016), Zambia and Ghana. Barry Sautman's 1994 article provides an insightful account of the experiences of African students across China in the 1980s, including the racist, violent attacks that were made on the students. Incidentally, that same Chinese attitude, despite the officially sanctioned rhapsodies about harmonious relations between the PRC and the African states, and how "the hearts of Chinese and African people are connected" (Xinhua 2015), seems to have found its way to Africa among the recent Chinese migrants.

Since culture matters, we have to grant that "culture influences the practice of policymakers by permeating their ways of thinking," as Liu argues (2013, 58). On that basis we cannot overlook the fact that Meles Zewani's policies were influenced by his local culture(s). This poses the question: is it enough to evoke Confucianism in an analysis of China's relations with Ethiopia, without a consideration (or better still, an understanding) of the latter's dominant socio-cultural values? This is, of course, a point that is equally relevant to Kenya, Ghana, and other African countries in their respective relations with China. It is also reminiscent of the old chestnut, or can of worms, if you will, found in Area Studies, and which Africanist scholars in particular love to sidestep, pretending Niger, for example, is understandable merely through its colonial history. Well, is it? Yet it seems the Chinese corporate employee who was selling green tea in Niamey went to Niger with nothing other than a body of Confucian values, bags of green tea, and a dash of French, the so-called official language of Niger to ply his trade, in total disregard for the local cultural values.

Certainly, this invokes a more basic question, one that has been raised by others. As Monson and Rupp (2013, 24) succinctly put it, does it make analytical sense "... to study a nation (China) in relationship with a continent (Africa)"? In addition, since African states have independent policymaking mechanisms, how is it that there are as yet only a few attempts among Africanists to understand the policymaking processes of African countries in their relations with China? In that respect, Africanists (both in the West and Africa) are no different from their Chinese equivalent mentioned above. More importantly, it is hardly scholarship to adumbrate the separate interests of the many African countries under the "shadow of Africa" in their relations with China. Moreover, is it right for researchers to just simply assume that the National Democratic Congress of Ghana, a social democratic party, has identical policies to those of the current political party in power in Ghana, the New Patriotic Party, a center-right and liberal conservative party, in regards to China?

Confucianism as the basis of China–Africa relations?

While the cultural imperative may be relevant to our understanding of the behaviour of the Chinese and Africans toward each other, it is perhaps not helpful to hide behind cultural essentialism to explain or obviate national policies. After all, our respective traditional norms and values, while not necessarily identical, are not in contention with each other; it is our circumstantial interpretations of each other's cultures that may suggest, or lead to, tension. Indeed, most of the values that are assigned to particular nations as their unique peculiarities have universal resonance. Thus, while Confucianism underscores the values of loyalty, filial piety, benevolence, courage, decorum, endurance, frugality, harmony, honesty, modesty, obedience, patience, respect, selflessness and sincerity, in the grand scheme of things these are values that are universally intelligible.

The Analects contains the following passage: "What is meant when the Documents say, 'when Gao Zong [a ruler of the Yin Dynasty] was in his mourning hut, he did not speak for three years'?" Confucius responded as follows: "Why insist on

Gao Zong? The men of antiquity all did likewise" (Confucius 1993, 51). Kwame Anthony Appiah (1992, 91) aptly notes that "Many African societies have as much in common with societies that are not African as they do with each other," for after all "what is distinctive about African thought is that it is traditional; there is nothing especially African about it." In African traditional behaviour apparently fathers cover up for their sons and vice versa, reminiscent of the following Confucian moral. In Chapter XIII of *The Analects*, Confucius remonstrates with Duke She who boasted that a man he knew served as a witness against his own father who had stolen a sheep (Confucius 1993, 51; Waley 1945, 175) because fathers and sons should cover up for each other instead of bearing witness against each other. Anyhow, Chan and Liu's references to Confucianism to explain Chinese attitudes to Africa may provide some insight into the evolving relations between China and the African countries. If so, then conceivably the application of African social norms and values may also be essential in our analysis.

Conclusion

On December 12, 2016 the popular BBC programme *University Challenge* included a bonus question about high-profile Chinese railway (infrastructure) projects in Africa, and the contestants got all three questions correct, an indication of how much purchase China's growing presence in Africa's development has already gained in the popular discourse.[4] At the same time, burgeoning economic relations between China and African countries has ignited a global political reaction that at once confirms Western anxieties about China's rise, and provokes despair on the part of Africa's traditional partners about losing control over Africa as articulated in the Africanists' discourse.

The elite African perspective on the topic, however, indicates the emergence of a counter-discourse to that popular in the West. Dambisa Moyo (2010) is among the elite African researchers and policymakers, the Sino-optimists, as Adem (2013) would say, who see China's role in Africa as implicitly positive, as opposed to the Sino-pessimists such as Adama Gaye (2008) and many African NGOs (French, 2014), who are forthrightly skeptical about China's economic relations with African countries and their political implications. Reflecting that dichotomy are the contrasting views of Western observers, amongst whom Bräutigam is cautiously optimistic about China's role in the economic development of Africa, with French at the other extreme. What is also evident in this configuration is the fact that among African observers (in Africa as well as in the diaspora), Sino-optimists dominate the discourse, while it is the other way around among Western observers.

The above discussion has reflected on how research about policymaking in Beijing and in the relevant African countries in regard to the growing relations between China and Africa would help to further reveal the nuances that define the evolving relationship. It would also help researchers to better understand and articulate the diversity of the interests of the African countries, and indeed of Beijing, within that framework.

Based on the conception that modernisation can only take place in the context of local social structures and cultural identity, it is interesting to note that Ghanaians love Chinese fried rice but in Ghana it is served with *shito*, a particular Ghanaian chilli oil, which is how the locals like it. Thus I have argued for more investment into mutual cultural understanding between Africans and Chinese.

I have suggested that the conception of Confucian values as unique to China and East Asia is a product of cultural essentialism based on a perceived difference, even misunderstanding, between elements of Chinese culture and African cultures. As indicated above, African thought and Confucian thought have a lot in common because they are both traditional: they are family-oriented, authority-centered and communal, and they glorify ancestor worship (Ampiah 2014). It would therefore be more useful to appreciate and emphasise in our analysis the commonalities rather than uniqueness. In essence, to ensure mutual respect beyond the anodyne convention of the "win-win" strategy, a one-sided didactic philosophy based on "Confucius say" will not help; rather, it will cause more confusion by exacerbating the cultural gap between Africans and the Chinese.

French (2014, 221) apparently encountered a Chinese official who recounted a popular Chinese view of Africans as people who "like to dance. That's their specialty. They may be poor, but they are very happy." This is nothing new! The Chinese might have got this idea from the Europeans who said that about Africans until the West became affected by happiness engendered by pop music and the accompanying style and dance. It is true that happiness is barely explored in Confucianism (Fraser 2013), but it is nevertheless the only form of success that counts in the grand scheme of things. The East Asian couple dancing semba in Rei Helder's video show how dancing ensures happiness.

Notes

1 Rei Helder, "Semba do Chinese," www.youtube.com/watch?v=ErsetXLvHQ8. Accessed March 20, 2017.
2 Pew Research Centre, "Global Attitudes and Trends: China's Image." July 14, 2014. www.pewglobal.org/2014/07/14/chapter-2-chinas-image/ (accessed November 28, 2016).
3 Beijing Statement of FOCAC Cultural Ministers Forum, CHINAFRICA www.china frica.cn/txt/2012-07/02/content_467390.htm (accessed December 14, 2016).
4 BBC University Challenge, Episode 22 www.bbc.co.uk/iplayer/episode/b085z105/uni versity-challenge-201617-episode-22 (accessed December 14, 2016).

References

Adem, Seifudein. ed. 2013. *China's Diplomacy in Eastern and Southern Africa.* New York: Routledge.
Alden, Chris, Daniel Large and Ricardo Soares de Oliviera. eds. 2008. *China Returns to Africa: A Rising Power and a Continent Embrace.* London: Hurst and Company.
Ampiah, Kweku and Sanusha Naidu. eds. 2008. *Crouching Tiger, Hidden Dragon? Africa and China.* Scottsville, VA: University of KwaZulu-Natal Press.

Ampiah, Kweku. 2014. "Who's Afraid of Confucius? East Asian Values and the Africans." *African and Asian Studies* 13(4): 385–404.

Appiah, Kwame Anthony. 1992. *In My Father's House: Africa in the Philosophy of Culture.* Oxford: Oxford University Press.

Bodomo, Adams. 2017. *The Globalization of Foreign Investment in Africa: The Role of Europe, China and India.*, Bingley: Emerald Publishing.

Bräutigam, Deborah. 2015. "5 Myths about Chinese Investment in Africa." *Foreign Policy*, 4 December. http://foreignpolicy.com/2015/12/04/5-myths-about-chinese-investment-in -africa/ (accessed November 1, 2016).

Bräutigam, Deborah. 2009. *The Dragon's Gift: The Real Story of China in Africa.* Oxford: Oxford University Press.

Chan, Stephen. 2013. "The Middle Kingdom and the Dark Continent: an essay on China, Africa and Many Fault Lines." In *The Morality of China in Africa: The Middle Kingdom and the Dark Continent*, edited by Stephen Chan, 3–43. London: Zed Books.

Confucius. 1993. *The Analects.* Oxford: Oxford University Press.

Ellis, Stephen and Gerrie ter Haar. 2004. *Worlds of Power: Religious Thought and Political Practice in Africa.* Oxford: Oxford University Press.

Fraser, Chris. 2013. "Happiness in Classical Confucianism: Xúnzǐ." *Philosophical Topics* 41(1): 53–79.

French, Howard W. 2007. "Commentary: China and Africa." *African Affairs* 106(422): 127–132.

French, Howard W. 2014. *China's Second Continent: How a Million Migrants Are Building a New Empire in Africa.* New York: Alfred A. Knopf.

Gaye, Adama. 2008. "China in Africa: After the Gun and the Bible: A West African Perspective." In *China Returns to Africa: A Rising Power and a Continent Embrace*, edited by Chris Alden, Daniel Large and Ricardo Soares de Oliveira, 129–142. London: Hurst and Company.

Halper, Stefan. 2010. *The Beijing Consensus: How China's Authoritarian model will dominate the twenty-first century.* New York: Basic Books.

Helder, Rei. "Semba do Chinese", available at www.youtube.com/watch?v=ErsetXLvHQ8 (accessed March 20, 2017).

Johnson, Chalmers. 1987. *MITI and the Japanese Miracle: The Growth of Industrial Policy, 1925– 1975.* Tokyo: Charles E. Tuttle Co. Publishers.

Kristof, Nicholas D. 1993. "The Rise of China." *Foreign Affairs* 72(5): November/ December: 59–74.

Lan, Shanshan. 2017. *Mapping the New African Diaspora in China: Race and the Cultural politics of Belonging.* London: Routledge.

Large, Daniel. 2008. "Beyond 'Dragon in the Bush': The Study of China-Africa Relations." *African Affairs* 107(426): 45–61.

Lee, Ching Kwan. 2009. "Raw Encounters: Chinese Managers, African Workers and the Politics of Casualization in Africa's Chinese Enclaves." *The China Quarterly* 199: 647–666.

Lee, Ching Kwan. 2017. *The Spectre of Global China: Politics, Labor, and Foreign Investment in Africa* Chicago, IL: University of Chicago Press.

Liu, Haifang. 2008. "China-Africa Relations through the Prism of Culture: The Dynamics of China's Cultural Diplomacy with Africa." *China Aktuell* 37: 9–44.

Liu, Jerry C. Y. 2013. "Sino-African Cultural Relations: Soft Power, Cultural Statecraft and International Cultural Governance." In *The Morality of China in Africa: The Middle Kingdom and the Dark Continent*, edited by Stephen Chan, 47–59. London: Zed Books.

Mbaye, Sanou. 2011. "Africa Will Not Put Up with a colonialist China." *The Guardian*, February 7, 2011.

Mohan, Giles and May Tan-Mullins. 2009. "Chinese Migrants in Africa as New Agents of Development? An Analytical Framework." *European Journal of Development Research* 24(4): 588–605.

Mohan, Giles, Ben Lampert, May Tan-Mullins and Oaphne Chang. 2013. *Chinese Migrants and Africa's Development: New Imperialists or Agents of Change?* London: Zed.

Monson, Jamie and Stephanie Rupp. 2013. "Introduction: Africa and China: New Engagements, New Research." *African Studies Review* 56(1): 21–44.

Moyo, Dambisa. 2010. *Dead Aid: Why Aid is not Working and How There Is Another Way for Africa.* London: Penguin Books.

Ohno, Kenichi and Izumi Ohno. eds. 2013. *Eastern and Western Ideas for African Growth.* London: Routledge.

Park, Yoon Jung. 2009. "Chinese Migration in Africa." SAIIA China in Africa Project Occasional Paper, January.

Procopio, Maddalena. 2015. "The Effectiveness of Confucius Institutes as a Tool of China's Soft Power in South Africa." *African East-Asian Affairs* 2 (June): 98–125.

Roy, Denny. 1994. "Hegemon on the Horizon? China's Threat to East Asian Security." *International Security* 19(1): 149–168.

Rupp, Stephanie. 2008. "Africa and China: Engaging Postcolonial Interdependencies." In *China into Africa: Trade, Aid, and Influence*, edited by Robert Rotberg, 65–86. Washington, DC: Brookings Institution Press.

Saavedra, Martha. 2009. "Representations of Africa in a Hong Kong Soap Opera: The Limits of Enlightened Humanitarianism in The Last Breakthrough." *The China Quarterly* 199: 760–776.

Sautman, Barry. 1994. "Anti-Black Racism in post-Mao China." *The China Quarterly* 138: 413–437.

Shambaugh, David. 1996. "Containment or Engagement of China? Calculating Beijing's Responses." *International Society* 21(2): 180–209.

Sishuwa, Sishuwa. 2015. "How China's Confucius Centres Affect African Culture." *New African Magazine*, May 21.

Strauss, Julia and Martha Saavedra. eds. 2009. *China and Africa: Emerging Patterns of Globalization and Development.* Cambridge: Cambridge University Press.

Su, Irene Yuan. 2017. *The Next Factory of the World: How Chinese Investment is Reshaping Africa* Boston: Harvard Business School Publication.

Sverdrup-Thygeson, Bjornar. 2017. "The Chinese Story: Historical Narratives As a Tool in China's Africa Policy." *International Politics* 54(1): 54–72.

Taylor, Ian. 2006. "China's Oil Diplomacy in Africa." *International Affairs* 82(5): 937–959.

Taylor, Ian. 2008. *China's New Role in Africa.* Boulder, CO: Lynne Rienner Publishers.

Thompson, Reagan. 2012. "Assessing the Chinese Influence in Ghana, Angola, and Zimbabwe: The Impact of Politics, Partners, and Petro." Stanford University Honors Thesis.

Waley, Arthur trans. 1945. *The Analects of Confucius.* London: George Allen & Unwin.

Xinhua. 2015. "Full Text: China's Second Africa Policy Paper." *China Daily*, December 5.

Zhang, Hongming. ed. 2014. *China International Relations: Review and Analysis (Volume 1).* Reading: Paths International Ltd.

Zhou, Min, Shabnam Shenasi and Tao Xou. 2016. "Chinese Attitudes toward African Migrants in Guangzhou, China." *International Journal of Sociology* 46(2): 141–161.

5

MEDIA AS A SITE OF CONTESTATION IN CHINA–AFRICA RELATIONS

Cobus van Staden and Yu-Shan Wu

In focusing on society and the structuring of social relations, the social sciences increasingly have to contend with media, which is more than a simple reflection of society, or a space where contentious social communication is amplified and accelerated. Media plays an active role in forging, defining and destroying links between distant states, governments and publics. China–Africa relations provide a potent example of this. It is an intensely mediated relationship, with governments, state and private broadcasters, and individuals in Africa, China, and the West using media to try and forge links, or weaken them. Both in the alliances and the contestation between China and Africa, media must be seen as a constant contributing, complicating factor.

This chapter explores two main components of China–Africa media relations: first, as a field of study, detailing and mapping the groundwork and avenues that have delineated the field up to the present and have involved different methodologies and disciplines of study. The second is a forward-looking section that includes discussion of media – namely social media – and its impact on the practice of China–Africa scholarship.

Context: China–Africa media relations

The interdisciplinary nature of scholarship on China–Africa relations makes it susceptible to the perpetual changes and reinterpretations that characterise the social sciences. Moreover, the study of the role of media in China–Africa relations is relatively new compared to other aspects of the relationship, such as China's engagement in infrastructure or diplomacy. This is despite the fact that media relations between the regions date back to the 1950s, when it existed mainly as a high diplomatic priority, for the purpose of establishing bilateral relations (see Wu 2012). Contemporary scholarship in this area thrived as a result of broader geopolitical shifts, in particular, China's "going out" strategy from the 1990s and the intensification of China–Africa links after the first FOCAC meeting in 2000. "Going out" can be disaggregated into two broad components. The first and practical aspect involved enhancing domestic economic development through the pursuit of natural resources and raw materials abroad, and thus the expansion of China's trade links and investments through the internationalization of China's state-owned enterprises (SOEs) (Friedberg 2006). Second, while Chinese media organisations were part of the larger trend of sending out SOEs, they were also part of a broader motivation: in the words of former President Hu Jintao, this was to "… help ensure that our country has a more friendly image, with greater moral appeal" (Chin and Thakur 2010, 121). It was thus also inextricably linked to the dissemination of intangible Chinese values and ideas. Another factor was the changing media landscape following the 2008 global financial crisis where traditional news players pulled back their activities, while emerging players expanded their external scope.

This context has provided a rich background for increased studies on China's "non-traditional" engagements. It has also, however, left lingering questions about issues such as the extent to which China has achieved its goals in promoting its international image in relation to its commercial presence abroad and, more broadly, how its role in "new" spaces has impacted the structures and ideas around media practises and reporting specifically, and African international relations more widely.

At the same time, the media topic has not only widened understanding of the very nature of China–Africa relations, it also happens to provide the instruments that can assist scholars in understanding the relationship. Media provides an additional window into public sentiments on the relationship, and transmits real-time information and deliberation across geographic borders. Social media, defined as "web-based tools and services that allow users to [comment], create, share, rate and search content and information" (Bohler-Muller and Van der Merwe 2011, 2), specifically functions as a kind of public sphere that allows users to articulate and discuss this relationship. However, such discussions neither have much power over political decision-making nor necessarily constitute the collaborative labour of identifying shared problems and articulating solutions as argued by Habermas (1989) in his account of the seventeenth- and eighteenth-century European public sphere. In this respect, we echo Nancy Fraser (1992) in basing our thinking on the model of a multitude of competing publics, all presenting and constructing themselves through debate, rather than a unified public sphere.

Thus the dual role of media as a topic in China–Africa studies, as well as an instrument *for* the very analysis and deliberation of the relationship, reflects the present complexities of the information age. More broadly, this has given scholars in the social sciences more platforms for the observation of society and more opportunities to become active participants in shaping the field. As observers of (and minor participants in) this area of study, our basic stance is that the role of Chinese media organizations in Africa *and* the emergence of communication technologies and social media platforms are neither positive nor negative for the field, but are progressively making it more multifaceted by including more voices, contexts, and disciplines.

Survey of China–Africa media research

The study of China–Africa media relations cannot be separated from larger conversations about the role of media in statecraft since the end of the Cold War. From the early 1990s, theorists pointed out that the globalisation of media is having geopolitical effects. These comments were couched either as a critique of the marginalization of non-Western cultural autonomy (Tomlinson 1991) or as an instrument to increase soft power. Joseph Nye's (1990) classic coining of soft power as including the kinds of perceptions created by media, and the resultant implied link between the power of corporate mass media and state power, created a narrative that inspired the construction of new apparatuses of state communication during the late twentieth and early twenty-first centuries. As an emerging global power, and one both risk-averse and constrained by its commitment to non-interference, China was receptive to the idea that media could smooth the way for its international relationships. In particular, this built on Chinese media outreach to the developing world (including Africa) during the Cold War, when media hardware and programming was supplied to poor countries as part of anti-American, anti-colonial and anti-Soviet campaigns. The analysis of media in the context of China–Africa relations, therefore, was built on wider investigations of Chinese global public diplomacy and the role of media in this outreach. These studies emerged from a few key concerns, and slowly expanded to cover a wider set of issues.

In the first place, early studies (which more recent studies on Chinese media and Africa have drawn upon) focused on the expansion, mechanics, and nature of Chinese soft power, focusing on the tools of state messaging as China expanded its global influence. These include Kurlantzick's (2006) work on China's public diplomacy in Southeast Asia and Ding's (2008) focus on China's soft power in the global South. This focus continues, exemplified by authors such as Fijałkowski (2011). These works tended to characterise China's outreach to the global South, and Africa in particular, as a form of soft power expansion, and located their discussion of media in the context of IR, rather than in cultural studies, media studies or related disciplines. With this focus came a particular framing of the issue as one of success or failure, gain or loss of power in the international arena, rather than simply mapping media flows. Throughout the development of this field, there were descriptive accounts of the mechanisms of global outreach and IR, for example d'Hooghe (2007) and Wang (2008). These accounts also drew close links between media and soft power.

Subsequent work by scholars like Sun (2010), Zhang (2010), and Zhu (2012) maintained the focus on soft power, but widened the analytical approach by including media studies methodologies like industry analysis and policy analysis to provide accounts of changes in the domestic Chinese mediasphere, as well as the mechanisms of the globalisation of Chinese media. Africa was a particularly salient case study in balancing the northern hemisphere bias of many accounts of Chinese media globalisation, because of China's long history of offering media infrastructure as part of its aid outreach to Africa. Scholars who paved the way for this new field include Farah and Mosher (2010), and Gagliardone, Repnikova and Stremlau (2010). These studies provided a base for the subsequent analysis of the role of media in China–Africa relations.

One of the key fields of research has been the expansion of Chinese official (Communist Party of China [CPC] affiliated) media in Africa. Most of these authors have linked this expansion with the Chinese government's attempt to address foreign publics directly. This analysis continues the above-described IR-influenced focus on media as a tool for the expansion of soft power, rather than as a flow of intellectual property, or as something consumed by key audiences. In other words, these accounts tended to focus on media as sender-directed. Less attention was paid to the consumption of the media. In other words, it focused more on what the (Chinese) senders intended with this media expansion, rather than whether any of these goals were actually achieved in the context of the (African) audience. One exception is Gorfinkel et al. (2014), who mapped audiences of Chinese state media in Kenya and South Africa. This survey was from a bird's eye view, however, and didn't include much use of focus groups, audience ethnography or other one-on-one methodologies. The research in this section tends to foreground media outlets like China Central Television (CCTV), Xinhua and China Radio International (CRI) as organs of state communication, rather than as corporate entities competing in the global news economy. Most of this writing dates from 2012, the year when CCTV (re-branded as China Global Television Network [CGTN] in early 2017) and Xinhua launched large new production facilities in Nairobi. It was also the year when the *China Daily* launched its weekly Africa edition on the continent. The striking overlap of the expansion of various Chinese CPC-affiliated media into Africa also drew attention to the ways that the Chinese government uses media, and the possible effects of this media expansion in Africa. Authors linking the expansion of this media with the public diplomacy of the state include Wu (2012), and a succession of special journal editions, some of which are related to projects dedicated to Chinese media in Africa (see, for example, Van Staden 2013; *Ecquid Novi: African Journalism Studies* 34(3): 2013; *Chinese Journal of Communication* 9(1): 2015).

The expansion of China's media abroad also drew the attention of the mainstream media, for example, Al Jazeera's article "Chinese Media Expands Africa Presence" (January 24, 2013), *The New York Times'* "Pursuing Soft Power, China Puts Stamp on Africa's News" (August 16, 2012) and the *BBC*'s "Chinese Media in Africa: What You Need to Know" (March 22, 2013). These accounts echoed the dual scholarly focus on media and soft power. A similar dual focus was visible in related enquiries into the modes of Chinese public diplomacy outreach, for example through university exchanges and curriculum expansion (Youngman 2014; King 2013) and the role of Confucius Institutes (Hartig 2015; Procopio 2015).

The second main approach tracks the coverage of China's presence in Africa as processed largely by African scholars through the African press. This approach moves away from the IR focus mentioned above, in favour of mixed quantitative and qualitative media studies methodologies. In particular, many of these studies have favoured content analysis and framing analysis. The studies have predominantly focused on using the coverage of China or China–Africa relations as a barometer of African popular or elite opinion about China. Authors who used framing analysis include Wekesa (2013), Wekesa and Zhang (2014), Jura and Kaluzynska (2013) and Finlay's survey of journalists' work (2013). These approaches have occasionally been aided through the use of electronic harvesting and interpreting of texts, and meditations on the mechanics of this methodology inform some of this writing. In most cases, the studies have tended to focus on a single country or compared small numbers of countries. In order to overcome logistical barriers to obtaining large numbers of newspapers, many have drawn on the online versions of print publications, which tend to locate these studies in larger African media markets like Kenya and South Africa. These studies skew toward news media, rather than entertainment or fiction. In this sense, this approach tends to echo the first approach in that it positions news media in the center of China–Africa discourse. Framing analysis also has the effect of implicitly supporting the sender bias seen in the first approach, while seemingly undercutting it. In other words, while framing the analysis's focus on interpretation seems to favour an audience perspective over sender strategies, the fact that this interpretation is left up to the analyst (or his/her algorithm) means that the actual audience doesn't get a word in. The field of China–Africa media studies arguably would benefit from more audience surveys, but that is frequently hampered by the cost and logistical difficulty of this kind of research in an African context.

A third approach has been to survey journalists involved in various fields of China–Africa coverage. This has included speaking with Chinese journalists covering Africa for Chinese state press (Li and Rønning 2013) and African journalists reacting to the growing influence of China in their field (Wasserman 2015). Wasserman combines this approach with an analysis of the political economy of media, especially his focus on shifts in reporting after Chinese investment in South Africa's Independent Media Group. This work is situated in the political context of South Africa's ruling African National Congress party's close relationship with China. Both from the Chinese and African sides, press freedom is one of the main preoccupations in these surveys. This is reflected in the use of political economy as a core methodology. This opens both government and commercial influence as fields of enquiry. It also positions the China–Africa debate in a normative scheme shaped by Western radical enquiry, and liberal rights-based discourse. However, these enquiries also include analyses of different traditions of journalistic practice. Different from a normative approach, this view asks whether there is only one "correct" (Western) approach to journalism, or different traditions, of which a Chinese "constructive journalism" approach might be one. It also allows for the possibility to question and redefine African journalistic tradition as distinct from

Western practice (Wan 2015). This approach arguably opens the door to more theorization of journalistic practice in the global South. More recently, the above-mentioned three themes have been discussed together, in separate chapters, in *China's Media and Soft Power in Africa* (Zhang et al. 2016).

A fourth, emergent approach focuses on the Internet as a vector of China–Africa relations. This includes a focus on the expansion of Chinese telecom conglomerates like ZTE, Huawei, TenCent (WeChat), and StarTimes into the African media economy (Cisse, 2012). Some authors have also focused on the actions of individual African governments in their interactions with network providers, frequently with a focus on state surveillance and the suppression of dissent (Human Rights Watch, 2014; Gagliardone, 2014). These accounts also frequently make clear that the enabling of surveillance is not unique to Chinese telecoms, but that Western telecoms are similarly complicit in the repression of Africa.

This field has also seen research on the role of social media in China–Africa relations. Wu (2013) compared the Chinese and South African public and governments' use of social media as a public diplomacy and communication tool. Kaigwa and Wu (2015) pointed out that Chinese companies' network provision has led to the development of thriving social media ecosystems in Africa, and these have also become the forum for the expression of African opinions about Africa's engagement with China, notably around issues like illegal wildlife trade and business. Lu and Van Staden (2013) traced the influence of Chinese chat forums in channeling migration and capital from China to Africa. Much of this work has been descriptive, constituting an initial delineation of the field, and more intense analysis is called for. However, these first approaches open the possibility to discuss the political role of social media beyond its usual theoretical limits of the liberal public sphere. Rather, it enriches this discussion by putting it in the context of the geopolitics and economics of the global South.

There are also some emergent and minor strands of research that should be noted. First, we have seen analyses of reactions in Western media to China–Africa relations. Emma Mawdsley's article "Fu Manchu Meets Dr Livingston in the Dark Continent" (2008) was an early example of this work. Other instances are Harper (2012) and Benabdallah (2015). Second, there is also historical research being done into the past mediation of China–Africa relations via official and entertainment media. With more entertainment media making its way from Asia to Africa, this is a key area of enquiry. The historicization of these flows has the potential to open up an entire new field in China–Africa media studies beyond the governmental sender bias that limits soft power-focused work. It also opens up a potentially rich new flank of audience research based on cultural studies, where the experiences of actual African audiences will receive more attention than is possible through framing analysis. Precursors to this work have used audience ethnography as a basis methodology (Joseph 1999). Much scope for similar ethnographic work exists. Newer scholarship has approached the same issue through a focus on distribution and exhibition networks that brought kung fu film to South Africa, mapped through archival industry analysis, film theory, and reception theory (Van Staden 2017). The increased engagement of other players such as Japan and South Korea, as reflected by their hosting of culture weeks and film festivals in Africa, means that this type of engagement could become increasingly salient in understanding the mediated relationship between Africa and Asia. Van Staden (2014) has historicised some of these trends in the context of Japan.

Looking forward

Having provided a brief snapshot of the important themes in China–Africa media research, our forward-looking section is organised into two main parts. The first recognises that media, especially social media, is not simply an object to be studied and will continue to exist as both a site and tool of deliberation and contestation in the China–Africa field. Having discussed this reality, the section will then move to exploring ways in which China–Africa media scholarship could advance.

China–Africa and media practices

The China–Africa field has evolved from foundational seminal studies by authors like George Yu or Philip Snow, and newer key works of scholarship, like Chris Alden (2007) and Deborah Bräutigam (2009). Since then different media platforms have expanded the China–Africa space, leading to a diffusion of authoritative voices and contributions from journalists like Tom Burgis (2015) or authors straddling academia and journalism (reaching a whole new target audience) like Howard French (2014). The field of study has thus expanded and its participants are more loosely defined, with a wider range of expertise than before.

The use of different media platforms has impacted the field in various ways. Prominent China–Africa scholars have extended their audiences with the use of online spaces such as the official blog of ambassador David Shinn and Deborah Bräutigam's blog *China in Africa: The Real Story*, which is now integrated into Johns Hopkins University's China–Africa Research Initiative (CARI).[1] The Centre for African Studies at Peking University also sends out a weekly newsletter, *PKU African Tele-Info*, in Chinese.[2] The same scholars are able to engage one another and share news stories and events on other online platforms (such as the Chinese in Africa/Africans in China network). Twitter has also become an important space for this engagement. The internet has even become a powerful organizing tool to store and categorise China–Africa publications and reports on university databases and other online portals, as well as research organizations' websites. This has been extended by entities such as the China–Africa Project, China House and the Wits Africa–China Reporting Project, who all play a dual role of producing original content while also disseminating China–Africa content to new audiences.[3] Finally, official actors such as the FOCAC secretariat have not only created their own website but are branding their articles on platforms such as *AllAfrica.com*, thereby opening a possible window for discussion and active engagement, beyond the official ambit and traditional audiences.

There is therefore a cross-pollination of works and topics over an increased number of platforms. Needless to say, this has not rendered physical interaction, including conferences, exchanges and collaborations, and the mediated analysis that comes from it, any less valuable. While online interaction has changed the pace at which news developments and discourse on the China–Africa relationship are shared, it is clear that the best means to move the relationship ahead is through dispelling myths and building trust between all sides, through increasing interaction that is still arguably limited to a handful of stakeholders. This is the reason that in 2016 Oxfam International and the African Union established the Africa–China Dialogue Platform to engage stakeholders on the relationship.

An important issue one should not lose sight of are the wider forces (in reality, and the governance of the internet) that affect China–Africa relations and engagement on issues of interest. Government internet policies have a direct influence on the degree of engagement between users across different regions. One example is the current legislative status of Facebook in China that has a knock-on effect on relations, as users are operating on different online platforms. (Perhaps this will change as the rising popularity of the Chinese mobile social media, WeChat, expands its reach into Africa.) There should be caution when measuring the impact of social media on the study of China–Africa, as shared knowledge on platforms such as Facebook or Twitter potentially excludes a wider Chinese audience and perspective (and even language), who in turn have their own platforms such as Sina Weibo and Renren. And while there are fewer such restrictions on the African side, actual internet access is not yet widespread (but rapidly changing). Consider, for example, that almost 70 percent of Kenyans use the internet (as of mid-2016), in comparison to 52.6 percent in South Africa and 1 percent in Eritrea, and in total, an average of 28 percent of the African population actually uses the internet.[4] The important question that this leads to, then, is the extent to which stakeholders on the China and African side are able to determine and contribute to how the discourse on the relationship moves forward.

Moreover, the potential increased "closeness" of people and societies needs to be further investigated, as it also means divisions, the obvious example being language, become more apparent. Specific to the media field is the closer proximity of differing identities, ideologies, uses and interests of China–Africa scholars and media practitioners (and even the differences within each group). A particular instance is the often-diverging agendas and views regarding through what framework the relationship should be approached. In other words, a real intellectual contestation over the meaning of credibility exists between scholars and practitioners. This is further highlighted by Hassid and Sun (2015, 3) who state "it is not an exaggeration to say that political scientists and media scholars may even have different understandings of what constitutes valid empirical data or worthy lines of inquiry and which theoretical models and paradigms are fashionable or out of date."

It is important to approach these divergences in a nuanced way, rather than simply issuing judgments regarding their relative validity. This is because although both researchers and journalists tend to have a preoccupation with good data, their professional obligations, timeframes and audiences are radically different. While there is frequently a tendency to pit the two against each other (both from within these camps and outside), the reality is that these are not monolithic groups. Various cultures of journalism exist, and professional parameters tend to shape the nature of the reporting produced (Hanitzsch 2007). To a certain extent, both sides have faced the same problem: the difficulty of getting credible information and numbers on which to base both reporting and theoretical exploration. At the time of writing this chapter, online platforms have pushed some of the newest China–Africa research to occupy a hybrid position between academic knowledge production and investigative journalism. A notable example of this process of seeking accurate numbers and statistics is the Johns Hopkins CARI, which collated data on Chinese aid and investment into a database, which will provide a resource for both scholars and journalists. This kind of initiative shows the potential for media platforms to house hybrid spaces that serve both scholars and journalists, and for new forms of collaboration and hybridization.

Future China–Africa media relations from a topic perspective

Having outlined the impact of media on the practice of scholarship and China–Africa relations, bringing together a range of experiences and voices into the fold, what is the future of China–Africa media related topics? Indeed as media follows particular news cycles to determine the relative importance of a topic, the evolution of the China–Africa topic will, naturally, remain dependent on international, continental and domestic developments. What have thus been identified are various issues that will likely move the area of study beyond its current juncture, although they are also dependent on the trends in specific disciplines of study and stakeholder interests, who together mediate the direction of the field.

Country-specific developments: As China's media relations unfold, the next step is to expand on countries' particular responses, moving beyond the over-arching surveys outlined above (see, for example, Zhang et al. 2016). The challenge here is to uncover the nuances of relations (such as the media's relationship with other sectors), and the particularities of specific constituents. Besides important content analysis of past media coverage, the exploration of how such dynamics could potentially impact China's broader media strategy, external relations and interests from a foreign policy perspective also matters. One instance is the reported 20 percent Chinese stake in South Africa's Independent News and Media, which was acquired by the Sekunjalo Independent Media Consortium, from previous Irish ownership (Lloyd 2013; Wu 2016). This development has not only raised the impact of respective governments' engagement in South Africa's media space but the fact that this consortium has intentions of an "African growth strategy," to increase its presence beyond South Africa (Wu 2016). This raises questions whether China and South Africa's respective public diplomacy and broader foreign policy interests can effectively operate jointly, through the instance of collective media engagement on the continent. Furthermore the rise of new corporate stakeholders (like South Africa's Naspers's stake in Tencent China); the widening of the range of content flowing to particular areas (from Chinese state broadcasting, print, and film) and the role of transcontinental networks, need to be considered in tandem to state engagement. This area will probably develop further in the short term, due to maturing research projects focusing on particular regional expansions of Chinese media.

Chinese media players in the present and beyond: The nature of China's media engagement also involves changing communication platforms. Consideration needs to be given to shifts in China–Africa media relations, as they are influenced by larger changes in geopolitics and the economics of media. While many scholars note the expansion of Chinese media in Africa since 2012, less is known about how Chinese media (this includes broadcast media, Chinese telecommunications firms, Xinhua News Agency, as well as newer players such as pay-TV operator StarTimes) have changed their practices, approaches, and outlooks due to wider shifts in the media field and local markets. This does not only relate to changes in how Chinese media outfits operate, but also how economic changes in the news business itself affect them. There also remains the need for more comparisons between China's media engagement between regions (such as its media links in Asia and Latin America, in comparison to those in Africa), highlighting whether its strategy is indeed overarching or not.

It is important that analysis of the nature of Chinese state media and beyond take note that its competitors are not monolithic either. Much has been written about the manner in which China's global media rise was a response to global criticism (particularly for its role in the Darfur crisis and Zimbabwe around 2008) and what Chinese officials called "irresponsible" reporting about China by Western news entities. Less is known, however, about how African respondents and local media players are impacting China's media practices and engagement. Moreover, the question remains how more established players, such as European and US news outlets, who remain long-standing strategic partners to Africa, and new players like Qatar's Al Jazeera, have responded to China's broader "charm offensive" on the continent. There is also the re-enforced engagement of other Asian partners, such as Japan and South Korea, who have hosted their own forums with Africa: the Tokyo International Conference of African Development, and the Korea–Africa Forum, respectively. The question remains whether these players will be influenced by China's media experience or offer new avenues for communicating and engaging African publics.

Using media to uncover trends in society: While it has been noted that Chinese and African social media users do not necessarily engage on the same online platforms, their respective social media usage and online discourse provide a view into the complexities and politics at the country level, including which strata of society engage online and around what issues. In addition, online media offers the chance to glimpse how Chinese and Africans talk amongst themselves about China–Africa relations, as well as how they communicate with each other. The online views of Kenyans and South Africans toward China were outlined in Kaigwa and Wu's

(2015) book chapter "#MadeinAfrica: How China–Africa Relations Take on New Meaning Thanks to Digital Communication." However, this could be broadened to understand the sentiments of other key players in other societies. A further step, indeed, would be to contrast how such online views contrast with street-level popular opinion and official rhetoric, and how they may impact bilateral relations. Social media is also an important window into the views of people and groups who are not confined by physical borders, for instance the important topic of how Chinese migrants organise themselves, as well as channel and facilitate economic migration to Africa (Lu and Van Staden 2013).

The official and unofficial interplay: Official and unofficial spaces are closer than ever before. This can be noted in the extent to which China's own policymakers are gauging and responding to Chinese social media sentiment. The media response to FOCAC VI also demonstrated a drive to bring civil society concerns (for example relating to wildlife and sustainable development) and the fostering of people-to-people relations, into the formal China–Africa relationship. China's engagement in Africa will also need to respond to continental developments such as a burgeoning youth demographic, urbanization and Africa's own uptake of communication technology (Kaigwa and Wu 2015). This also links to the urgent issue of the relative agency of African citizens. Whether social media platforms can change state–citizen relations became a crucial issue in the case of Tunisia and Egypt, and even in China. The power of citizens will also, in turn, impact countries' external relations.

Beyond media: Further study of non-conventional public diplomacy instruments is needed in order to determine to what extent China is keeping to traditional ideas of public diplomacy. How do physical network initiatives such as President Xi's 2013 vision of closer cross-regional economic and political links via the "Belt and Road Initiative" (BRI) that includes an overland route running from China to Eastern Europe and a complementary maritime route that stretches from Southeast China across the Indian Ocean to Africa and to Europe, relate to media expansion? In particular there is discussion amongst Chinese officials regarding a "Digital Silk Road," as part of BRI. In addition, little is understood about the way African states can leverage public diplomacy and media in promoting their own interests in China and beyond, and what challenges might lurk in this approach.

Current and potential study areas, moreover, provide empirical evidence of the broader issue of China in IR, and how it challenges traditional conceptions of power. Much debate has taken place about the role of China's public diplomacy instruments, namely the media, and assessments of their potential to aid China's soft power. It is true that Chinese state broadcasters have faced challenges in penetrating the African media market, particularly as a latecomer in relatively robust civil society spaces, such as Kenya and South Africa. The increased number of local and external voices means that players are required to adopt competitive strategies to gain audience attention. However, it is also worth unpacking what China itself views as successful engagement, and looking at these strategies in relation to its own interests, rather than in comparison to other players. For instance, China is not only seeking to win popularity through image building but also to gain support for, and justify, specific long-standing economic and geopolitical interests in Africa.

A lesser-reported story, in fact, is the ways that China continues to build its attractiveness through its economic and infrastructure engagement in Africa, engagement that is linked to media. In particular, how Chinese telecommunications firms mediate and structure the relationship between China and Africa raise further lines of enquiry. This influence is exercised on a few different levels. First, Chinese telecommunications companies are instrumental in setting up new internet and mobile telephone networks across the continent. Second, they are also increasingly collaborating with Western companies to tailor services to particular African regions. Third, companies like Huawei are also central to hardware provision, both in terms of networks and consumer handsets. Fourth, allegations have been raised that Chinese companies are enabling or facilitating government surveillance in countries like Ethiopia (Human Rights Watch 2014). Of course there is also the political and economic implications of the entrance of the Chinese pay TV company, StarTimes, that offers more affordable packages than established players and the role of Chinese media interests in local media (and vice versa), through content and strategic direction, as exemplified by the Independent News and Media in South Africa and Naspers in China.

An important additional question is whether China's relationship with stakeholders could change. As the continent likely continues to "leapfrog" toward wider communication technology access, China may find itself in a position where it is required to expand the spaces in which it engages (which is arguably already happening, with Chinese media entities setting up Twitter accounts). Nye (2002, 6) argues that gaining soft power through information requires access to multiple channels of communication, which can only aid in further influence on how particular issues are framed. The social media debate is then an indicator of the implications for China's soft power, often described as a government instrument. New communication technology has elevated public voices and interests, leading to questions regarding whether China can continue to favour official government links (such as official media forums and specific stakeholder exchanges, for example between officials and specific journalists) without engaging opposition groups and the broader society. What is apparent is that China's rise and power will, as the media space attests, become increasingly mediated.

Conclusion

Monson and Rupp (2013, 24) warn that the study of "China–Africa" is in danger of being analyzed in isolation and researchers should not omit or diminish the understanding of global and national contexts under which such relations exist. Rather than a binary partnership, they see relations as "a flexible and emerging process that links two areas of the globe – including their people, business, governments, ideas and networks – in ways that are both fundamentally rooted in historical processes and unfolding before our eyes" (2013, 28). Some of these aspects have received more scholarly attention than others. As we showed in this chapter, the focus on Chinese media expansion in Africa has tended to foreground government-level concerns about soft power and international expansion of state media over the level of audience engagement. We have also so far seen more attention paid to news media than to other forms of media, despite the rapid expansion of Chinese telecom companies into the African internet space. On a wider level, research has leaned toward a China-centric approach (perhaps inadvertently), reinforcing a view of Africa as the party that is acted upon. More research into African agency in relation to Chinese media is needed, both on the institutional and audience levels.

China–Africa relations have been intensely mediated from the start, and the field has been shaped by the fact that it developed while two of its constitutive arenas – academia and journalism – have themselves been changing rapidly. What is certain is that whatever shape the field takes in the future, it will be even more mediated than it is now. Media is increasingly both the object of and the forum for research on China–Africa relations. The mediation of these relations is sure to shape them in the future, and it is incumbent on researchers to trace the role of politics and power in this mediation.

Notes

1 See http://davidshinn.blogspot.co.za/ and http://www.chinaafricarealstory.com/.
2 This newsletter is said to reach 6000 readers on five continents interested in African studies. For more information, see Li, 2014.
3 See, respectively: www.chinaafricaproject.com; www.facebook.com/chinahousekenya; and, http://china-africa-reporting.co.za.
4 As expressed in: *Internet World Stats.* "Internet Usage Statistics for Africa," June 2016, www.internetworldstats.com/stats1.htm, accessed September 26, 2016.

References

Alden, Chris. 2007. *China in Africa*. London: Zed Books.
Benabdallah, Lina. 2015. "Political Representation of China–Africa: The Tale of a Playful Panda, or a Threatening Dragon?" *Africa Review* 7(1): 28–41.
Bohler-Muller, Narnia and Charl van der Merwe. 2011. "The Potential of Social Media to Influence Socio-political Change on the African Continent, Africa Institute of South Africa." *Africa Institute of South Africa Briefing 46*.
Bräutigam, Deborah. 2009. *The Dragon's Gift: The Real Story of China in Africa*. Oxford: Oxford University Press.
Burgis, Tom. 2015. *The Looting Machine*. London: William Collins.
Chin, Gregory and Ramesh Thakur. 2010. "Will China Change the Rules of Global Order?" *The Washington Quarterly* 33(4): 119–138.
Cisse, Daouda. 2012. "Chinese Telecom Companies Foray into Africa." *African East-Asian Affairs* 69: 16–23.

Ding, Sheng. 2008. "To Build a 'Harmonious World': China's Soft Power Wielding in the Global South." *Journal of Chinese Political Science* 13(2): 193–213.

d'Hooghe, Ingrid. 2007. *The Rise of China's Public Diplomacy*. Hague: Clingendael Institute.

Farah, D. and A. Mosher. 2010. *Winds from the East: How the People's Republic of China Seeks to Influence the Media in Africa, Latin America, and Southeast Asia*. Washington, DC: Center for International Media Assistance, September.

Finlay, Alan. 2013. "Tracking the Effects of a 'Soft Power' Strategy on Journalism in China." *Ecquid Novi: African Journalism Studies* 34(3): 155–160.

Fijałkowski, Łukasz. 2011. "China's 'Soft Power' in Africa?" *Journal of Contemporary African Studies* 29(2): 223–232.

Fraser, Nancy. 1992. "Rethinking the Public Sphere: A Contribution to the Critique of Actually Existing Democracy." In *Habermas and the Public Sphere*, edited by Craig Calhoun, 109–142. Cambridge and London: MIT Press.

French, Howard. 2014. *China's Second Continent*. New York: Alfred A. Knopf.

Friedberg, Aaron L. 2006. "'Going Out': China's Pursuit of Natural Resources and Implications for the PRC's Grand Strategy." *National Bureau of Asian Research* 17(3): 5–34.

Gagliardone, Iginio. 2014. "New Media and the Developmental State in Ethiopia." *African Affairs* 113(451): 279–299.

Gagliardone Iginio, Maria Repnikova and Nicole Stremlau. 2010. "China in Africa: A New Approach to Media Development?" report on a workshop organised by the Programme in Comparative Media Law and Policy and Stanhope Centre for Communications Policy Research. Oxford: University of Oxford.

Gorfinkel, Lauren, Sandy Joffe, Cobus Van Staden and Yu-Shan Wu. 2014. "CCTV's Global Outreach: Questioning the Audiences of China's 'New Voice' on Africa." *Media International Australia, Incorporating Culture and Policy* 151: 81–88.

Habermas, Jurgen. 1989. "The Public Sphere: An Encyclopedia Article." In *Critical Theory and Society: A Reader*, edited by Stephen Bonner and Douglas MacKay Kellner, 136–144. New York and London: Routledge.

Hanitzsch, Thomas. 2007. "Deconstructing Journalism Culture: Towards a Universal Theory." *Communication Theory* 17(4): 367–385.

Harper, Stephen. 2012. "The Chinese are Coming!: Representations of Chinese Soft Power in a BBC Television Documentary." In *China and the West: Encounters with the Other in Culture, Arts, Politics and Everyday Life*, edited by Lili Hernandez, 33–44. Newcastle: Cambridge Scholars Publishing.

Hartig, Falk. 2015. "Communicating China to the World: Confucius Institutes and China's Strategic Narratives." *Politics* 35(3–4): 245–258.

Hassid, Jonathan and Sun, Wanning. 2015. "Stability Maintenance and Chinese Media: Beyond Political Communication?" *Journal of Current Chinese Affairs* 44(2): 3–15.

Human Rights Watch, 2014. "'They Know Everything We Do': Telecom and Internet Surveillance in Ethiopia" www.hrw.org/report/2014/03/25/they-know-everything-we-do/telecom-and-internet-surveillance-ethiopia.

Huynh, Tu T. 2012. "What People, What Cultural Exchange? A Reflection on China– Africa." *African East-Asian Affairs: The China Monitor* 2, November: 3–16.

Joseph, May. 1999. "Kung Fu Cinema, Frugality and Tanzanian Asian Youth Culture: Ujamaa and Tanzanian Youth in the Seventies." In *SportCult*, edited by Randy Martin and Toby Miller, 41–63. Minneapolis, MN: University of Minnesota Press.

Jura, Jaroslaw and Kaluzynska, Kaja. 2013. "Not Confucius, nor Kung Fu: Economy and Business as Chinese Soft Power in Africa." *Africa East-Asian Affairs* 1: 42–69.

Kaigwa, Mark and Wu, Yu-Shan. 2015. "How China-Africa Relations Take on New Meaning Thanks to Digital Communication." In *Africa and China: How Africans and Their*

Governments are Shaping Relations with China, edited by Aleksandra W. Gadzala, 149–170. Lanham, MD: Rowman & Littlefield.

King, Kenneth. 2013. *China's Aid and Soft Power in Africa*. Oxford: James Currey.

Kurlantzick, Joshua. 2006. "China's Charm: Implications of Chinese Soft Power." Carnegie Endowment for International Peace Policy Brief 47 (June).

Li, Anshan. 2014. "Soft Power and the Role of Media: A Case Study of PKU African Tele-Info (Draft)", www.cmi.no/file/2655-.pdf, accessed 25 September 2016.

Li, Shubo and Helge Rønning. 2013. "Half-orchestrated, Half Freestyle: Soft Power and Reporting Africa in China." *Ecquid Novi: African Journalism Studies* 34(3): 102–124.

Lloyd, Libby. 2013. "South Africa's Media 20 Years After Apartheid." Center for International Media Assistance, July 17.

Lu, Jinghao and Cobus van Staden. 2013. "Lonely Nights Online: How does Social Networking Channel Chinese Migration and Business to Africa?" *African East-Asian Affairs* 1: 94–116.

Mawdsley, Emma. 2008. "Fu Manchu versus Dr Livingstone in the Dark Continent? Representing China, Africa and the West in British Broadsheet Newspapers." *Political Geography* 27(5): 509–529.

Monson, Jamie and Stephanie Rupp. 2013. "Africa and China: New Engagements, New Research." *African Studies Review* 56(1): 21–44.

Nye, Joseph S. Jr. 1990. "Soft Power." *Foreign Policy* 80: 153–171.

Nye, Joseph S. Jr. 2002. "Hard and soft power in the global information age." In *Re-ordering the world*, edited by Mark Leonard, 2–10. London: Foreign Policy Centre.

Pan, Su-Yan. 2013. "Confucius Institute Project: China's Cultural Diplomacy and Soft Power Projection." *Asian Education and Development Studies* 2(1): 22–33.

Paradise, James F. 2009. "China and International Harmony: The Role of Confucius Institutes in Bolstering Beijing's Soft Power." *Asian Survey* 49(4): 647–669.

Procopio, Maddalena. 2015. "The Effectiveness of Confucius Institutes as a Tool of China's Soft Power in South Africa." *African East-Asian Affairs* 2: 98–125.

Rawnsley, Gary D. 2009. "China Talks Back: Public Diplomacy and Soft Power for the Chinese Century." In *Routledge Handbook of Public Diplomacy*, edited by Nancy Snow and Philip M. Taylor, 282–291. New York and London: Routledge.

Sautman, Barry and Yan, Hairong. 2009. "African Perspectives on China–Africa links." *The China Quarterly* 199: 728–759.

Sun, Wanning. 2010. "Mission Impossible? Soft Power, Communication Capacity, and Globalisation of Chinese Media." *International Journal of Communication* 4: 54–72.

Tomlinson, John. 1991. *Cultural Imperialism: A Critical Introduction*. Baltimore, MD: Johns Hopkins University Press.

Van Staden, Cobus. 2013. "Editor's Introduction: Electric Shadows: Media in East Asian/African Relations". *African East-Asian Affairs* 1(1): 4–15.

Van Staden, Cobus. 2014. "Moomin/Mūmin/Moemin: Apartheid-Era Dubbing and Japanese Anime." *Critical Arts* 28(1): 1–18.

Van Staden, Cobus. 2017. "Watching Hong Kong martial arts film under apartheid." *Journal of African Cultural Studies* 29(1): 46–62.

Wan, James. 2015. "Propaganda or proper journalism? China's media expansion in Africa." *AfricanArguments*. August 18.

Wang, Yiwei. 2008. "Public Diplomacy and the Rise of Chinese Soft Power." *The ANNALS of the American Academy of Political and Social Science* 616(1): 94–109.

Wasserman, Herman. 2015. "China's 'Soft Power' and Its Influence on Editorial Agendas in South Africa." *Chinese Journal of Communication* 9(1): 8–20.

Wekesa, Bob. 2013. "The Media Framing of China's Image in East Africa: An Exploratory Study." *Africa East-Asian Affairs* 1: 42–69.

Wekesa, Bob and Zhang Yanqiu. 2014. "Live, Talk, Faces: An Analysis of CCTV's Adaptation to the African Media Market." CCS Discussion paper, Stellenbosch University, Centre for Chinese Studies Discussion Paper 2.

Wu, Yu-Shan. 2012. "The Rise of China's State-led Media Dynasty in Africa." SAIIA Occasional Paper no. 117, available at www.saiia.org.za/occasional-papers/the-rise-of-chinas-state-led-media-dynasty-in-africa.

Wu, Yu-Shan. 2013. "The Political and Diplomatic Implications of Social Media: The Cases of China and South Africa." *African East-Asian Affairs* 1: 70–93.

Wu, Yu-Shan. 2016. "China's Media and Public Diplomacy Approach in Africa: Illustrations from South Africa." *Chinese Journal of Communication* 9(1): 81–97.

Youngman, Frank. 2014. "Engaging Academically with China in Africa – the Institutional Approach of the University of Botswana." *African-East-Asian Affairs* 3 (October 6).

Zhang, Xiaoling. 2010. "Chinese State Media Going Global." *East Asia Policy* 2(1): 42–50.

Zhang, Xiaoling, Herman Wasserman and Winston Mano. eds. 2016. *China's Media and Soft Power in Africa*. London: Palgrave Macmillan.

Zhang, Xiaoling, Herman Wasserman and Winston Mano. 2016. "China's Expanding Influence in Africa: Projection, Perception and Prospects in Southern African Countries." *Communicatio: South African Journal for Communication Theory and Research* 42(1): 1–22.

Zhu, Ying. 2012. *Two Billion Eyes: The Story of China Central Television*. New York: The New Press.

6

"CHINA IN AFRICA" IN THE ANTHROPOCENE

A discourse of divergence in a converging world

Ross Anthony

Within the context of China's growing integration into the global economy, mainstream Euro-American voices, emanating from the realms of politics, media, think-tanks and, to a certain extent, academia, have portrayed the opposite: namely, isolating China from this integration. Callahan (2012) refers to champions of China's rise as being subject to "China Exceptionalism," viewing China as an almost timeless entity to be viewed differently from the rest of world-historical time. Instead of articulating China as part of a "network-based logic of globalisation that ties us all together," it asserts "a sharp geopolitical vision of the world" (Callahan 2012, 50). Critics of contemporary China highlight its state driven model of market engagement, authoritarian government, perceived threat to global stability and pressure on global resources and environmental sustainability as threatening aspects of its "rise."

Nowhere do these anxieties manifest themselves more than in Euro-American discourses on the Chinese engagement in Africa. Former US president Barack Obama, for example, stated in 2014 that China has a "need for natural resources that colours their investments in a way that's less true for the United States" ("Barack Obama talks to *The Economist*: An Interview with the President." *The Economist,* August 2, 2014). Former British Prime Minister David Cameron was quoted as being "increasingly alarmed by Beijing's leading role in the new 'Scramble for Africa'" and "warned African states over China's 'authoritarian capitalism' ... claiming it is unsustainable in the long term" (Hirono and Suzuki 2014, 444), while during her tenure as US Secretary of State, during a visit to Africa in 2012, Hillary Clinton stated: "America will stand up for democracy and universal human rights even when it might be easier to look the other way and keep the resources flowing" (Smith, 2012) – a barb widely believed to be aimed at China. In this context, Africa functions as a site of contestation in which Euro-American ideologies exert influence through championing, in many regards, their difference from China.

The construction of a worldview that is essentially binary is both a powerful and popular one. Today, both nationalism and transnational-capitalism are normative modes of engagement in international affairs, and it is often tempting to deploy nationalist representations that betray an increasingly complex and interlinked world. In focusing on the Western fixation on China's influence within Africa, this chapter highlights the growing inappropriateness of such a model within the current global economic system. Against the backdrop of what is increasingly being dubbed the Anthropocene, understood as human-based alteration of both climate and eco-systems exacerbated by a hyper competitive economic model focused on consumption and GDP growth, I interrogate the Euro-American discourse on China–Africa. In highlighting how this discourse seeks to separate China's behaviour from its own behaviour, and ascribe binary values to these engagements, the discourse signals a broad-scale failure to engage in one of the largest challenges of the twenty-first century: namely how to grapple with the contradiction which is the spread of post–Cold War market capitalism to large swathes of the globe (a system both historically and currently endorsed by Western powers), with the planetary damage that such a spread is accelerating.

Due to space constraints, this chapter focuses primarily on the Euro-American dimension of the China–Africa discourse. It should be noted at the outset that China itself is also often at pains to point out its own exceptionalism. The Chinese government often asserts a moral position vis-à-vis Western engagement, highlighting the latter's colonial past in Africa and its solidarity with Africa in terms of its own subjection to Western imperial incursion. Within China, there exists a very different China–Africa discourse, couched in a positively charged language of historical struggles against western imperialism and the championing of development assistance. China's integration into the global market economy, and its "going out" policy over the past few decades, however, entail that this exceptionalism is at odds with China's growing role in world affairs. In this respect, the criticism laid against the Euro-American account of "China-Africa" is increasingly applicable to China itself. Within Africa, discourses around Chinese interaction are far more complex, due predominantly to the kaleidoscope of different countries involved, as well as significant differences in state and civil society accounts of the engagement. Additionally, both China and Euro-America influence African discourses through their comparatively powerful media. Further research into the convergences and disparities between these different versions of the China–Africa discourse is needed.

The Euro-American construction of "China in Africa"

Beginning in the nineteenth century, with the work of Friedrich Nietzsche, a focus on the conditions under which scientific, historical, ethical, and political norms are produced has influenced generations of prominent thinkers. Nietzsche's *On the Genealogy of Morality* (1887), for instance, argued how values, such as the concept of "good," are essentially rooted in how the nobility class came to stand as a general norm against which the marginalised defined themselves (2006: 13). Karl Marx and Friedrich Engels developed this line of thinking in *The German Ideology* (1846) insofar as they assert that large portions of the history of humankind has constituted mainly the history of elite, often literate classes who wrote history by themselves and for themselves (1968: 17). Later theorists, such as Antonio Gramsci, used the notion of hegemony to explain how the cultural norms of ruling elites extend over entire populations (Bates 1975, 353). A more recent thinker who interrogated how "truth" discourses and power relate to each other is Michel Foucault, whose books frequently focus on denaturalizing taken-for-granted institutions, such as the prison, the insane, or sexuality, then tracing shifting social practices and often idiosyncratic changes in their histories (genealogies). Heavily indebted to Nietzsche, Foucault argues for a knowledge–power axiom in which discourse plays an important role in structuring human subjectivity, something which has been subsequently taken up in critical theory studies of power in relations to race, gender, and other naturalised classifications.

The emphasis on discursive formations was subsequently to have implications for re-assessing the West's construction of other parts of the globe, particularly the Western colonial period. A pioneer in this regard is Edward Said, whose notion of "Orientalism" demonstrated how scholarly, "scientific" work on regions which were being colonised (Said focuses on the French construction of the Levant) effectively produced a body of knowledge that presented itself as neutral scientific fact but was, in reality, a discourse that reflected social norms of contemporary French society. Said's work is indebted to Foucault's focus on how institutions produce knowledge and how this process is influenced by pre-existing power structures. Thus, knowledge is not neutral; it categorises the world in ways that align with the interests of the originators of knowledge practices, reflected in Foucault's term "power/knowledge."

The emergence of the contemporary China–Africa discourse, which has arisen more or less in tandem with China's rise as a global economic actor, is in some ways akin to Said's use of the term. It attempts to articulate a series of objective facts "out there," while simultaneously functioning as a product of the social and political conditions under which it has been constructed. These underlying conditions are important because they play a role in structuring the way in which the world in conceived, and thus influence the kinds of knowledge produced. Within the Euro-American construction of the China–Africa discourse, a key marker of this is the way in which China's engagement in Africa, and its economic rise more generally, is treated with suspicion. This bias is widespread within mainstream media[1] and has filtered into high-level political discourse. Miwa Hirono and Shogo Suzuki argue that the political science literature on Chinese foreign policy in Africa is "heavily influenced" by notions of a China threat (2014), the likes of which are echoed in the mass media.

This critique can be situated within a broader literature that examines how traditionally powerful actors construct representations of the other that suit them, which Luttwak calls "Great Power Autism" (2012). The Euro-American sphere has been a site of particular focus in this regard due to its several hundred years of dominance in the global arena (Frank 1998; Wallerstein 2011). It is thus not only a discursive construction of China that is relevant here, but equally so, a particular depiction of Africa. Within the context of the Euro-American relationship with Africa, Mbembe argues that "Africa as an idea, concept, has historically served, and continues to serve, as a polemical argument for the West's desperate desire to assert its difference from the rest of the world" (2001: 2). In this reading, Africa serves as a binary against which Europe constructed concepts of civilization and modernity, achieved through institutional interventions in religion, schooling, administration, language and many other spheres of social life.

Dubbed "neo-colonialism" by Ghana's first president, Kwame Nkrumah, it has been argued that Africa's post-colonial situation functions, in certain cases, as an extension of this earlier period (Yates 1996). From this perspective, Western powers are responsible for its perpetuation through the likes of engagement during the Cold War period and structural adjustment reforms of the 1980s and 1990s, the latter of which, it has been argued, contributed toward the erosion of democracy (King 2003) and opening up markets to devastating foreign competition (Bond 2004). From this *longue dureé* standpoint, there is a sense that Western powers have viewed Africa as its own preserve, and that the rise of China and other "emerging powers" are unsettling this relationship. The emergence of the BRICS countries and others who have taken an active economic interest in Africa has now begun to challenge this monopoly in a way not evident since the Cold War era. These new interactions have given rise to a discourse that asserts the West is "losing out" to countries such as China in Africa, which for decades had been dubbed "the Lost continent," a continent that had little to do with the new challenges of globalisation (Ferguson 2006: 26).

Within the context of new emerging powers, particularly China, there has arisen a discourse that has been referred to as "the new scramble for Africa" (Carmody 2011). Referring to Thomas Packenham's original colonial "scramble for Africa" (1992) amongst European powers, now this is involving China and other developing world countries, with the stakes focused on natural resources as opposed to territorial claims. This process has put Africa back on the map of the broader geopolitical imagination, with mainstream publications such as *The Econo-mist* now portraying Africa as the "hopeful continent" ("Africa Rising: A Hopeful Continent." February 28, 2013) with "1.2 billion opportunities" ("Business in Africa: 1.2 billion opportunities." April 14, 2016). It is particularly the notion of Africa's under-development, that is to say, the vast future *potential* in terms of resource extraction, consumer markets, agricultural development, and other hitherto untapped markets, which underlies the current discursive rise of Africa in the West, and particularly "China in Africa." Conversely, this carries the implication that Africa is, in effect, the continent least integrated into networks of global capital; with the exception of certain urban, resource and eco-friendly enclaves, Africa continues to exist largely in the "shadow" of the neoliberal order (Ferguson 2006).

China (and the world) in Africa

This antagonistic discourse toward the rise of China, as well as China's counter-discourses, assert themselves today, in practice, primarily through the mechanism of the global market. Cerny argues that we should understand the market as facilitating a process whereby states increasingly function as "quasi enterprise" associations within the context of political globalisation (Cerny 1997, 251). In this vein, anxiety around China is expressed, for instance, in the likes of China's manufacturing advantage, its implications for job loss, the debt the United States owes China, issues around intellectual property appropriation, protection of strategic industries from Chinese influence (and vice versa), and accusations of protectionism. Within Africa, the discourse reiterates the aforementioned "scramble" narrative, in which blocs of power vie for commercial influence in Africa, with the West highly critical of China's general style of market engagement (including accusations of labour abuses, environmental degradation, and shoddy workmanship).

Nevertheless, while the notion of competition of nation states, as well as broader multilateral groupings, easily lend themselves toward thinking in terms of potentially antagonistic global imaginaries, they equally have the potential to foster a more complex understanding of the world. While contemporary globalisation may, on the one hand, exacerbate nation state competition, at another level, such globalisation initiates numerous transnational forces (i.e. commodity and population flows) that complicate the simplicity of competitive sovereign states. From within both the Euro-American and Chinese positions, there are alternative proponents that tend to accept China's increasingly normative position within the global market economy. In this vein, China's integration is embedded within much larger global circulations of finance, goods, and people. Within China's broader economic engagement, this complexity is evident in new institutions such as the Asian Infrastructure Investment Bank, initiated by China and including most major countries (but not Japan or the United States), which will fund development projects on a global scale. Even the BRICS grouping, despite rhetoric aiming to challenge the existing global economic order, declares in its Fortaleza Declaration that it "will supplement the efforts of multilateral and regional financial institutions for global development" (BRICS Ministry of External Relations, 2014).

This more integrative view of China's "rise" is also increasingly evident with regards to its African engagements. Within Africa, China is involved, directly and indirectly, with numerous international actors. Within Sudan's oil industry, for instance, Sudanese, Indian, Malaysian and Chinese companies extract oil in the form of a multinational joint venture; in Uganda's Lake Albert region, China is involved in a joint multi-national venture in which the China National Offshore Oil Corporation holds a stake along with (Anglo-Irish) Tullow and France's Total. Chinese companies engage with various other partners, for instance in the case of Italian Eni striking a deal with China to procure natural gas off the coast of Mozambique (Economy and Levi 2015, 46–47). Within the context of infrastructure, huge projects such as the Lamu Port Southern Sudan Ethiopia Transport Corridor (or LAPSSET), include a host of different investors and contractors, in which the Chinese are but one of a number of actors, including the Dutch, Korean, and Japanese companies, the Kenyan and Ethiopian governments and other bodies such as European Union and the African Development Bank. Additionally, it must be kept in mind that commodities extracted in Africa often feed into a number of materials and processes whose end result is consumable goods in other parts of the world. For instance, Chinese oil extracted within Africa is sold on international markets, rather than hoarded domestically. There is relatively poor co-ordination between the various national oil companies in Africa with the Chinese presence expanding rather than contracting the amount of oil available on global markets (Downs 2007, 47).

This more integrative understanding of China's presence in Africa serves as a counter-discourse that has been promoted by both Western and Chinese political classes. Even some who are suspicious of Chinese engagements in Africa simultaneously grasp their role as a market actor. Barrack Obama, commenting on the Chinese presence in Africa, said "the more the merrier" and that the Chinese should be "welcomed" (*The Economist*, August 2, 2014). On the Chinese side, this pragmatism is also visible. The former Special Representative of the Chinese Government on African Affairs, Zhong Jianhua, argued that "China's job, our responsibility, is to try and help Africa compete with us" (Africa Research Institute 2013), while academics from the Chinese Academy of Social Sciences, like He Wenping, argue that China is "glad to see … benign competition" and "growth of African business environment" ("Sino-African Ties Demonstrate Cooperation, not competition or exploitation." *Global Times,* December 7, 2015). Additionally, within Africa itself, there are a multiplicity of discourses on China, with numerous governments championing, and occasionally chastizing, the China–Africa narrative.[2]

A number of researchers in the field have tried to overcome this disconnection through relativizing China as one amongst many other actors on the African continent. Barry Sautman and Yan Hairong, for instance, argue:

> The main problem with the China-in-Africa discourse is not empirical inaccuracies about Chinese activities in Africa, but rather the decontextualization of criticisms … Some analyses positively cast Western actions in Africa compared to China's activities; others lack comparative perspective in discussing negative aspects of China's presence, so that discourse consumers see a few trees, but are not given a view of the whole forest. Such analysis reflects Western elite perception of national interests or moral superiority as these impinge on "strategic competition" with China.
>
> *(2008: 27–28)*

Sautman and Yan assert that while China is part of the "world system," their style of market engagement is tempered due to China's past experience as a semi-colony, its socialist legacy and developing country status which make its engagement "less injurious to African sensibilities about rights" than Western counterparts. This position echoes the work of C.K. Lee who, in a comparison of Western and Chinese-run copper mines in Zambia, argues that there are "two varieties of capital" (2014, 35) at work, the former, for instance, more at ease with disinvesting during market down-turns, as opposed to Chinese companies' longer-term commitments despite market fluctuations. Nevertheless, this latter model offers "no capacity to undermine the prevailing neoliberal order, nor any interest in replacing it" (Lee 2014, 64). Similar critiques, although ones that put China more on an even plane with Western actors, are advanced by authors such as Mohan and Power, who argue that rather than "south-south cooperation," China's interest in Africa might be the more familiar and hegemonic "north-south relationship" (2009, 7), in which China is equated both domestically and internationally as subject to a process of neoliberalism. Ian Taylor argues more broadly of "state capitalism" evident in the BRICS economies:

> It is quite obvious that the state capitalist emerging economies are structurally integrated into the ongoing world order under the hegemony of neoliberal capitalism. They do not represent a different or alternative order, other than one where these activities are incorporated as notional equals.
>
> *(2014: 345)*

Additionally, scholarship has emerged that has reversed the notion of a passive Africa being host to the agent, China. Authors such as Corkin (2013) and Gadzala (2015) have demonstrated how African actors in Angola, Equatorial Guinea and Ethiopia, function as savvy operators who drive hard bargains with companies, be they Chinese or others, in terms of resource and infrastructure deals. Their active engagement often entails the accumulation of stupendous wealth and power at the expense of the broader populace.

We're all in this together: global capital in the Age of the Anthropocene

The persistence of a China–Africa discourse that is binary and ascribes positive and negative values to actors is not only empirically inaccurate but, more crucially, signals a failure to locate an adequate representation of the world in an age of economic globalisation. The current paradigm, which echoes an earlier Cold War imaginary of a globe divided between good and evil, seems increasingly ill fitted to the current context. This issue is most pressing as we enter what scientists have now officially categorised as the Anthropocene, an ecological era in which human-based planetary alteration will be inscribed in the fossil record. The Anthropocene is a chronological geological term that dates evidence of human activities in terms of their significant global impact on the earth's ecosystems. Sometimes dubbed "the sixth extinction" (Kolbert 2014), the event shares with the other five great extinctions on earth, the following traits: the catastrophic loss of global biodiversity; the rapid speed of this loss relative to evolutionary and geological time; a non-random extinction which effects entire taxa of species, while other groups remain largely unaffected; survivors which are often not previously dominant groups (Sodhi et al. 2008). [3]

While there are debates as to when, precisely, the Anthropocene began, it is almost universally recognised that capitalism, and particularly contemporary global capitalism, is playing a key role in its acceleration. Jason Moore makes a forceful argument for the inextricable link between the rise of the Anthropocene and the historical rise of European capitalist economies. Drawing on the Marxist theory of "value in motion," he argues that a world view begins to emerge which must perpetually accumulate "nature's free gifts" through commodity production (Burkett 1999; Moore 2015). As increasing numbers of nation state elites promote integration into the global economic system, the issue of ecological and climatological crises has, unsurprisingly, been propounded by the expansion of this principle to various parts of the world. Regions such as Africa, and other areas of the global South, viewed as new "frontiers" in terms of commodities and trade, also possess, for example, the majority of the planet's biodiversity. In the current context, the rise of multiple market actors expanding into new markets, coupled with the desires of recipient nation states' need for development and economic growth, increasingly come into conflict with the limits of "cheap nature."

Within this context, the following points stand out. First, the expansion of global market competition over the past few decades is exacerbating the shift into the Anthropocene; China's integration into this system has accelerated the problem. Second, the acceleration of this process entails a far greater degree of entanglement between Chinese and other actors, in which flows of capital, commodities and labour form part of global commodity chains. As a result, separating "Chinese" actors from other actors becomes increasingly difficult. In this vein, consumption patterns in one part of the world can be linked to extraction processes in other parts. For instance, with relevance to the China–Africa debate, a common criticism in the Euro-American sphere is that China is destroying global forestry reserves in regions such as Southeast Asia and Africa (in countries such as Mozambique and Gabon). While it is very hard to trace these kinds of chains, a survey report by Forest Trends argues that a key driver of China's illegal logging in these regions is the Euro-American sphere itself: the growing demand for affordable furniture in the United States and the European Union, imported from China, is a "key driver" in these broader global forest product commodity chains (Sun et al. 2005, 16).

A world thus increasingly interconnected, coupled with an emerging realization of increased ecological destruction propelled by the economic and political ideologies underlying this interconnectivity, makes the dominant principle of territorial, nation-state sovereignty increasingly difficult to sustain. As Biermann and Dingwerth argue, "global environmental change increases the mutual dependence of nation states, thereby further undermining the idea of sovereignty as enshrined in the traditional Westphalian system" (2004, 2). This is not least because planetary ecology long pre-dates human territorial constructions and does not acknowledge their borders. If we situate the dominant Euro-American discourse on China–Africa relations within the context of national and regional territorial competition, it follows that, in light of the current global environmental crises, this paradigm is increasingly ill-suited for the challenges that lie ahead. In many respects, China's integration into the global market system over the past few decades has signalled a new era of "emerging powers," a world in which regions once dominated by the West now compete with them predominantly through market means. The persistence of the "new scramble" narrative, the notion of the "Beijing consensus" and the likes of a "New Cold War", suggest a reactionary conceptualization of the current global context. In many respects, the rapid influence of China on continents such as Africa and South America is symptomatic of this new global reality, and thus reactions to this within the Euro-American sphere serve to highlight how new forms of global circulation are subordinate to notions of national and regional competition.

In a general sense, the issues raised here are vast and open up a host of concerns around re-thinking global political economy in the context of the Anthropocene. An emerging body of literature examines issues of ethics of the Anthropocene, and how it relates the re-thinking of issues such as the acceleration of consumerism and capitalism more generally, as well as the role of territory, scarcity, and inter-species relations (Zylinska 2014). The issue of China's rise is but one extension within this broader context – but an important one. The sheer scale of China's population, and its shift toward a consumer-based society, is significant, as is the fact the Euro-American sphere now faces significant global market competition. Issues such as the China–Africa question can be viewed as an extension of this process, highlighting a number of ethical questions relevant to the Anthropocene in general, such as the sustainability of neoliberal capitalism, as well as issues of historical equity, justice, and the right to development. Within the context of Africa, one of the least integrated regions within the global market economy while simultaneously uniquely rich in resources, these issues are particularly pertinent. It is for this reason that considerably novel phenomena, such as China's economic rise on the continent over the past two decades, may benefit from analyses that take the Anthropocene into account.

Research implications

Within this context, the China–Africa discourse, as characterised here, is concerning insofar as it articulates the issue from a perspective which, in many respects, runs counter to these challenges. In many respects, criticisms implicit in the China–Africa discourse (cut throat competition, environmental degradation, corruption, "primitive" accumulation and exploitation) can be attributed to underlying mechanisms of global capitalism as a whole. In this vein, both the Euro-American and Chinese discourses, which tend to pitch the engagement in terms of a moral choice which Africans should make, misses, in Sautman and Yan's parlance, "the forest for the trees" (Sautman and Yan 2008). If, hypothetically, China were still deeply immersed in Maoist-era Communism today, it would be more plausible that Africa indeed faced an "authentic" choice between traditional western partners and China. Today capital flows frequently transcend the boundaries of nation states and entail collaborations across international scales. Take, for instance, the multi-billion-dollar Chinese company, Queensway Group, with significant dealings in Angola, Guinea, and Zimbabwe, described in the *Financial Times* (Tom Burgis, "China in Africa: How Sam Pa Became the Middle Man." August 8, 2014):

> The group is in business with BP, Total and the commodity trader Glencore; it boasts interests stretching from Indonesian gas and oil-refining in Dubai to luxury apartments in Singapore and a fleet of Airbus jets; it is active in North Korea and Russia. It comprises a web of private and offshore companies underpinning two main enterprises: China Sonangol, which is principally an oil company (although it also own the former JP Morgan building opposite the New York Stock Exchange on Wall Street) and China International Fund, an infrastructure and mining arm, whose flag flies above the entrance to Luanda's golden skyscraper.

This description of a Chinese actor – a very wealthy one, admittedly – captures a network in which Africa is merely a part of a much broader set of relations. In doing so, the notion of "Chinese engagement" is grasped in increasingly integrative, rather than excised, terms. Within academia, new approaches within the social sciences, such as actor-network theory, trace networks of flows of peoples, finances, commodities, and other forms of material life so as to better describe empirical reality. Within the spirit of this approach, numerous detailed accounts have emerged which follow the likes of pipelines (Marriott and Minio-Paluello 2012), roads (Harvey and Knox 2015), commodity chains (Hughes and Reimer 2004), and the transnational exchanges of migrant communities (Chu 2010). Within the China–Africa research community, work of a similar forensic nature exists, particularly in research into the oil industry (Patey 2014; Corkin 2013). From this perspective, the roles of China, and other actors tend to emerge from the tracing of various relations, rather than beginning one's analysis with China.

Nevertheless, the network perspective must still contend with dominant representations of China–Africa relations, in which authors can sometimes perpetuate the very issues they are arguing against. For instance, numerous research pieces (including ones by this author), embark by setting their arguments up *against* the accusation that China's engagement in Africa is largely negative. Terms such as "rogue aid," "neo colonial," issues of "human rights" and "labour abuses" and "environmental degradation" are frequently raised, often to be dispelled or argued against.[4] Given the large-scale negativity around China's image in Africa as portrayed in the media, such an approach seems justified. However, such a perspective nevertheless risks perpetuating territorially based, binary, value imbued thinking. Thus, even the best research in the field is in some sense "structured" by this broader discursive formation. It echoes what Stuart Hall argues regarding the constricting powers of discourse more generally; they enable the construction of a topic in a certain way but also "limits the other ways in which the topic can be constructed" (2006, 165). The China–Africa scholarship of the future will have to negotiate and interrogate such a limit.

It is thus imperative that the rise of China in Africa be approached within the broader context of the spread of market capitalism to a number of developing regions in the world and the increasing interconnectivity this produces. This is in the interests of making empirical, on the ground, investigations more accurate, which are at risk of being obscured when an obsessive focus on China brackets out other actors. More broadly, within the context of the rapid ecological changes that are being accelerated by the spread of market economies across the globe, analyses dwelling almost exclusively on the role of China fail to identify the broader context and its challenges. The global spread of an economic and political ideology based on profit and growth beyond the confines of the western world increasingly heralds a crisis of global proportions, not only insofar as it increasingly threatens species survival but also because none of the actors in question have offered any serious economic and political alternatives to the current model, other than mitigation and further market-based solutions. The persistence of Cold War and colonial-era binary conceptualizations in an increasingly singular world-system tends to deflect addressing some of the major problems facing the twenty-first century.

Notes

1 See, for examples, Andrew Malone, "How China is Taking over Africa, and Why the West Should be VERY Worried." *The Daily Mail.* July 18, 2008; CNBC's "Recolonizing Africa: A Modern Chinese Story" (Esposito, Tse, Al-Sayed 2014), *New York Review of Books'* "The Chinese Invade Africa" (Johnson 2014).
2 In countries such as Zambia, "China bashing" has been mobilised in the interests of electoral politics (Sautman and Yan 2014) while in South Sudan, at least at the level of civil society, there is lingering resentment against Chinese support for Khartoum during the Sudanese civil war (Anthony and Jiang 2014).
3 Dubbed "The Big Five": The Cretaceous-Paleogene; the Triassic Jurassic; Permian-Triassic; the Late Devonian; and the Ordovician Silurian.
4 A few randomly chosen quotes from introductory paragraphs and abstracts include: "Discussions on the politics of Chinese engagement with African development have been marked by increasing concern over Chinese use of aid in exchange for preferential energy deals" (Mullins, Mohan and Power 2010, 857); "The West has been critical of China's rise in Africa, pointing to a 'Chinese resource grab', support for dictatorial and corrupt regimes under the guise of non-interference, and its use of non-transparent practices to 'corner the African market' (at the expense of Western economic interests) in the context of a 'new' scramble for Africa's resources and markets" (Cheru and Obi 2011, 72); "At this critical juncture in our history, it behooves the masses of the African people, political leaders, intellectuals, politicians and lovers of Africa to earnestly determine what the Chinese really want from Africa? What is the hidden motive of these Chinese in Africa?" (Insaidoo, 2016).

References

Africa Research Institute. 2013. "Ambassador Zhong Jianhua: China's Special Representative on African Affairs, on Trade, Aid and Jobs," August 6, www.africaresearchinstitute.org/newsite/publications/ambassador-zhong-jianhua-on-trade-aid-and-jobs/.
Anthony, Ross and Hengkun Jiang. 2014. "Security and Engagement: The Case of China and South Sudan." *African-East Asian Affairs* 4: 78–96.

Appadurai, Arjun. ed. 1988. *The Social Life of Things: Commodities in Cultural Perspective.* Cambridge: Cambridge University Press.

Bates, Thomas. 1975. "Gramsci and the Theory of Hegemony." *Journal of the History of Ideas* 36(2): 351–366.

Biermann, Frank and Dingwerth, Klaus. 2004. "Global Environmental Change and the Nation State" *Global Environmental Politics* 4(1): 1–22.

Bond, Patrick. 2004. *Against Global Apartheid: South Africa Meets the World Bank, IMF and International Finance.* Cape Town: University of Cape Town Press.

BRICS Ministry of External Relations. 2014. "Sixth BRICS Summit: Fortaleza Declaration," http://brics.itamaraty.gov.br/media2/press-releases/214-sixth-brics-summit-fortalez a-declaration.

Burkett, Paul. 1999. "Nature's 'Free Gifts' and the Ecological Significance of Value." *Capital and Class* 23: 89–110.

Callahan, William A. 2012. "Sino-speak: Chinese Exceptionalism and the Politics of History." *The Journal of Asian Studies* 71(1): 33–55.

Carmody, Padraig. 2011. *The New Scramble for Africa.* Cambridge, MA: Polity Press.

Cerny, Philip. 1997. "Paradoxes of the Competition State: The Dynamics of Political Globalization." *Government and Opposition* 32(2): 251–274.

Cheru, Fantu and Obi, Cyril. 2011. "De-Coding China–Africa Relations: Partnership for Development or (Neo) Colonialism by Invitation?" *The World Financial Review*, 25: September.

Chu, Julie. C. 2010. *Cosmologies of Credit Transnational Mobility and the Politics of Destination in China.* Durham, NC: Duke University Press.

Chun, Lin. 2013. *China and Global Capitalism: Reflections on Marxism, History, and Contemporary Politics.* Basingstoke: Palgrave Macmillan.

Comaroff, Jean and Comaroff, John L. 2011. *Theory from the South: Or, How Euro-America is Evolving Toward Africa.* London: Routledge.

Corkin, Lucy. 2013. *Uncovering African Agency: Angola's Management of China's Credit Lines.* London: Routledge edition.

Dews, Fred. 2014. "8 Facts about China's Investments in Africa." The Brookings Institute. May 20. www.brookings.edu/blogs/brookings-now/posts/2014/05/8-facts-about-china-investment-in-africa.

Dickson, Bruce J. 2007. "Integrating Wealth and Power in China: The Communist Party's Embrace of the Private Sector." *The China Quarterly* 192: 827–854.

Downs, Erica S. 2007. "The Fact and Fiction of Sino-African Energy Relations." *China Security* 3(3): 42–68.

Economy, Elizabeth C and Levi, Michael. 2015. *By All Means Necessary: How China's Resource Quest is Changing the World.* Oxford: Oxford University Press.

Edward Said. 1979. *Orientalism.* New York: Vintage.

Easposito, Mark., Tse, Terence and Al-Sayed, Merit. 2014. "Recolonizing China: A Modern African Story?", December 30. www.cnbc.com/2014/12/30/recolonizing-africa-a-mode rn-chinese-story.html.

Ferguson, James. 2006. *Global Shadows: Africa in the Neoliberal World Order.* Durham, NC: Duke University Press Books.

Ferguson, Neil and Schularick, Mrtiz. 2007. "'Chimerica' and the Global Asset Market Boom." *International Finance* 10: 215–239.

Foucault, Michel. 1990. *The History of Sexuality, Vol. 1: An Introduction.* Translated by Robert Hurley. New York: Vintage (Reissue edition).

Frank, Andre Gunder. 1998. *Re-Orient: Global Economy in the Asian Age.* Berkeley: University of California Press.

Gadzala, Aleksandra W. ed. 2015. *Africa and China: How Africans and Their Governments are Shaping Relations with China*. Lanham, MD: Rowman & Littlefield.

Gallagher, Mary Elizabeth. 2001. *Contagious Capitalism: Globalization and the Politics of Labour in China*. Princeton, NJ: Princeton University Press.

Halakhe, Abdullahi Boru. 2014. "China's Rise Meets America's Decline in Africa." *Aljazeera*. May 27. http://america.aljazeera.com/opinions/2014/5/china-africa-unitedstatestra deinvestmentdiplomacy.html.

Hall, Stuart. 2006. "The West and the Rest: Discourse and Power." In *The Indigenous Experience: Global Perspectives*. Edited by Roger C.A. Maaka and Chris Anderson, 165–173. Toronto: Canadian Scholars Press.

Halper, Stefan. 2012. *The Beijing Consensus: Legitimizing Authoritarianism in Our Time*. New York: Basic Books.

Hardt, Michael and Negri, Antonio. 2011. *Empire*. Cambridge, MA: Harvard University Press.

Harvey, Penny and Knox, Hannah. 2015. *Roads: An Anthropology of Infrastructure and Expertise*. Ithaca, NY: Cornell University Press.

Hirono, Miwa and Suzuki, Shongo. 2014. "Why Do We Need 'Myth-Busting' in the Study of Sino–African Relations?" *Journal of Contemporary China* 23(87): 443–461.

Hughes, Alex and Reimer, Suzanne. 2004. *Geographies of Commodity Chains*. London: Routledge.

Insaidoo, Kwame. 2016. *China: The New Imperialists and Neo-colonialists in Africa?* Bloomington, IN: Author House.

Jacques, Martin, 2012. *When China Rules the World: The End of the Western World and the Birth of a New Global Order*. London: Penguin Books.

Johnson, Ian. 2014. "The Chinese Invade Africa", September 25. www.nybooks.com/articl es/2014/09/25/chinese-invade-africa/.

Kermeliotis, Teo. 2011. "Is the West Losing Out to China in Africa?" CNN. September 9.

King, Stephen J. 2003. *Liberalization against Democracy: The Local Politics of Economic Reform in Tunisia*. Bloomington: University of Indiana Press.

Klein, Naomi. 2015. *This Changes Everything: Capitalism vs. the Climate*. London: Penguin.

Kolbert, Elizabeth. 2014. *The Sixth Extinction: An Unnatural History*. New York: Henry Holt and Co.

Latour, Bruno. 2007. *Reassembling the Social: An Introduction to Actor-Network-Theory*. Oxford: Oxford University Press.

Lee, Ching Kwan. 2014. "The Spectre of Global China." *New Left Review* 89 (Sept–Oct): 29–65.

Luttwak, Edward. 2012. *The Rise of China vs the Logic of Strategy*. Cambridge, MA: Bellknap Press.

Marriott, James and Mino-Paluello, Mika. 2012. *The Oil Road: Jorneys from the Caspian to the City of London*. London: Verso.

Marx, Karl and Friedrich Engels. 1968. *The German Ideology*. Moscow: Progress Publishers.

Mbembe, Achille. 2001. *On the Postcolony*. Berkeley: University of California Press.

Mitchell, Timothy. 2001. *Colonising Egypt*. Berkeley, CA: University of California Press.

Mohan, Giles and Power, Marcus. 2009. "Africa, China and the 'New' Economic Geography of Development." *Singapore Journal of Tropical Geography* 30(1): 24–28.

Moore, Jason W. 2015. *Capitalism in the Web of Life: Ecology and the Accumulation of Capital*. London: Verso.

Nietzsche, Friedrich. 2006. *Nietzsche: 'On the Genealogy of Morality' and Other Writings*. Cambridge: Cambridge University Press.

Nest, Michael. 2011. *Coltan*. Cambridge: Polity.

Nkrumah, Kwame. 1965. *Neo-Colonialism: The Last Stage of Imperialism*. London: Thomas Nelson.

Ong, Aihwa and Collier, Stephen J. eds. 2004. *Global Assemblages: Technology, Politics, and Ethics as Anthropological Problems*. Oxford: Blackwell.

Pakenham, Thomas. 1992. *The Scramble for Africa*. New York: Avon Books.

Patey, Luke. 2014. *The New Kings of Crude: China, India, and the Global Struggle for Oil in Sudan and South Sudan*. London: Hurst.

Ramo, Joshua Cooper. 2004. *The Beijing Consensus: Notes on the New Physics of Chinese Power*. London: Foreign Policy Centre.

Republic of Togo. 2013. "Togo Urges Africa to Emulate Chinese Development Model." September 29. www.republicoftogo.com/Toutes-les-rubriques/In-English/Togo-urges-Africa-to-emulate-Chinese-development-model.

Rose, Nikolas and Miller, Peter. 2008. *Governing the Present: Administering Economic, Social and Personal Life*. Cambridge: Polity.

Sautman, Barry and Yan Hairong. 2008. "The Forest for the Trees: Trade, Investment and the China-in-Africa Discourse". *Pacific Affairs* 81(1): 9–29.

Sautman Barry and Yan Hairong. 2014. "Bashing 'the Chinese': contextualizing Zambia's Collum Coal Mine shooting." *Journal of Contemporary China* 23(90): 1073–1092.

Smith, David. 2012. "Hillary Clinton Launches African Tour with Veiled Attack on China." *The Guardian*. August 1. www.theguardian.com/world/2012/aug/01/hillary-clinton-africa-china.

Smith, David. 2013. "Kenya Misses Out as Obama's Africa Tour Plays Catch-up." *Mail and Guardian*. May 22. http://mg.co.za/article/2013-05-22-kenya-misses-out-on-obamas-africa-tour.

Sodhi, Navjot S., Barry W. Brook and Corey J.A. Bradshaw. 2009. "Causes and consequences of species extinctions." In *The Princeton Guide to Ecology*, edited by S.A. Levin, S.R. Carpenter, H.C.J. Godfrey, A.P. Kinzing, Michel Loreau, J.B. Losos, B. Walker and David S. Wilcove, 514–520. Princeton, NJ: Princeton University Press.

Strange, Susan. 1998. *States and Markets*. London: Continuum.

Sun, Xiufang, Cheng Nian and Canby Kerstin. 2005. "China's Forest Product Exports: An Overview of Trends by Segment and Destinations." *Forest Trends*. January. www.forest-trends.org/publication_details.php?publicationID=162.

Tan-Mullins, May., Mohan, Giles and Power, Marcus. 2010. "Redefining 'Aid' in the China–Africa Context." *Development and Change* 41(5): 857–881.

Taylor, Ian. 2014. "Emerging Powers, State Capitalism and the Oil Sector in Africa." *Review of African Political Economy* 4(141): 341–357.

Tsing, Anna Lowenhaupt. 2005. *Friction: An Ethnography of Global Connection*. Princeton, NJ: Princeton University Press.

Wallerstein, Emmanuel. 2011. *Historical Capitalism and Capitalist Civilization*, London: Verso.

Whalley, John. 2011. *China's Integration into the World Economy*. Hackensack, NJ: World Scientific.

Yates, Douglas. 1996. *The Rentier State in Africa: Oil Rent Dependency & Neocolonialism in the Republic of Gabon*. Trenton, NJ: Africa World Press.

Zylinska, Joanna. 2014. *Minimal Ethics for the Anthopocene*. Ann Arbour, MI: Open University Press.

7

DOING ETHNOGRAPHY BEYOND CHINA

The ethic of the ignorant foreigner

Gabriel Bamana

Anthropologists have struggled with issues of ethnographic representation and the production of grand narratives about (alien) societies. Such persistent issues were located not only in the politics of writing ethnography (Clifford and Marcus 1986) but also in the dynamic encounter between the ethnographer and the ethnographic Other in an ethnocentric context defined by West–Other power relations. The issue becomes even more problematic when the ethnographic encounter takes place between a native ethnographer (with a Western educational background) and his/her own people (Kuwayama 2003; Ndaya Tshiketu 2016). The matter is even more problematic when the ethnographer, not a native or Western, is rather an individual traditionally identified as the ethnographic Other, yet he/she uses tools from a Western intellectual tradition to process and mediate knowledge in the universe traditionally identified as that of the ethnographic Other. In the latter case, as an honorary Western-self, not only the ethnographer confronts his/her own tradition but also answers questions regarding the inherited intellectual tradition through which knowledge is processed. As a matter of fact, in such contexts, as it is in the case of the encounter between Africa, China, and beyond, local traditions do not necessarily align with Western rationality and knowledge protocol, although it remains the dominant framework of the anthropological enterprise.

Henrietta Moore (1999, 5–9) succinctly discussed some of the issues connected to ethnographic representation. Accordingly, Anthropology's grand theories and generalization seemed "exclusionary, hierarchical and homogenizing." In fact, the questioning of anthropological practices in the wake of postmodernist critiques (Trouillot 1991) regarding "the institutionalised practices and relations of power" extending to new historical conditions of performance insists on the partiality of anthropological interpretation and its relevance to a particular "electoral politics." Anthropological assumptions are but "partial truths" (Clifford 1986).

Even a retreat into the ethnography of the particular still raised questions about how narratives inadequately represent the ethnographic Other as essentially different. The ethnographic turn (Moore 1999, 6) still engaged the ethnographer in a continuous dialogue with the Other, and did not solve the issue of essentialist representations of the ethnographic Other. The basic assumptions of the discipline value cultural difference, assuming difference mediates the interpretation of meanings in an effort to link personal experiences to social processes. Anthropology, as Moore (1994, 1–2) suggests, largely ignores sameness as a significant component in the process.

Notwithstanding discussions around the crises of representation, the politics of knowledge production and the questioning of Anthropology's core concepts, such as culture (Abu-Lughod 1991), the ethnographic encounter with the different Other still fuels the ethnographic imagination. Indeed, even when the ethnographic encounter takes place in a familiar context of the home, one is still concerned with the politics of representation that underscores difference as the substance of the story to be told.

Two major contexts have emerged from the above assessment: the West and the Rest; the ethnographic Self and the Other. Both contexts involve the "political unconscious" (Limón 1991) in dealing with issues of ethnographic representation. The West and the Rest contexts deal with representation as a political dilemma, while the second context deals with representation as a moral, if not a personal dilemma. Nevertheless, both contexts remain located in the same historical background of the hegemonic ethnocentrism where Anthropology mediates knowledge in a comparative model of inquiry.

The necessity for Anthropology to remain engaged in the possibility of critical politics, as well as critical ethics linked to the ethnographic encounter and to processes of knowledge production, still stands as an ongoing task. There have been tremendous changes as Anthropology has critically engaged with these unresolved issues to find a way out in its pretention to enquire about cultural differences and social systems, and how such differences and systems are embedded in hierarchical relations of power (Fox 1991; Moore 1999, 2). These issues are finding new ground and elements of expression as African and Asian encounters intensify.

In this chapter, I intend to revisit some of these issues indirectly, from a different perspective. My discussion takes into account an ethnographic encounter that puts into play individuals that were historically ethnographic identified as Others (power dynamics) and yet who use conventional (mainly Western) methodological and epistemological assumptions that are still subject to the unresolved political and historical issue of representation and knowledge production. The choice not to bear the burden of the political and moral dilemma of the ethnographic encounter is a return to basics, which finds its way out in thinking ethnography as "remembrance" (Fabian 2006) of an encounter and dialogue between two mutual beings in a social context governed by a cultural protocol of hospitality.

Such a choice, my choice indeed, remains tributary to efforts to locate the ethnographic encounter in a lineage of human encounters that enhance our understanding of meanings of human identities in their particular social environment. This choice in doing ethnography remains within a heterodox tradition that has usually sought to open Anthropology to a new direction that values other forms of rationalities and knowledge traditions. Engagement in such a direction reveals the above-mentioned issues as primarily issues in methodology and epistemology.

The present discussion is mostly inspired by my experience as an African ethnographer in Mongolia. The country was home to me from 1995 for a little more than a decade when I embarked on field research on the meaning of tea practices in social processes. I was fluent in the Mongolian language, and my initial encounters with herders in rural Mongolia were not a series of encounters with academic entries and exits – *à la* Malinowski (Pratt 1986, 37–38). I will specifically discuss three main topics: the ethnographic encounter, the ethic of the ignorant foreigner and lastly, research conversations and remembrance as techniques of knowledge process.

The ethnographic encounter

I conducted research in Mongolia at different times between 2007 and 2010. The subject of my research was meanings of tea practices in this society. The study involved a large range of people not necessarily identified as research participants. In a context of familiarity and a multi-sited practice, my objective was to break away from common sense knowledge about tea practices and acquire an informed experience. This locates the ethnographic encounter within everyday social encounters and yet distinctively distinguishes such an encounter from the routine through its purpose.

Since my subject of study was mostly confined to the home, and most specially to what has conventionally been identified as the woman's domain of activity (Rosaldo 1974, 23), my research mainly took place inside people's homes. My position in most homes I entered for research was that of a guest, not primarily an ethnographer. Even when I entered a home announced as an ethnographer, I was welcomed as a guest who happened to be an ethnographer. Being an ethnographer was a quality of a guest.

This situation located the ethnographic encounter in a social context where, paraphrasing a discussion by Ndaya Tshiketu (2016), the encounter process, ways of communicating, social identities of participants, topics of discussion during the encounter as well as the authority over the encounter process, were matters subjected to a particular hospitality protocol rather than being pre-established in an off context tradition.

Considering the experience a posteriori, I realised that I struggled and juggled positions to find my place as an African ethnographer in Mongolia. There were two basic registers of power relations in my encounters with friends and herders in rural Mongolia: ethnographer–research participant versus guest–host. The encounter took place in either of these registers, although not in a sequential process.

In the first register, as an ethnographer, I came in the home with the historical assumptions of the anthropological discipline. I assumed to be in charge of the protocol that would translate these friends' actions and discourses into "data" for a particular knowledge production project as well as transform my friends into research participants. The hierarchical power rapport this protocol establishes was far from involving the reciprocity I expected in a social encounter.

As Ndaya Tshiketu (2016) writes, "research participants" do not expect to be interviewed about certain subject matters at a particular time. Not only are topics of discussion somehow imposed onto them but the whole process of the ethnographic encounter and its protocol are imposed on the participants who find themselves on the defensive end; their position of power, as host, becomes undermined as they have to answer rather than ask questions, or at least reciprocate. Indeed, I recall that whenever I interviewed learned individuals (e.g. academics), there often was resistance in the process as they would not want to submit entirely to the research protocol and rightly so because of the inherent power dynamics. My suspicion grew even stronger as I understood that my research protocol apparently operated in conditions that put me in a higher position. However, my personal identity did not necessarily entitle me to such a position. As for me, an African, research participants expected me to be different and not an heir to an established Western tradition and part of the lineage of Western researchers they had the chance to meet.

Retrospectively, doing ethnography was part of encountering the other (not the ethnographic Other) but the other as a stranger with whom I may share a similarity in identity and meanings about life although in different social and ecological environment. Sameness, as Moore (1994, 1–2) writes, is part of the process. Furthermore, it became my understanding that it is the encounter with the other that mediates moments that produced knowledge awareness. Thus, as Fabian (2002) writes, ethnology remains essentially dialogical.

One of my frustrations was that in the ethnographer–research participant register, there hardly was a mutual encounter I expected. I was in people's homes to learn about a particular subject matter and had previously defined myself as an ethnographer, leaving them with no choice but to welcome me as such. This register was alien to the actual social environment of Mongolia, although some individuals could connect me to other researchers they had had a chance "to serve." This register placed both myself and my friends into a context of established ethnographic endeavour and was thus denied its primary substance as a human encounter committed to finding out about meanings of human identity and the social processes these are embedded in.

Initially, I did not question my choice of assuming anthropology's unresolved issues of methodology and epistemology in the framework of the historical West–Other encounter. All along, as a "familiar stranger" or an "outside insider," I struggled to define a new identity in line with my Anthropological training yet in line with the social conditions of the people I was working with. The resolution to resort to the guest–host register came from tensions, especially about interviewing my friends and long-time acquaintances. I had to abandon interviewing research participants because the idea of interviewing my friends was practically awkward and did not involve reciprocity. Moreover, interview situations did not provide social conditions that bring about breakthroughs in communication (Bourdieu 2003).

The tension was so strong that during my field research I noted:

> There exist "ethnographic moments" when I am aware of being there as a researcher which implies observation and writing down notes. This requires particular attention. At other moments, being there is instead "my life" and I do not ask why I do what I am doing or why they do what they are doing. All is part of the routine, and I am part of this routine too. To be an ethnographer, I need to create some distance from my identity as a member of this community in order to participate methodologically. However, this is not a "real" participation because my status as researcher becomes important. In real life, participation is existential, and that makes the difference.
>
> *(May 5, 2007)*

I thus argue that an ethnographic encounter without the existential component presents problems. As Fabian (2006, 142) explains, "field work is carried out through communicative interaction mediated by language and that whatever objectivity we can hope to attain must be founded in inter-subjectivity." Accordingly, doing ethnography without any existential moment was similar to encountering the other essentially as a performer of social practices. Social processes would be a theater. People were not busy implementing a script or pretending to perform some actions (Bourdieu 1977, 16–22). People meant what they did as well as they did what they meant.

In the guest–host register, it was about the ethnographer as a guest. The host controls a reciprocal protocol of hospitality in his/her home as this transforms one from being a stranger into becoming an acquaintance and a friend. In the particular case of my research in Mongolia, the hospitality protocol included the exchange of greetings and the offer of a bowl of tea and a conversation that generally sought to establish sameness through an inquiry about my social identity. Herders inquired if in my country people would be herding the same five kinds of animals; they asked if we had similar practices and even spoke the same Mongol language.

Humphrey (2012) writes that in the hospitality protocol, the guest comes from without into the stable realm of the home. The guest is the unknown stranger who may even become a danger in the stable realm of the home. Therefore, the series of greeting exchanges seeks to find out about the identity of the guest. The offer of the hospitality tea that is imbued with the host's best wishes for the guest, and the obligation for the guest not to pass on such offer seeks to establish a relationship essential to the encounter where hierarchical beings chose to communicate (or not) about certain matters. The guest, like the ethnographer, coming from without, usually brings about the matter that is the purpose of his/her visit. A conversation is usually the mode of communication.

In this register, it is the guest that is "the Other" and not the other way around. It is about finding common ground for communication breakthrough in an encounter of mutual beings yet in hierarchical social relations of power. The hos-pitality protocol in the ethnographic encounter engages actors into inter-subject communication and when such communication is about knowledge production, agents engage in what I call a research conversation. Of course being the "Other" put me in a vulnerable position that defined the ethical dimension of my research.

The ethic of the ignorant foreigner

My research was tremendously enriched by encounters with several people who contributed their experience, insights, and actions to the construction of narratives. One ethical dimension of this research surrounded the quality of my relation with the people I encountered. My position was constantly shifting. Any pretension to control the process would portray my hosts as passive participants subordinated to my research protocol.

Most people looked at me as an "ignorant foreigner" as they welcomed me into their homes as a guest.[1] These two qualifications constituted the framework of my ethical attitude and power balance in the process. It mainly implied two elements. First, as a foreigner, I did not have practical knowledge of my subject matter (tea practices), as well as knowledge of the Mongolian tradition (*ulamjlal*) according to which actors adopt certain forms of behaviour and carry out those practices. Second, as mentioned earlier, I was allowed into people's homes as a guest (which means an outsider subject to the hospitality protocol), and it was from this position that I carried out my research.

Being both a guest and a foreigner somehow obliged my friends to instruct me on a simple subject matter such as tea practices in everyday life processes. In normal circumstances, "ignorance" such as mine would be inexcusable. Tea practices were so ordinarily entangled in everyday life that people expected everyone to have at least a practical knowledge of them. It is the assumption that I was an ignorant that stimulated my hosts to engage in conversation and eventually instruct me on this particular subject matter. I definitely was the "Other ignorant."

As a long-term guest with substantial knowledge of the language and familiarity with the context, I also qualified as an insider. Whenever there were other guests in a home, I was usually treated as an insider. It is this position as an insider, after being filtered through the hospitality protocol, that provided a trustful connection for sharing knowledge. Indeed, I came to understand that knowledge of tea prac-tices was both a life skill (practical knowledge) and a privilege (theoretical knowl-edge). I also realised that people did not always want to share information with an individual with a hybrid identity or, simply, whose social identity they did not know. In this social context, knowledge sets social boundaries and hierarchy. People who shared similar knowledge belonged together.

Because of my position in the local kinship system, my status remained some-how ambiguous. I continued to be an outsider with the privilege of a long-term guest position. This was a hybrid identity. Consequently, I had the privilege of accessing information through conversations with my hosts because the hospitality protocol had established my (fluid) identity and the appropriate power balance. Apparently, the power balance I experienced put me in a vulnerable position and went against the ethnocentric assumption of the stereotypical Western ethnographer.

As for my particular subject of study, I realised that since tea practices were entangled in the domestic process, being a guest provided me with a privileged vantage point. However, this privilege does not imply there were no difficulties. There were various "gray zone" situations where I was rather treated as an outsider in a relation that involved power games because of the other aspects of my identity.

In reality, although talking about tea and tea practices was an easy subject of conversation, there were instances in which some of my acquaintances preferred to remain friends rather than become research participants. As a guest in their homes, I was not allowed to change the quality of this relationship because such change would introduce a different power balance in our relationship. The hospitality protocol requires that a guest be a recipient and a passive participant. This situation made my participation into some activities difficult.

The above-mentioned hospitality protocol implies that although women had substantial practical knowledge about tea practices, men presented themselves as the people to have the (theoretical) knowledge about tea practices and thus they would initially engage into conversation with me. As a guest, I was subject to male discourse as per hospitality protocol according to which it is generally a male responsibility to perform the discursive aspect of hospitality (Humphrey 2012).

All the above constitute the social conditions of my ethnographic encounters in Mongolia. It would be misleading to suggest that historical ethnocentrism writes off these circumstances. It is rather necessary to account for how the ethnographer, taking into account these conditions, establishes a rapport for engagement into a process that brings about knowledge awareness and eventually knowledge production. Such a process would distinguish routine conversations from research conversations intended to construct knowledge in a particular discipline tradition.

As in my case, gray zones included the fact that I was able to move beyond the guest protocol because of my established status as an ignorant foreigner. Several times, I went beyond the boundaries of the gendered hospitality protocol to hold conversations with women. Other gray zone situations included restrictions in participation because of my gender and status as a guest.

As a foreign guest, I had to enjoy the privilege of my (hierarchical) social position and the subsequent hospitality protocol, which kept me away from everyday processes. In the latter case, my male hosts treated me to streamlined information stripped of social contingencies.

It was, after all, the hospitality protocol and the position I was given as a guest ignorant foreigner that set the tone for research conversations. This does not mean the ethnographer does not come with his/her assumptions. Rather the nature of the hospitality protocol converts both the guest and the host into related individuals able to engage in different activities yet using established positions and identities. As a guest/ignorant foreigner, I was able to participate in research conversations with my hosts and discuss meanings of the practices that constructed the identity of a few individuals in that society.

Research conversations and ethnography as remembrance

The strategy I used to learn from my friends was engaging in what I call research conversations (Mong: *yum yarikh,* talk over a matter). In the register of the ethnographer as a guest, I had personally opted for (research) conversations that transpire in writing as reflexive remembrance. The focus in such conversations is communication – not an interrogation – with research partners, leaving room for multiple rationalities, including symbolic knowledge.

As I mentioned earlier, in the case of the ethnographer as a guest and in my particular experience in Mongolia, I, the ethnographer was the Other because I came from the outside into the realm of the home. My host wanted to learn about my identity and set a pattern of relation appropriate for communication according to our momentously established identities. This provided the social conditions for meaningful communication. I was thus the Other in the sense of being an "inter-subjective Other."

Ndaya Tshiteku (2016) experienced the same awkwardness researching at home in the Democratic Republic of Congo and resorted to conversation as a suitable means of communication in that context. She insisted that conversation (Ling: *masolo*) allows social and cultural conditions of communication and of the social encounter to transverse and influence the ethnographic encounter with participant partners open to the rationality of their mutual understanding of meanings of human identity in the particular context of their discussion. *Masolo* in the Democratic Republic of Congo and *yaria* in Mongolia refer to a conversation within particular traditions of hospitality and communication and these traditions would be included in the specificities of the (ethnographic) encounter between Africa, China, and beyond.

Ndaya Tshiketu (2016) intentionally omits to tell us that the practice of *masolo* originally takes place in the cultural backdrop of the hospitality protocol according to which a guest is usually welcomed in the community and in the family as he/she sits in the *ingomba* (meeting place) and is asked "*masolo nini oyeli biso?*" (what is the news?) since the last meeting. If the guest has come for the first time, the host introduces the *masolo* (conversation) by inquiring about the guest's identity and the intention of his/her visit.

It is only then that the host may decide, in openness to the rest of the community, to welcome the guest and establish communication. These conditions favour trustful communication, while the reliability of information is checked against community approval. Community control on the hospitality process implies that communication should not involve anything that may harm the community and individual reputation and wellbeing. The format of communication during this encounter remains a conversation. Nevertheless, this pattern does not exclude hierarchical relations of power. Power is ubiquitous with social relations, says Foucault

(1982). In this context, the difference is that the hierarchy is not pre-established before the encounter as it seems to be the case in the conventional approach.

The social protocol of hospitality establishes the social conditions of communication including a hierarchy of power in an inter-subjective encounter. Further, the inter-subjective other pattern implies a search for meanings yet in a process of communication open to the mutual understanding of identity as embedded in practices (including discourses). A research conversation is about communication of respective meanings of identities as agents engage in an exchange of information and explanation of those meanings. In my case, it was the meanings of tea practices I was most concerned about. Learning about these meanings opened up to the identity of the people who constructed such meanings to deploy them in social processes.

I use the expression "inter-subjective Other" to express the experience of alterity in the research process. Fabian's (2002, 2006) discussion about how anthropology constructed its object (the Other) is most helpful to understand my suggestion. Indeed, Fabian (2006, 145) suggests that "for the ethnographer, there is a kind of experiencing the other" and writing ethnographies is about remembering these experiences of alterity. The experience of alterity includes seeking to understand both difference and sameness. The process remains comparative and aims to establish in a specific language, what two human beings do differently to construct particular identities and deploy these in social processes in order to achieve specific objectives.

A research conversation is thus a technique that requires openness to different rationalities referring to lived experiences and to the meaning of living-those-experiences. Research conversations are existential because the shared experiences reflexively engage both partners. In a research conversation, there is, on the one hand, a guest ethnographer sharing the anxiety of his/her quest for understanding and, on the other hand, the host participant sharing the experience and knowledge of the experience that answers – or not – the quest of the researcher.

Throughout field research, I was a student of the subject matter and, as such, vulnerable to answering questions that explained my ignorance about this particular subject matter. Research conversations were primarily learning processes involving reciprocity. The power balance in this conversation was always shifting and subject to different criteria beyond my control. From this point, the process moves into converting these conversations into narratives that focused both on the features meanings of particular identity and the practices that socially constructed these identities and the subsequent cultural ideologies as well as the social processes of their deployment in hierarchical relations of power.

In the process of such knowledge production and the presentation of the acquired knowledge, there exist significant tensions such as selection of representative ethnographies, representative ethnographic moments and relevant information, and how to present such information in a way that does justice to the experience of alterity acquired in research conversations.

In writing about tea practices in Mongolia, I mostly relied on the concept of ethnography as "remembrance" (Fabian 2006) to avoid the unresolved political and historical issue of representation. I think of ethnography as a remembrance of learned experiences. Ethnographic remembrance implies as Fabian (2002) would say, the idea of coevalness with and recognition of the host participants in the presentation of the ethnography.

Throughout the text, the ethnographer presents remembrance of the ethnographic encounters, the conversation with the host that included an experience of alterity. Ultimately, ethnography is about the presentation of meanings. Of course, meanings imply a representation of meanings of individual identities and social networks involving the author's own subjectivity as being part of the process that stimulated knowledge awareness and production. Ethnography comes out as a remembrance of inter-subjective encounters. The challenge in writing ethnography as remembrance in regards to the assumption of the anthropological disciplines would be similar to what Pratt (1986, 31–33) points out: "subjective experience is spoken from a moving position already within or down in the middle of things, looking and being looked at, talking and being talked at."

The ethnography of tea practices in Mongolia as I wrote it, recalls those existential encounters which provided me with a personal experience of life in Mongolia and that became a catalyst in my interpretation of tea practices and local knowledge. I struggled to present memories of personal encounters with different people as "alter ego" in Mongolia.

An anthropological ethnocentrism hinders the production of multiple voices and rationalities that would do justice to the diversity of practices and historical conditions under which different human beings construct identities and pattern of social processes and cultural ideologies to sustain such processes. An inherited ethnocentrism becomes even harder to shake off as one is constantly concerned with the discipline of orthodoxy.

Toward a new direction in doing ethnography

When I engaged in research in Mongolia, I had more than a few assumptions. My task was mainly to document tea practices and find out about meanings that local herders attributed to these practices. To some extent, these meanings were assumed to be there as raw materials for collection. Engagement with herders in rural Mongolia allowed me to understand that meanings, similar to the identities they represent, are fluid and subject to interpretation according to the individuals engaged in the conversation. Meanings were tuned to a variety of objectives (what would be the added value?) and intentionality (why would participants engage?) as well as geared to actions (what are the outcomes and consequences?).

A process of understanding these meanings required an engagement with individuals constructing these meanings in inter-subjective encounters that allow an awareness of such meanings through an interpretative process. In this particular case, it is the hospitality protocol that governs the encounter between the guest ethnographer and the host participant and thus determines the power balance and the process of the conversation between the two. Hospitality thus defines the social conditions of the ethnographic encounter (not the opposite), and more precisely, it clarifies the identity of the unknown guest as an individual with whom one can share the information or not.

These social conditions, as well as the sort of knowledge every society produce, are sometimes subordinated to the historical and political issues inherent in the process of the anthropological enterprise. Whether it is a choice or an inheritance, anthropological ethnocentrism remains a hindrance to the diversity of rationalities that would enrich the discipline in its query to understand meanings of human identity in its diversity.

The issue remains mainly methodological and epistemological and it finds new grounds for discussion in African, Chinese and encounters beyond these. Indeed particular African and Asian traditions define protocols of social encounters, of communication and exchange of information. These protocols channel knowledge production, which is sometimes embedded in a symbolic language. It is these local traditions of knowledge that are left out as the anthropological enquiry responds to a particular mode of enquiry that is an outcome of a Western intellectual tradition and which has an essentialist and universalist pretension because of its historical circumstances.

Local traditions of knowledge are suitably equipped with processes of inquiry and of representation of meanings of identities and social processes in local categories that certainly would serve in a dialogical anthropological project. Such a dialogical endeavour would bring to shore difference and sameness in each tradition, thus not having one alienating the other through the construction of the ethnographically different Other as an object of scientific inquiry.

As encounters between Africa and Asia intensify, chances are that African ethnographers will realise, as I did, that the intellectual traditions they carry with them in the field may be but a re-appropriation that portrays them as individuals in between traditions which is a liminal position and a hybrid identity. The encounter with Asian traditions (Chinese and Mongol included) serves as a filter to help African researchers sort out what is their intellectual tradition against what is an African heritage of a Western discourse. Such filtration and sorting are hard to achieve in the African–West encounter as similarities in the intellectual tradition may be blinding differences.

Afro-Asian encounters do not necessarily exclude the West. However, old paradigms such as the West considering the rest as its object and thus setting out to defend Africa from a Chinese "invasion" hardly advance the emergence of an alternative rationality and intellectual tradition. China, if not Asia, also uses a Western framework, a Western discourse about Africa as well as a Western power dynamic to meet the people of Africa and their tradition would generate a missed opportunity for such an innovative intellectual encounter. Afro-Asian dynamics could help the West to sort out its historical burden and move forward in the encounter with other traditions, especially Afro-Asian.

Afro-Asian encounters are the encounters of particular traditions, and should not be encounters of traditions of disguise. As much as I experienced, Mongolian herders have always wanted to discuss my African identity in the communication protocol of their tradition. These conversations have often pointed to meanings of who we are and the sort of things we did in our respective communities to sustain our identity. These, I should admit, are legitimate matters of an inclusive anthropological inquiry. Indeed, why would ethnographers from Africa and Asia assume the historical and unresolved assumptions of their disciplines that have constructed stereotypes that have sometimes misled our understanding of the local communities both in Africa and Asia? Much energy has been spent on the continent to prove sameness with the West while the West insisted on difference. The present historical moment allows for renewed discussions on the methodology and epistemology involved in the ethnographic encounter as the locus for the construction of knowledge in an inter-subjective dynamic.

Note

1 Very often, I was greeted with the phrase: *Khöörkhii amitan, yu ch medekhgüi*, "Poor one, he doesn't know anything."

References

Abu-Lughod, Lila. 1991. "Writing Against Culture." In *Recapturing Anthropology: Working in the Present*, edited by Richard Fox, 137–162. Santa Fe, NM: School of American Research Press.

Bourdieu, Pierre. 2003. "Participant Objectivation." *The Journal of the Royal Anthropological Institute* 9: 281–294.

Bourdieu, Pierre. 1977. *Outline of a Theory of Practice*. Cambridge: Cambridge University Press.

Clifford, James and George Marcus. eds. 1986. *Writing Culture. The Poetic and Politics of Ethnography*. Berkeley: University of California Press.

Clifford, James. 1986. "Introduction: Partial Truth." In *Writing Culture: The Poetic and Politics of Ethnography*, edited by James Clifford and George Marcus, 1–26. Berkeley: University of California Press.

Fabian, Johannes. 2006. "The Other Revisited. Critical Afterthoughts." *Anthropological Theory* 6(2): 139–152.

Fabian, Johannes. 2002. *Time and the Other: How Anthropology Makes Its Object*. New York: Columbia University Press.

Foucault, Michel. 1982. "The Subject and Power." *Critical Inquiry* 8(4): 777–795.

Fox, Richard, ed. 1991. *Recapturing Anthropology: Working in the Present*. Santa Fe, NM: School of American Research Press.

Humphrey, Caroline. 2012. "Hospitality and Tone: Holding Patterns for Strangeness in Rural Mongolia." *Journal of the Royal Anthropological Institute* 18(s1): 563–575.

Kuwayama, Takami. 2004. *Native Anthropology: The Japanese Challenge to Western Academic Hegemony*. Melbourne: Marston.

Kuwayama, Takami. 2003. "'Natives' as Dialogic Partners: Some Thoughts on Native Anthropology." *Anthropology Today* 19(1): 8–13.

Limón, Jose. 1991. "Representation, Ethnicity, and the Precursory Ethnography: Notes of a Native Anthropologist." In *Recapturing Anthropology. Working in the Present*, edited by Richard Fox, 115–135. Santa Fe, NM: School of American Research Press.

Moore, Henrietta L. ed. 1999. *Anthropological Theory Today*. Cambridge, MA: Polity.

Moore, Henrietta L. 1994. *A Passion for Difference: Essays in Anthropology and Gender*. Bloomington, IN: Indiana University Press.

Ndaya Tshiteku, Julie. 2016. "L'Anthropologie et ses méthodes: Le dilemme d'une Ethnographe chez soi." In *Eprouver l'altérité: Les défis de l'enquête de terrain*, edited by Vincent Legrand and Clementine Gutron, 221–238. Louvain-la-Neuve: Presses universitaires de Louvain.

Ortner, Sherry. ed. 1999. *The Fate of Culture. Geertz and Beyond*. Berkeley: University of California Press.

Pratt, Marie Louise. 1986. "Fieldwork in Common Places." In *Writing Culture. The Poetic and Politics of Ethnography*, edited by James Clifford and George Marcus, 27–50. Berkeley, CA: University of California Press.

Rosaldo, Zimbalist. 1974. "Woman, Culture, and Society: A Theoretical Overview." In *Woman, Culture, and Society*, edited by Zimbalist Rosaldo and Louise Lamphere, 17–42. Stanford, CA: Stanford University Press.

Strathern, Marilyn. 1995. *Partial Connections*. Savage: Rowman and Littlefield.

Trouillot, Michel-Rolph. 1991. "Anthropology and the Savage Slot: The Poetics and Politics of Otherness." In *Recapturing Anthropology: Working in the Present*, edited by Richard Fox, 17–44. Santa Fe, NM: School of American Research Press.

8

GLOBAL AFRICAN STUDIES AND LOCATING CHINA

Jamie Monson

What is "China–Africa" studies? Is it different from, or the same as, "Africa–China" studies? What is its relationship to African Studies, Asian Studies, or Area Studies more generally? Is it something we should even consider to be a field of study, and if so, what constitutes a field? Is it transnational and therefore innovatively boundary-crossing? Or rather, as critics have said about Area Studies in the past, is it merely an "empty category" that reacts to shifting world strategic political and economic affairs without sharing any of the foundational characteristics of a field, for example a body of theory and a methodology?

To begin to address these questions that are in contention among today's scholars, in this chapter I lay out the forgotten roots of the historical relationship between African Area Studies and China–Africa studies. In doing so, I trace the relationship between US-based Area Studies and the development of African Studies in Chinese institutions of higher education, for these have an entangled history. This chapter will draw on my own personal autobiography as well as the archives of the Ford Foundation, an institution that played a critical role in the post-1980 development of African Studies in China by linking US and Chinese Africanists in a fifteen-year scholarly collaboration.

As earlier literature reviews have shown, the body of scholarship now widely described as "China–Africa" gained momentum immediately following the Beijing Summit of 2006, as western scholars in particular (and their governments) responded to anxieties about expanding political–economic relationships between Chinese and African partners (see Alden 2007; Manji and Marks 2007; Ampiah and Naidu 2008; Large 2008; Taylor 2009; Monson and Rupp 2013). These debates framed concerns that have been foundational to Area Studies all along, and raise provocative questions about the degree to which Area Studies can be said to have been primarily about place. For we might have expected an emergent literature on China–Africa engagement to share characteristics with global studies or cultural studies, both trajectories of post–Area Studies scholarship that were unfolding in dynamic ways by 2006. And there is now certainly much important new scholarship emerging along these lines, especially among a new generation of scholars.[1] Yet in its disciplinary focus (political–strategic) and its ideological orientation (pursuing national interests) the China–Africa field from its inception strongly resembled familiar frameworks for Area Studies and resulted in similarly contentious arguments about the changing world order. The emergence of "China–Africa" and these ensuing debates have therefore illuminated key and enduring underlying principles of Area Studies – as their political–strategic foundations have framed them as much as their geographical boundaries.[2]

Moreover, it is critically important to recognise that Chinese academic institutions have their own deep histories and contemporary traditions in African Studies. For while African Studies in China has many strong connections to its Western counterparts, for example its relationship to national priorities in political and economic policy making, there are also differences, including limitations until recently in training in African languages and cultures,[3] and a focus on library and documentary rather than fieldwork methods.[4] Chinese African Studies has not been isolated from US counterparts – far from it, as over the last three decades there has been significant exchange and dialogue despite current geopolitical anxieties about competing US and Chinese interests (Sylvanus 2016, 170). Few remember the critical dialogue and exchange that took place among Chinese and US-based Africanists over a period of fifteen years from 1983 to 1998, through the leadership of Professor George Yu from the University of Illinois, and funded by the Ford Foundation. These exchanges produced important ideas and framed a lasting conversation among Chinese and US-based academics that helped to shape African Studies in China as university programs were revived there following the Cultural Revolution in response to China's "opening up" to international engagement under the leadership of Deng Xiaoping (Yu and Xia, 1983). The purpose of the Ford funded program was to "strengthen teaching and research in African Studies in China" and it supported both senior and junior scholars with mentorship, opportunities for research and study in Africa as well as joint conferences. The African Studies model, the same model that came under critique so heavily from the late 1990s and continues today, therefore played a contributing role in the structure of academic engagement in Chinese African Studies.

I start this chapter with my own autobiography, from my years as a graduate student at UCLA in the 1980s (when we hosted the first of the Ford Foundation–funded delegations of Chinese Africanists to visit the US) to more recent experiences as a visiting scholar in Chinese African Studies academic institutions. I move on to bring in a view from the archive, to engage a much-needed historical perspective on these critical questions. And lastly, I conclude with reflections from my current vantage point as director of one of the largest and most enduring federally funded US-based African Studies Programs, Michigan State University, where like our peers we continue to debate the past, present, and future of Area Studies. And as we look toward the future, I pose a different question: if scholars today were to ask a major foundation like the Ford Foundation to fund a new fifteen-year global research initiative on Africa and China, what would it look like, and why?

Autobiography

The field of Africa–China studies has been entangled with my own autobiography since the early 1980s. I was still a graduate student at UCLA in May 1983 when the first Ford-funded delegation of five Africanist scholars from China visited our African Studies Center. Among them was a senior scholar from the Chinese Academy of Social Sciences, Dr. Wu Shenxian, who was at that time a researcher at the Institute for West Asian and African Studies (IWAAS) in Beijing and a respected Africa specialist on Egypt, Sudan, and agricultural development. I was working in those years for the UCLA African Studies Center and in that capacity I supported our director, Michael Lofchie, with making arrangements for the visit and hosting the Chinese guests. My own special assignment was to offer hospitality and assistance to Ms. Wu (the only woman member of the delegation), whom we addressed respectfully as "Madame." Dr. Wu and I soon developed a warm rapport and she ended up staying on at UCLA that year as a visiting scholar.

This opportunity to be engaged in a historic international scholarly exchange with Chinese partners was something remarkable, and I was especially touched by my personal connection with the now late Wu Shenxian. Over the years, however, my memories of this early connection between Chinese and US Africanists drifted away as my career as a historian of East Africa took me in different directions. It was only twenty years later in 2003, when I found myself invited to give a lecture at IWAAS in Beijing about my research on the TAZARA railway, that strong memories of that earlier connection returned.[5] When I entered the building of the IWAAS complex there were photographs of Dr. Wu on the wall alongside those of other scholars and directors.[6]

And my conversations with experienced Chinese Africanists like Dr. Yang Lihua over lunch that day drifted back to those first connections between our US and Chinese African Studies programs. We lamented, in 2003, the low levels of funding and interest that we experienced as Africa specialists within our respective academies; these institutions seemed to prize other continents and regions more highly. I recall being asked in 2003 by at least one Chinese university if I couldn't offer a lecture on American Studies instead of African Studies. Little did we know then that by the time of my next visit to Beijing in 2007, the relationship between Africa and China would have become a hot topic and we would all be swept up in the tide of "China–Africa."

Several years later, I was invited to spend a semester as a visiting researcher at the recently established Institute for African Studies located at Zhejiang Normal University in Jinhua. This institute was founded as Chinese interest in Africa was expanding rapidly, especially in this region of southeastern China. My office at the Institute was next to that of now retired (yet very active) scholar and diplomat, Ambassador Shu Zhan, who had spent much of his career in Africa working for China's

Ministry of Foreign Affairs after completing his degree in linguistics at the London School of Economics. He and I spoke often and we developed a collegial friendship; his wife, Liu Ping (also a retired diplomat) took me on outings to show me the city. During one of our conversations Ambassador Shu mentioned to me that he himself had taken part in the Ford Foundation program to foster scholarly engagement between US and Chinese Africanists. He had attended the second joint Sino-US African Studies conference in 1991 in Los Angeles, and stayed on for three months at George Mason University, where he was hosted by Professor John Paden.

I had left UCLA by the time of the Sino-US conference and we did not meet there. But as Ambassador Shu and I shared notes during the months that I stayed in Jinhua, it was fascinating to remember those historic tours and to learn the significance of the exchange program from his perspective. In the end, the initiative lasted for a total of fifteen years and brought fifty-two Chinese scholars together with thirty-one Americans, through exchange visits and two joint conferences. Many of the Chinese experts had stayed for several months at US African Studies Centers and also received funding support to visit African universities to do field research. The program was the outcome of an extraordinary commitment and vision on the part of Professor George Yu, who was at the time in the Political Science department at the University of Illinois and also had strong links to UC Berkeley.[7]

During the past several decades, my personal career in African Studies has intersected at multiple times and in multiple places with those of Chinese scholars. And over a period of some thirty years the focus of our engagement has shifted from sharing our experiences as fellow (sometimes lonely, always underfunded) Africanists during the 1980s and 1990s, to participating today in the fast-developing field of "China-Africa." The senior generation of scholars in China and in the US has now retired, but many have remained active and their students have become leaders of their departments and institutes. At a recent conference on Chinese infrastructure projects in Africa held in Washington, DC, I was present when Ambassador Shu Zhan and his mentor during the Ford Foundation program, Professor John Paden, were reunited. The field of China–Africa has therefore rekindled old friendships and relationships as it has provided new and different momentum for scholarship.

Archives

Documents held in the archives of the Ford Foundation in upstate New York give testimony to the strength and duration of the personal relationships that developed among scholars that participated in the fifteen years of exchange among Africanists in China and in the United States between 1983 and 1998. The archives also provide important information about the ways that African Studies in both the United States and in China was developing at the time. The project was firmly embedded in the disciplines of Political Science and International Relations, and was one of a set of US foundation–funded exchanges in China focused on developing these fields after 1980.[8] Participants from both the US and the Chinese side came from programs closely linked with national foreign affairs institutions, including on the US side the Council on Foreign Relations, the US House of Representatives, the National Intelligence Center, and the Carnegie and Ford Foundations. In China the collaborating partners were IWAAS (and its parent institution, the Chinese Academy of Social Sciences), the Chinese Association for International Understanding, and the China Institute of Contemporary International Relations.

The exchange program was also based firmly in the methodological and strategic frameworks that prevailed in Area Studies programs in the US in the 1980s and 1990s – with a special focus on the crisis and transition that was then taking place in southern Africa over the fifteen-year period (see Masao and Harootunian 2002). The three themes of the second Sino-US Conference on African Studies held in California in 1991 were Africa's economic potential; "great power" involvement in Africa; and the political future of southern Africa. Not only was the disciplinary base focused on political affairs, but there was also a strong developmentalist paradigm that extended simultaneously toward Africa and China: to become professional African Area Studies experts, Chinese scholars were invited to follow an American model to develop a cohort of "area experts." Area expertise included language and cultural study, in particular familiarity with the African context. This particular aspect of Area Studies, knowledge developed through the cultural and linguistic immersion that was required for reputable "field research," was something that was lacking in the Chinese academic institutions.

Yet the Chinese scholars clearly held their own during these visits and in the symposia. There were polite yet clear divergences in the points of view of the participants. For example, there were conversations between Professor Richard Sklar of UCLA and Professor Zhu Zhonggui about the potential for multi-party democracy to provide a pathway to development in post-independence Africa. While Sklar was enthusiastic about the new democratic traditions breaking out on the continent at the time, Zhonggui argued that the one-party state model was better suited because of its potential for generating national unity, political stability and development. At the core of both positions was the question of whether democracy was a Western institution imported into Africa, or based on African initiative.[9] Yet these debates took place without any official participation by African academic or other institutions.

While there were no African institutional partners in this exchange (though two US-based African academics participated, Michael Chege from the Ford Foundation and C.E.D. Halisi of the University of Illinois), there was an effort with funding provided to create opportunities for Chinese academics to engage in field work study in selected African countries, including Zimbabwe, Tanzania, and Kenya. Professor Yu made personal visits to African universities to make the necessary agreements for the hosting of the Chinese academics, some of whom had never traveled to Africa before. The developmentalist and pedagogical expectations of the exchange were evident in many of the reports, for example in the report from the

1991 conference, which quoted one US participant as observing that the US-based exchange committee could "take pride" in the intellectual growth of the Chinese Africanists, whose development they had not anticipated "would grow so quickly" through their support.[10]

The first grant of just over $20,000 was awarded in April 1983 to support the visit of five Africa specialists from the Chinese Academy of Social Sciences to visit US African Studies Centers. Their itinerary lasted for two weeks, from May 2 to May 19, starting in Los Angeles where I served as one of the hosting staff. According to the itinerary held in the archive, in addition to high level consultations the delegates heard talks in Los Angeles by leading Africanists on such themes as American political science and African affairs (delivered by James Coleman); human rights and ethnicity (Leo Cooper); teaching African Studies (Gerald Bender at University of Southern California); and development in southern Africa (Carol Thompson). At UC Berkeley and Stanford University they were invited to attend seminars and panels, coordinated by Carl Rosberg at Berkeley and David Leonard at Stanford. George Yu then hosted the visitors at the University of Illinois and they heard a talk on agricultural research in Africa by Earl Kellogg, befitting the land grant university model. The delegation then visited Northwestern University where John Paden led a series of conversations with Africanists from Northwestern and from Wisconsin who addressed broad themes of the role and organization of African Studies in the university.

At the Center for Afro-American and African Studies at Ann Arbor, Michigan, the delegation heard a talk on Africa and the African Diaspora focusing on the question, "Why a Center for Afro-American and African Studies?" by Niara Sudarkasa, while historian Frederick Cooper spoke on a panel in the afternoon on the theme of Marxist theory in the study of African history, and Ernest Wilson later discussed energy policy and politics in Africa. In Washington, DC, they visited the Library of Congress and at Howard University the group met with Robert Cummings before going into conversations with Helen Kitchen of the Center for Strategic and International Affairs at Georgetown University, followed by a meeting with Congressman Howard Wolpe who was then Chairman of the House Subcommittee on African Affairs for the House Foreign Relations Committee. The delegates then traveled to New York where they were invited to the Social Science Research Council (SSRC) for talks with Sally Falk Moore, Jane Guyer, and Sara Berry. The final day before their departure they managed to squeeze in a visit to the African American Institute, the United Nations, and a dinner with the National Committee on US–China Relations, which had played a key role in organizing the visit.

I have provided this description of the inaugural Chinese delegation visit in some detail because it demonstrates that the Chinese scholars were able to meet with a high level group of African Studies experts during their seventeen-day visit – not only in academia but also in the Washington and New York offices of organizations like the SSRC that were key players in academic research and policy in US–Africa relations in the 1980s. Their itinerary comprises a veritable "who's who" of leading Africanists from that era. Handwritten notes taken by Peter Geithner, and a tentative guest list, show that luncheon guests included additional representatives from the American Committee on Africa; the Rockefeller and Carnegie Foundations; Crossroads Africa; the Phelps-Stokes Fund; and the Council on Foreign Relations among others. The evaluation reports and other notes that circulated after the tour were uniformly positive. The next step in the exchange was for a delegation from the US institutions to visit China, an event that took place in 1984 and in 1985 a committee was created to facilitate ongoing exchanges, which George Yu helped to form and then served as chair of the US–China African Studies Exchange Committee, a group headed by the same leaders who had led the first visits. The ongoing Ford-funded exchanges included both junior and senior Chinese Africanists, joint conferences and seminars.

In March 1983, George Yu circulated an important paper to the Ford Foundation, and presumably to other audiences, co-authored by himself and Xia Jisheng of Beijing University (and at that time also a visiting professor at the University of Illinois), on the topic "The Development of International Studies in China: The African Area." In the opening introduction to the paper they defined Area Studies as "a concept for the development of comprehensive and particular knowledge with regard to a world region and/or country, the 'whole culture' approach." This concept – and its development – had faced many difficulties in mid-twentieth-century China before the end of the Cultural Revolution allowed the beginnings of renewal in China's universities. And Area Studies was one of the important targets for the development of expertise in International Development. For many in US African Studies institutions, this was surely their first exposure to the background to Chinese African Studies. The paper explained China's interest in African Studies as stemming from its Third World policy orientation, and cited the developments in African Studies from the late 1950s, including the teaching of a course on African History at Peking University in 1962 followed two years later by the founding of the Institute of Asian and African Studies there, creating a "serious beginning" for African Studies in China at several levels. African Studies, like other academic areas of study, experienced heavy setbacks during the Cultural Revolution years, and was only revived again afterwards, especially with the creation of the Institute for West Asian and African Studies of the Chinese Academy of Social Science in Beijing in 1977. Since then, as the work of Li Anshan has shown, there has been an increasingly active group of scholars including graduate training both at universities and in the Academy of Science.

Area Studies to global studies? Navigating the twenty-first century

Chinese African Studies has shared much in common with Western approaches in terms of its interest in policy and international political analysis over the last several decades. Yet until recently there were still not as many Chinese academics doing research on the continent when compared with western Area Studies experts. This explains in part why former ambassadors like Ambassador Shu Zhan, who had several years of experience living in Africa, remain critically important in Chinese foreign affairs today – and why they hold appointments at universities and think tanks where they are mentoring new generations of specialists. Graduate students on the other hand are increasingly spending time doing Africa-based archival and field research, as are younger scholars such as Xu Liang at Peking University and Xu Wei at Zhejiang Normal University. There have been 152 doctoral theses on African topics in Chinese universities between 1981 and 2005, half of which focus on South Africa (Chen Hong and Zhao Ping, 2006, cited in Li 2005). Chinese African Studies centers are expanding scholarly inquiry further into the humanities, for example Zhang Yong, a recently hired scholar of African film at Zhejiang Normal University, and hosting experts in the arts among other fields. PhD candidates from Beijing University have specialised in fields from history to International Relations, and training in African Studies and Africa–China Studies programs is diverse in Beijing where students come together from all parts of the globe in bilateral exchange programs.

Still, as they shared with me personally, Chinese Africanists through the late 1990s perceived their status as marginalised within their own institutions. At this same time, US and other Western-based Africanists began our own period of introspection and rethinking of the Area Studies enterprise, as the Cold War came to an end and the rationale for knowledge production of world areas and regions came into question. This was the era of "perpetual crisis," as Paul Zeleza (1977) put it, especially in the United States where African Studies had a particularly problematic legacy of exclusion that has continued to undergo critique and recalibration yet has not become inclusive of its diaspora nor resolved its fundamental tensions.

It was in the midst of this "perpetual crisis" that scholars and their institutions began to consider repositioning Area Studies into new configurations of transdisciplinary and comparative global studies. And it was during this same decade that the global political economic landscape shifted with the rise of the BRICS economies and new circuits of capital accumulation and expansion. Therefore, the study of China–Africa has emerged in a critical moment of rethinking and repositioning Area Studies globally, so that "South–South" and China–Africa studies are emerging both because first, area studies are being restructured into different varieties of global studies and bridges among these programs are one way that people are doing this; and second, because of the shifting global context and the rise of "the South" politically and economically.

The decade of the 2000s brought with it an expansion of Chinese political and economic engagement in Africa and a corresponding resurgence of scholarly interest in African Studies in China, and in the study of Africa–China relations globally. This shift in global studies was marked early on by a conference at Cambridge University organised by the editors of this volume, Chris Alden and Daniel Large, together with Ricardo Soares de Oliveira, resulting in a subsequent edited book titled *China's Return to Africa*. The book's abstract stated that despite deepening engagement between China and Africa post-2006 (the year of the Beijing Summit), the relationship had been "relatively neglected in academic and development policy circles." The volume marked an important transition taking place in the global context. Since that time scholarship and public discourse has grown exponentially on the topic of "China–Africa."

What I am arguing here, therefore, is that the initial five years or so of scholarship in China–Africa studies in the West seemed at least epistemologically to mark a resurgence of the African Studies paradigm and to provide new life blood to some of its expert players and its dominant institutions; indeed frequent references to the Cold War or to some new form of superpower engagement and rivalry in Africa are repeatedly emphasised both in the academy and in the media. Even a qualitative analysis of the literature on the China–Africa relationship from 2001 to 2011 framed its research question on the binary theme of "win-win" or "win-lose propositions," focusing on investment, human rights and international relations.

So if this was the case for Western institutions, what has been taking place in the Chinese academy? There seem to be several different trends, and they are changing as the global landscape is also shifting. First, there has been an increase in funding, support and training for Chinese African Studies, as reflected in the establishment of new university centers and think tanks (for example at Zhejiang Normal University) and state support for research and analysis. At the same time, one could say that these programs continue to reflect a national agenda, as they focus on themes of interest to the Chinese state (agricultural development, economic investment and development). These state interests have been pivoting increasingly toward the study of the relationship between China and Africa.

And at the same time, as China has been expanding its own economic interests and political relationships in Africa, scholars are generating ties that would lead to a flourishing of bilateral academic exchanges between China and Africa; publications in a broadening range of disciplines; and the strengthening of China's own associations for the study of Africa by 2015 (see Li Anshan, in this volume). Africa experts are increasingly being trained through short-term fieldwork, language study, graduate programs and think tank–based research that could support China's expanding engagement with the continent.

The way forward: a new initiative?

If a major donor such as the Ford Foundation were to be asked to support a new fifteen-year initiative today, what would it look like? Who would be the players, what would be the critical themes and topics? Any initiative that would be launched today would certainly be Africa-led, in contrast to the earlier exchanges. African institutions of higher education were conspicuously absent from the 1980s–1990s Ford-funded program, except as the hosts for visiting Chinese scholars who were beginning to gain fieldwork experience on the continent. African scholars and African universities would be the center rather than the margins of a future initiative, and this would also include the broader global African diaspora or as Zeleza put it, "many Africas."[11] A core primary starting point of inquiry for such a program would be epistemological: how is knowledge constructed, by whom and to what ends? And lastly the past, present, and future of Area Studies would be a critical element in moving forward to post–Area Studies scholarship.

The way forward may look much like the University of Minnesota African Studies Initiative (ASI) mission statement: "It is with these complex processes and notions in mind – processes and notions that transcend traditional categories of area, region, and nation-state, but also traditional demarcations of academic fields and disciplines – that ASI approaches African Studies." Other newly formed Afro-Asian collaborations, for example the AFRASO ("Africa's Asian Options") program at Goethe University in Frankfurt, offer similar possibilities. This move would go in at least two important directions: it would acknowledge the importance of depth that has characterised Area Studies, in particular the critical role of language, cultural and social science fields that are place-based and grounded. At the same time, however, this approach would relocate those spaces in the actual geographies where they reside. As we have known all along, these "Africas" and "Asias" are and have always been global in their connections and mobilities.

A new paradigm would therefore not simply reformulate the Area Studies framework into a transregional one that, while appearing to share characteristics with new approaches to globalisms has in fact returned to many of the intellectual underpinnings of Area Studies. Given the dynamic rethinking of place, and the reinsertion of theory (for example the Comaroff's *Theory from the South*, and new approaches to twenty-first-century global capitalisms) there is potential to go much further. The replacement of "North–South" paradigms by "South–South" paradigms is also in question, in productive ways, as what initially seemed to be a promise of a different global order with the so-called "rising powers" has led to inquiry into the power inequalities of formations such as BRICS (Taylor 2017). Forums have queried the inequalities of global capitalism and questioned whether the "South–South" space is one of equality of negotiation or one of unequal positions.

Through a new initiative, African Studies has the potential to become an Africa-led field that is truly global – indeed it can only become truly global – through the collaboration of interdisciplinary and global partners. We not only need new paradigms for the study of global Africa but we also need new institutional models that are collaborative and consortial across national and regional boundaries. A new generation of emerging scholars is starting to form along these lines, and many of these scholars not only carry out their research but also have mentors on multiple continents. For this kind of new innovation to flourish, it requires the support of a new vision of global African Studies through cross-disciplinary, cross-regional and inter-generational dialogues and the formation of the robust consortia that will sustain them.

Notes

1 Huang Mingwei, Cheryl Schmitz, Vivian Wu, Roberto Castillo, Derek Sheridan, Xu Liang, Hezron Makundi, among others.

2 A recent article by Rita Abrahamsen argues the opposite, i.e. that it has been geo-graphical specificity rather than production of theory that has given Africanists legitimacy in the field of International Relations (Abrahamsen 2017, 128–129).

3 With the exception of institutions for foreign studies, in particular the Beijing Foreign Studies University where seven African languages are taught.

4 At the time of writing, the Institute of African Studies at Zhejiang Normal University was due to launch its new "Anthropology Research Center of African Studies" by investigating how "Chinese anthropology will further develop" in a multi-cultural and multi-national context (IASZNU, "Call for Papers," 2017).

5 The TAZARA railway was constructed to link the Zambian copperbelt with the Indian Ocean port of Dar es Salaam in Tanzania with Chinese assistance between 1968 and 1975.

6 I am grateful to Professor George Yu for facilitating my first visit to IWAAS (and therefore to China), when he was supporting the attendance of Chinese Africanists to the African Studies Association at the time. His introduction allowed me to meet Yao Guimei at the 2002 meeting and she invited me to give a talk at IWAAS, beginning a long-term friendship between myself and IWAAS researchers.

7 Ford Foundation Archives.

8 The Ford Foundation funded international affairs exchanges between China and the US after 1979 that focused on the "nexus between international political, economic and security issues" and geographic area competence. These efforts were also supported by the Rockefeller Foundation, the John D. and Catherine T. MacArthur Foundation and the Henry R. Luce Foundation.

9 The paper authored by Zhu Zhonggui (1992) argued that multi-party democracy was not suited for African countries after independence because it was not consistent with African traditional values of consensus in decision-making. Zhonggui stated that the one-party state model was better suited because of its potential for generating national unity, political stability and development. Richard Sklar (1992) argued on the contrary that Africans were generating new forms of democracy that combined tradition and modernity.

10 Ford Foundation documents: Report on the Second Conference at Malibu.

11 Jama Adams' research on African migrants in China is one example of the importance of including diasporic perspectives; research on the history of overseas Chinese and on the histories of Asian indenture and the African slave trade also demonstrates the significance of this perspective (Adams, 2015).

References

Abrahamsen, Rita. 2017. "Research Note: Africa and International Relations: Assembling Africa, Studying the World." *African Affairs* 116(462): 125–139.

Adams, Carlton Jama. 2015. "Structure and Agency: Africana Immigrants in China." *The Journal of Pan African Studies* 7(10): 85–108.

Alden, Chris. 2007. *China and Africa*. London: Zed Books.

Ampiah, Kweku and Sanusha Naidu. eds. 2008. *Crouching Tiger, Hidden Dragon? Africa and China*. Scottsville, SA: University of Kwa Zulu-Natal Press.

A'Zami, Darius. 2015. "'China in Africa': From Under-researched to Under-theorised?" *Millennium: Journal of International Studies* 43(2): 724–734.

Comaroff, Jean and John L. Comaroff. 2012. *Theory from the South: or, How Euro-America is Evolving toward Africa*. Boulder, CO: Paradigm Publishers.

Institute of African Studies at Zhejiang Normal University. 2017. "Call for Papers."

Large, Daniel. 2008. "Beyond 'Dragon in the Bush': The Study of China-Africa Relations," *African Affairs* 107(426): 45–61.

Li, Anshan. 2005. "African Studies in China in the Twentieth Century: A Historiographical Survey." *African Studies Review* 48(1): 59–87.

Manji, Firoze and Stephen Marks. eds. 2007. *African Perspectives on China in Africa*. Oxford: Fahamu.

Masao, Miyoshi and Harry Harootunian. eds. 2002. *Learning Places, the Afterlives of Area Studies*. Durham and London: Duke University Press.

Monson, Jamie and Stephanie Rupp. 2013. "Introduction: Africa and China: New Engagements, New Research." *African Studies Review* 56(1): 21–44.

Sklar, Richard. 1992. "Problems of Democracy in Africa." In *Papers Presented at Second Sino-US Conference, January 11–15, 1991*, edited by George T. Yu. Urbana, IL: US-China African Studies Exchange Program.

Sylvanus, Nina. 2016. *Patterns in Circulation: Cloth, Gender, and Materiality in West Africa*. Chicago and London: The University of Chicago Press.

Taylor, Ian. 2009. *China's New Role in Africa*. Boulder, CO: Lynne Rienner Publishers.

Taylor, Ian. 2017. *Global Governance and Transnationalizing Capitalist Hegemony: the Myth of the "Emerging Powers"*. London: Routledge.

Yu, George and Xia Jisheng. 1983. *A Report on The Development of International Studies in China: The African Area*. Urbana, IL: US-China African Studies Exchange Program.

Zeleza, Paul. 1977. "The Perpetual Solitudes and Crises of African Studies in the United States." *Africa Today* 44(2): 193–210.

Zhu, Zonggui. 1992. "A Probe into Democracy in Africa." In *Papers Presented at Second Sino-US Conference, January 11–15, 1991*, edited by George Yu. Urbana, IL: US-China African Studies Exchange Program.

Views from downstairs: ethnography, identity, and agency

9

CHINESE PEANUTS AND CHINESE *MACHINGA*

The use and abuse of a rumour in Dar es Salaam (and ethnographic writing)

Derek Sheridan

There were Chinese people selling peanuts on the streets of Dar es Salaam. This was a story a Tanzanian friend and colleague told me before I even arrived in Tanzania. The Europeans and Indians at least, he continued, had respected boundaries. The Chinese, on the other hand, would compete in even the "lowest" occupations, even selling peanuts. The young Tanzanian men in this occupation walk around town carrying nothing more than trays over their shoulder; arranged with peanuts, candies, and cigarettes; advertising by clanging loose change in their free hand. The notion that Chinese would come all the way to Tanzania to compete against them exemplifies the extremity and novelty of the idea of "China-in-Africa," all the more extraordinary when placed against the image of a "rising China" and its investments in African infrastructure. It is also inaccurate. I could find no such individuals while conducting 17 months of ethnographic research among Chinese migrant entrepreneurs in Dar es Salaam, and I was ready to categorise it as yet another "myth" about the Chinese in Africa (cf. Sautman and Yan 2012).

The appearance of "China–Africa," or "Africa–China,"[1] as a field of knowledge production has been accompanied, after all, by falsifiable claims, persistent tropes, and overdetermined narratives; and the *raison d'etre* of an emerging network of scholars has largely been positivist-empiricist "mythbusting" (Bräutigam 2009, 2015; Hirono and Suzuki 2014). In this literature, a clear line is drawn between the genre of China–Africa myths and an empirically grounded "real story." Myths are understood as deficits of information; at best, benign ignorance correctable through research and better public communication; and at worst, malign defamation promoted by actors argued to gain politically from promoting anti-Chinese sentiment. In this latter case, myths hide not only "real stories," but also "real motives."

In this chapter, I take a different perspective on myths about Africa and China, examining their social lives in ordinary discourse among those supposedly closest to the "real story," ordinary Africans and Chinese themselves. In showing that even the Chinese in Africa sometimes reproduce myths about the Chinese in Africa, I demonstrate the benefit of an ethnographic stance in the study of Africa and China. An ethnographic stance toward myths is concerned less with empirical verification than with intelligibility; why particular stories *make sense* to people, and what that *sense* can tell us about the social and cultural realities in which people live. The premise is that even empirically problematic statements still say *something* about social reality, and that they might be more sympathetically read as forms of political and moral critique. Myths, like conspiracy theories, can be a "key mode of cultural knowledge that seeks to reveal and to make locally intelligible the hidden forces and estranging dynamics of modern social experience" (Boyer 2006, 327).

This ethnographic stance is potentially in tension with currents in Africa–China studies that may be wary about the academic reproduction of already debunked claims. The dilemma of an ethnographic stance is that even if anthropologists read myths not as propositions (statements which are either true or false), but as indices; statements "pointing" to social realities which can be marshaled for arguments about either the people who articulate them, or a conceptual "third," ethnographies themselves (re)produce propositions about the world. However nuanced the genre of writing, ethnographers may have limited control over the interpretants of their arguments. This is true in any context, but in the case of popular interest in "China–Africa," conventional disclaimers about the partiality of knowledge are challenged by invitations to expertise; the expectation that such researchers will be able to say something definitive about their research sites. In other words, what does an ethnographer with an agnostic stance toward the ontology of myths do when confronted with interlocutors who make claims, which the ethnographer could, on good authority given the "mythbusting" scholarship, argue to be incorrect?

Myths pose a challenge for ethnography as a genre assumed to share an elective affinity with demonstrating the "agency" of ordinary African and Chinese actors. The call to recognise *agency* in the study of African–Chinese interactions is based on the assumption that ethnographies would provide "real stories" about how ordinary actors shape Africa–China relations "on the ground." In addition to these forms of *residential agency*, however, there is also the question of *representational agency*; how ordinary actors themselves shape narratives of Africa–China relations, including narratives of their own *agency* in those relationships. I borrow these terms and their distinction from Kockelman (2007), who defines *representational agency* as the capacity "to refer, predicate, and infer ... what we are representing, how we are representing what we are representing, and what other representations were used to infer this representation (or were inferred from this representation)" (p. 384).

In what follows I examine the politics of knowledge production about "China-in-Africa" from the perspective of Chinese and Africans directly involved. I explore how claims are produced, validated, contested, and circulated in different social contexts, including academic research. In particular, I focus on *casual hyperbole*, and its use as ethnographic data. *Casual hyperbole* refers to how "myths" about China–Africa, at the same time as they are being debunked by scholars, are also being circulated by the same potential "research subjects" themselves. These myths afford reflection and evaluation of both actually existing and speculative forms of Chinese presence independent of their empirical referents. In this chapter, I focus on a single claim in one location: that there are Chinese in Dar es Salaam, Tanzania, who engage in "petty trade." I then explore the multiple uses to which this claim is applied. I argue that the claim is shorthand for describing the presence of Chinese wholesale traders who do not, by and large, engage in retail trading. Nonetheless, the claim indexes a set of discontents among certain Tanzanians about the role of Chinese in wholesale trading in the first place, and expectations of industrialisation. For expatriate Chinese as well, the figure of the petty trader challenges desired representations of Chinese presence. The "truth" of the claim is less significant than the rhetorical uses to which it is applied.

Seeing like an ethnographer: peanuts as indices

I had already forgotten the story of the Chinese peanut seller when a Tanzanian shoe retailer told me one afternoon that if I went downtown, I would see Chinese driving taxis and selling peanuts. The retailer worked inside a crowded complex in Kariakoo, Dar es Salaam's wholesale market, selling shoes purchased from the Chinese wholesalers who occupied the well-located shops along the street outside. I had never seen Chinese taxi drivers or peanut sellers in the city, so I later asked Ali, another shoe trader whom I knew. He told me the man was a "liar" (*mwongo*, in Swahili). "If you hate someone," Ali explained, "you will tell lies about them." People disliked the Chinese, he explained, because they rented shops, raising the cost of rent and forcing smaller traders into more inaccessible locations. It was okay to operate warehouses, but did they have to open shops? Ali argued the claim about Chinese peanut sellers indexed discontents about the visibility of Chinese traders in Kariakoo, and discontents that they were not investing in factories. In other words, the Chinese were not respecting desired divisions of labour.

These are standard interpretations with an elective affinity for the genre of ethnographic writing. Wholesale distribution was a major site of investment for Chinese migrant entrepreneurs who began arriving in Tanzania in the early to mid-2000s. The most recent estimate is that there are 100 shops distributed between shoes, plumbing, electronics, and other products (Sun et al. 2017). Despite the diversification of Chinese investments after 2010, despite the fact that, in absolute terms, Chinese traders are outnumbered in Kariakoo, and despite the falling profitability for wholesaling, the presence of Chinese traders remains iconic. While these Chinese have predominantly restricted themselves to wholesale only, it has become a matter of speaking for Tanzanians to claim that Chinese do "petty trade." The claim is more accurately a statement that there are some who engage in forms of "low-end globalisation" (Matthews 2011), competing directly against African wholesalers for control of the distribution of goods between China and East Africa.

Petty traders themselves have benefited from Chinese wholesalers opening shops, as have consumers for whom Chinese production has lowered costs. At the same time, Chinese shops were frequently invoked by Tanzanians to be symptomatic of the absence of both local industry and quality consumer goods. There has in fact been Chinese investment in the construction of factories, but their effects have been limited. Whenever I mentioned there were factories on the edge of the city, Tanzanian traders would insist these were not "real" factories, but secret warehouses for doing trade.

A language of surfaces and hidden realities accompany discussion of the Chinese presence, and this is reinforced by the spatial practices of the marketplace itself. The earliest Chinese traders presented samples on the streets of Kariakoo of their industrial products before opening shops. Chinese employees working for wholesale shops circulate daily through the marketplace visiting customers or looking for new ones. Some carry samples in their bags or backpacks, making them look, especially to Tanzanian customers who tease them, like *wamachinga*, the term for the young Tanzanian men with low capital who informally sell retail on the street. The term *wamachinga* is the plural form for *machinga*.

The epistemological stance I have taken so far is that the figure of the Chinese *machinga* is a *proximate* truth insofar as it indexes the position of Chinese wholesalers in Kariakoo and the semiotics of their work. Chinese traders in Kariakoo during the time of my research (2013–2016) largely restricted themselves to wholesale trade, but even this was considered by many Tanzanian traders to be too low in the trade hierarchy. As I argue next, however, claims like these can operate in discursive practice as truths rather than as simply indices for secondary order interpretations. As Kockelman (2007) argues, things are "true" insofar as an agent, "however distal, sundry, or unsuspecting – takes up one's claim and thereby presumes it in subsequent actions" (p. 384).

Seeing like a Chinese manager: peanuts as possibility

According to Chinese expatriates with whom I spoke, the image of Chinese traders in Kariakoo is as evocative as it is misleading for being a privileged site of ethnographic research. In the same conversation, a Chinese expatriate who I had just met around the city advised me that if "I wanted to study the Chinese, I should go to Kariakoo," but then he advised me that the real story was not to be found there, but in "big projects" like roads. The figure of the Chinese trader is problematic for the image of China in Tanzania because it exemplifies the lowest common denominator of China–Africa comparative advantage: Chinese industrial overcapacity and African industrial under-capacity. In multiple interviews, Chinese investors and managers argued the presence of wholesalers would diminish and eventually disappear as Chinese investment developed.

During an interview with a manager for a Chinese state-owned enterprise, he described wholesaling as a "stage one" enterprise undertaken by people with lower capital and lower "cultural level." They did not represent the "mainstream" of Chinese presence. This is perhaps why he accepted the story of the peanut seller as a possibility. I shared the story with him because he had been talking about misperceptions of the Chinese presence in Africa, and I shared my own example in the interest of building rapport. His response surprised me. Although he did not believe Chinese migrant entrepreneurs engaged in such activities as a primary source of income, he did not dismiss the story entirely. He speculated that a Chinese businessman might have temporarily encountered financial difficulties. The reason for these difficulties could range from slow business to delayed overseas shipping. He might have set up a stand to sell peanuts for a day. Somebody saw this, he continued, and falsely concluded that Chinese were coming to Tanzania in order to sell peanuts. This was an example, he argued, of how misperceptions are created on the basis of limited experience.

Questioning the reality of the referent, but not the reality of the signifier, he constructed a speculative narrative on the basis of a fact that may or may not have been "true." The plausibility of the story indexes ideological notions of the Chinese work ethic, willing to take on even simple tasks, and, in a common Chinese idiom, 吃苦 *(chīkǔ)*, or "eat bitterness," meaning to endure hardship in the pursuit of economic well-being and accumulation. In the context of the larger interview, however, the manager wanted to argue that wholesale and retail trading was only a temporary phenomenon that reflected an early stage of economic engagement between China and Africa. The sector would pass from the scene as Chinese increasingly become involved in industrial investment.

The argument that Chinese traders are non-representative of the Chinese presence is present in other situations where the myth of the "Chinese *machinga*" appears in ordinary discourse. My next example is a passing reference in the discussant comments of a Tanzanian government official attending a presentation on "Chinese engagement in Tanzania: Is it considered positive or negative by Tanzanians?"[2] The value of attending to an ephemeral reference in a larger conversation is to illustrate the significance of ordinary hyperbole, and the way rumours can become signified in distinct contexts without these contexts themselves being explicitly about the truth-value of the rumour in question. The explication of such ephemera, however, requires first explaining the context in which they appear.

Seeing like a Tanzanian diplomat: who does the peanut speak for?

The presentation was conducted by REPOA (Policy Research for Development), a development think tank in Dar es Salaam, in cooperation with Afrobarometer, a public opinion survey project in operation since 1999. In attendance were representatives of the Chinese embassy, representatives of other embassies, journalists and representatives from Tanzanian government agencies. Also in the audience were myself and two Chinese interlocutors whom I had invited.

The main conclusion of the presentation was that Tanzanians, as randomly surveyed (n=2386), had a positive opinion concerning the "influence" of Chinese investment. Eighteen percent of respondents, however, did not have an opinion. The two discussants invited to respond were both diplomats who had previously been posted to China. Their comments were structured in the form of a recognizable narrative about Sino-Tanzanian friendship and cooperation over the *longue durée*. The survey, the ambassador explained, is about perceptions, and "perceptions are a way of talking about others, but those perceptions may or may not align with reality. In this case, however, they align."

The discussants considered themselves authorised to evaluate the "reality" of claims about the Tanzania–China relationship. This was evident both in how they framed points of critique regarding the presentation, and in how they framed their own agency in the processes described. It was in this context that the figure of the "Chinese *machinga*" was invoked. The ambassador took issue with the use of the word "influence" in the presentation:

> "Influence" is a strong word, coming from the idea of a "sphere of influence." China will never build a sphere of influence in Africa. It is about cooperation. At FOCAC [Forum on China-Africa Cooperation], there was a recognition of the need to change relationships from being just state-to-state to being people-to-people. China has encouraged the development of people-to-people exchanges ... China practices non-interference. *It will never interfere, well maybe the Chinese petty traders interfere with the machinga, but that's not state level.* States go outside because they can't fulfill their interests at home. Countries have their own interests. That is fine. [Italics Added for Emphasis]

He later added that a random survey could not provide an accurate portrait of Sino-Tanzanian relations because not everyone in the survey had the same kind of direct experience with the Chinese. He stated that the survey is "0.05 percent of the country. Only 2000 people is too small a sample, and what period are we talking about? 50 years is a long time, but the Chinese *machingas* (sic) are recent. When I was in China, I encouraged Chinese to come. We created FOCAC because we've been talking about 'South-South' cooperation, but how do we do it?"

According to the ambassador, the Sino-Tanzanian relationship was primarily the product of state agency, and the people most authorised to speak about these relationships were officials like himself who had been involved in negotiations over perhaps the largest periodic diplomatic ritual of Sino-Tanzanian relations and Sino-African relations more generally, the Forum on China-Africa Cooperation. The figure of the "Chinese *machinga*," however, disrupts his argument. It appears first as the exception to a central trope of state-diplomatic discourse about China. "Non-interference" usually refers to the claim that while China engages *economically* with foreign states, it does not "interfere" with domestic politics in the fashion of traditional Western neo-imperial powers. In this case, however, Chinese involvement in a "low" economic sector is characterised as a form of "interference," but quickly disavowed as "not state level" even though the same logics of capital expansion applied to the Chinese state would also be an apt description of the motivations of Chinese traders who "go outside because they can't fulfill their interests at home." The claim is further complicated by the ambassador's assertion of agency ("I invited them to come"). In the larger context, he is referring to "Chinese investors" (whether public or private is not clear), and by taking responsibility for that action, he is also challenging the assumption that Chinese investment in Tanzania is primarily authored by Chinese leaders, a common assumption of the "neo-imperialist" narrative. However, whether wholesale traders could be considered "investors" is controversial to many Tanzanians, and explains part of the ambivalence in the ambassador's comments.

The ambassador's comments cannot be understood without considering the larger discursive field. His sensitivity to the use of "influence" is important because the term bears the traces, intentional to the report's authors or not, of the narrative of Chinese "neo-imperialism." The production, validation, and critique of claims about "China–Africa" is hard to disentangle from the geopolitical field that frames them. The anticipatory (and somewhat defensive) tone of counter-narratives is evident in the fact that even though the Afrobarometer survey suggested a generally positive portrait of Tanzanian perceptions, I noticed multiple questions and answers during the Q&A that magnified on challenging those elements which could be construed as "negative."

For example, survey respondents were also asked to list issues that they personally believed contributed to negative views of Chinese. A small group of respondents had criticised the quality of Chinese goods. The Chinese interlocutor I brought with me asked a question about this response, publicly challenging her "Tanzanian friends" that the responsibility for low-quality goods in the marketplace was due to the preference of Tanzanian traders and consumers, rather than Chinese manufacturers and suppliers. She had spoken to me before about an anecdote from a friend who worked in Kariakoo: a story of an archetypical Tanzanian consumer who could not be prevailed upon to spend a little extra for a higher quality auto part when given the option of a cheaper low-quality product. The identification of "African consumers" as responsible for market demand is understood to relieve, to a certain extent, the popular association of China with "low-quality" or "fake." Interestingly, however, some Chinese expatriates *did* blame Chinese traders for selling low-quality goods, and being responsible for hurting the image of China in Tanzania. These were sometimes connected with discriminating social judgments about the kind of Chinese who engaged in trade. Tanzanians, on the other hand, tended to blame their government for not enforcing quality control in customs. This was connected with broader domestic political discontents rather than evidence of anti-Chinese sentiment. The diversity of options for assigning blame in this situation shows how the identification of agency and responsibility is an exercise that requires simplifications of otherwise entangled processes.[3] The ambassador's comments about Chinese *machinga* as both *not* the product of state agency, but also potentially the product of "invitations" to economic interaction, exemplifiy this contradiction.

The actions of individual Chinese and Tanzanians very often get scaled up to geopolitical agency and responsibility. The exercise can also be seen at work when another audience member asked a question about ivory. They phrased the question in a manner referring to the earlier discussion of "interference": "China practices non-interference and expects others to do the same, but I wonder if there is any way to influence the Chinese public. For example, poaching in Tanzania is encouraged by high demand in China. Maybe because China is generous, one is embarrassed to raise these questions."

A representative from the Tanzanian Ministry of Foreign Affairs in the audience had the next question, but he first wanted to respond to the "accusations" of his colleague, continuing that "the idea that the poaching of elephants is championed by China is not true. I went to China with [President] Kikwete, and he signed an accord on wildlife protection. China has addressed this issue." The ambassador, in his own comments, added that "these are Chinese *criminals* and Tanzanian *criminals*. That is not official trade, so you should not count that. You can't call undercover Chinese business official because that is not official."

There is a lot happening in this exchange. A comment about influencing domestic demand for ivory in China in order to protect African elephants is glossed by a foreign affairs official as an "accusation" directed at "China" (specifically the Chinese government) and therefore an accusation that must be refuted. The fact the Chinese government does not promote the ivory trade, or that China does not condone corruption is taken to be evidence that the critique is unfounded. If it is not "official," it should not count.

I have found this is a pattern of argumentation not unique to this Tanzanian audience, but also found in discussions among both Chinese and foreign academics who study China–Africa. Because in the global media, discussions of "Chinese" and "Africans" in the abstract very quickly become discussions of "China" and "Africa" as agents whose moral responsibility is premised on their intentionality.

In the case of public diplomacy, whenever Tanzanian and Chinese officials challenge the veracity of "Western" accusations, entire categories of individuals are excluded as relevant agents, because the relevant agents are primarily the state rather than the wider assemblage of actors involved in these relationships. In the situation described here, the ambassadors serving as discussants themselves "uncover African agency" (cf. Corkin 2013) but insofar as they position themselves as authors of these China–Africa initiatives. This excludes other possible agents. The ephemeral reference to Chinese *machinga* in the course of this larger debate is casual, but encapsulates a wide range of contradictions in both the China–Tanzania relationship and the politics of knowledge around making claims about this relationship. And all this regardless of whether such a claim is "true."

Seeing like a peanut

It was an ironic surprise when late in my fieldwork, when talking again to my Tanzanian colleague, he insisted that the story of the peanut seller was indeed true. His uncle had eaten the peanuts. The man would set up his stand every evening, and his uncle would occasionally purchase them. There was a special spice the man must have used, because his uncle said the peanuts were very delicious. When I finally asked my colleague's uncle directly, I learned there had indeed been a Chinese man selling peanuts, but that was in the 1970s! At the time, in addition to 30,000 Chinese experts working on the Tanzania–Zambia (TAZARA) railway, there was a small community of Chinese living in Dar es Salaam who had come there from Guangdong early in the twentieth century. The uncle did not know from which group the man had come. I found Mr. Wang, a Zanzibari-born Chinese who had lived in Dar es Salaam since the 1970s, and asked him. He did remember such a man, but he told me the man had come from "mainland China" and Mr. Wang knew neither where he came from nor his eventual fate. It was over 40 years ago.

There was indeed a referent, after all, but it was distinct from that which its contemporary signifiers referred. The trace had its own distinctive histories, a trace which in a new context, could take on meanings and associations far removed from the initial conditions of its production. On the other hand, one might argue the trace of a forgotten Chinese presence hidden under the shadow of grandiose projects like the TAZARA railway echoes the semiotic marginalization of Chinese traders today in the face of the imagined futures to which China–Africa aspires. The thematization and characterization of this as an "echo," however, is itself a product of the genre of knowledge production which I am engaged in here, the production of an ethnographic text. That is, it is my own argument.[4]

Conclusion

In this chapter, I have used the rumour of a Chinese peanut seller, and the related figure of the "Chinese *machinga*," in Tanzania to examine how claims about the China–Africa relationship are produced, validated, contested, and circulated in the context of differing political economies of epistemological value. In doing so, I have tried to flatten the conceptual distinction between vernacular and expert forms of knowledge production by showing traffic, in the form of casual hyperbole, between them. My approach to these forms of myth and hyperbole has been different than approaches that aim to uncover a "real story," or to critique the "real motives," behind them. Instead, I have traced how myths and hyperbole work in rather quotidian ways in ordinary discourse. An "ethnographic stance" (Keane 2014) provides an alternative perspective to the study of Africa–China knowledge production than either "myths" or "real stories."

I examined a story that a Chinese individual had been selling peanuts in the streets of Dar es Salaam. It is the kind of story that lends itself readily to the ethnographic interpretation that the myth indexes popular Tanzanian discontents with Chinese migrant entrepreneurs who place themselves too low in the trade hierarchy. The story is closely related to the more prevalent claim that Chinese engage in "petty trade," or can even be called *machinga*, the term for the young Tanzanian men with low capital who sell retail on the street. Chinese traders in Kariakoo largely restrict themselves to wholesale trade, but even this is considered by many Tanzanian traders to be too low in the trade hierarchy. More than a manner of speaking, however, the casual hyperbole of Chinese trade as "petty trade" can become an un-extraordinary "fact" that serves particular functions in other discourses.

I showed how the myth appeared in the context of an event where Tanzanian officials commented on the claims of a social science research survey of popular Tanzanian attitudes toward the Chinese in the country. Although the survey revealed generally positive attitudes, it is perhaps a learned defensiveness and propensity to mythbusting that led these discussants to emphatically exclude certain types of subjects (Chinese "petty traders" and ivory traders) as unauthorised to represent "China–Africa" because they are not the "official" agents (the Chinese and Tanzanian governments). In the process, they ironically reproduced, in a casual fashion, the very rumour of a Chinese *machinga*. The final irony is that the story of the peanut seller is partially true, demonstrating the importance of considering the histories of the signs of myths and rumours themselves as having their own particular logics exclusive of the claims made on them by either ethnographic interlocutors or the ethnographers who marshal them to produce academic work.

Myths are not simply "misinformation" to be debunked; they are the mobilization of resources and imaginative projects, including the very field of "China–Africa" itself. The Chinese presence in Africa semiotically exceeds the materiality of people and things Chinese, beyond even the dominance of Chinese-manufactured goods in the market. Part of the reason for this is that the diplomatic staging of investments and aid projects produces dozens of announcements each year about projected investment pledges and projects. These announcements are sometimes collected uncritically by international scholars and used to produce inflated accounts of Chinese commitments, turning the potential into the accepted reality of China–Africa. These can in turn produce new mobilizations.

For example, a European university several years ago obtained funding for a multidisciplinary study of a reported Chinese investment project in southwest Uganda, but after scouting the area, research assistants reported back that there was in fact no such project (Joseph Maiyo, Personal Communication: July 2013). Pedersen and Nielsen (2013) propose the notion of the "trans-temporal hinge," defined as "any configuration of socio-cultural life that is imbued with the capacity for bringing together phenomena that are otherwise distributed across disparate moments in time" (p. 123). Applied to Chinese infrastructure projects in Mozambique, the authors observe contemporary mobilizations of people, resources, and knowledge claims "couched in a vocabulary defined by the properties of events that had not yet occurred" (p. 130). In much of the writing on China–Africa, in fact, the present has been situated as the prologue to various futures. The irony of the "mythbusting" stance is that it promises the deconstruction of the very premise of having a "field" of China–Africa studies. The story of the Chinese peanut seller is distinct because it represents the extremity of claims about China–Africa, the unprecedentedness, which is itself a source of fascination that exceeds the situation itself.

Ethnographic methods provide a privileged vantage point for studying the interaction between "myths" and "real stories." Ethnographic fieldwork requires extended immersion among the people one is researching, and the development of long-term and multifaceted relationships. Immersive fieldwork creates space for contradiction and the unexpected. Ethnographers don't just record interviews, but everything they see, hear, touch, taste, and even feel. It is a "deliberate attempt to generate more data than the researcher is aware of at the time of collection" (Strathern 2004, 5–6; Strauss 2013). Ethnographers don't just capture singular stories, but a sometimes discordant symphony of actions, discourses, and forms of social life. Ethnographies are "thick descriptions" (Geertz 1973) precisely because they endeavour to connect threads of social, cultural, political, and economic realities; but they do so from particular locations. This is why it would be a mistake to consider ethnography simply as a method for generating empirical data about a "real story" in the "downstairs" room of Africa–China. Ethnographies are "partial truths" (Clifford 1986) shaped by the specificities of the social encounter between the ethnographer and their interlocutors.

On purely empirical grounds, one benefit of ethnographic research beyond just interviews is that researchers may discover that informants may act differently in practice than they say or imply in interviews. This is not necessarily about deception. How people behave and how people speak may be shaped by their immediate social context. In conducting fieldwork, ethnographers are placed in a variety of contexts. Ethnographers who have worked with Chinese expatriates in many African countries, for example, might recognise the broad outlines of a contrast between "frontstage" and "backstage" discourse, or "official" and "unofficial transcripts" (Goffman 1959; Scott 1985). Frontstage discourses may resonate with state-diplomatic narratives of friendship and development cooperation, whereas the unofficial transcript of Chinese expatriate talk entails discontents about corruption, robbery, and inefficiency, much of it racialised in the everyday language of 黑人 [*hei ren*, or black person]. These two registers are not necessarily in contradiction, and it cannot be assumed that the latter is more "sincere" than the former. Complaining with fellow expatriates is just as much a socially contextual speech genre as official-ese. Ethnographers are neither mind-readers nor investigative journalists, but the benefits of immersion are witnessing how different forms of everyday knowledge claims are articulated in relation to each other, and the relation between these claims and ordinary practice and experience. Myths, hyperbole, and contradictions provide ethnographers insight into the messy politics of everyday knowledge production.

This is true in any ethnographic research setting, but in the case of China–Africa, these politics are complicated even further by the policing of China's "image" in Africa. In the context of contending global narratives of "imperialism" and its denial, ethnographers have to reflect on how different kinds of Chinese informants, in particular, represent themselves to researchers, and how they themselves in turn represent their informants in their writing. Many Chinese informants, aware of the way international media organizations have reported on people like themselves, will be self-conscious about how they present themselves to researchers, and it will matter where these researchers come from; if those researchers are perceived as friendly or not to their own interests and reputation. In more informally structured settings like ethnographic research, one's informants will not always be aware that what they say may potentially be converted into "ethnographic data" for academic publications. Formalised procedures of "informed consent" are designed to mitigate these ethical/legal dilemmas, but anyone who has done ethnographic fieldwork would recognise that the boundaries between research and not-research are porous.

The cases I discuss here raise important questions about how ethnographic data is used and, in turn, how much people have control over their representations of the world, and their own experiences. Compared to distant readings of China–Africa encounters, ethnography does highlight agency, but what I have tried to demonstrate here is that ordinary people themselves reproduce the myths and dominant narratives about them, which scholars are trying to debunk. This is why attention is warranted to what Kockelman (2007) calls "representational agency," the degrees of control people have over the production and circulation of their own interpretations of their own experiences. The ethnography of Chinese–African relations is necessarily reflexive, not simply as a standard exercise of anthropological claims-making, but as a recognition of the overdetermined interest in the topic itself. Understanding ordinary perspectives depends on understanding the ecology and political economy of discourses and stances within which people make sense of and communicate their experiences.

Notes

1 Scholars like Adams Bodomo (2012) and Bob Wekesa (2017) have advocated for and used the term "Africa–China" to challenge the Sinocentric appellation "China–Africa" which privileges Chinese agency and intentions. I deliberately use the terms interchangeably, but defer in this chapter to the ordering preferred by the editors for the volume.

2 See "Citizens Perception on Chinese Engagement in Tanzania," including the presentation, at www.repoa.or.tz/highlights/more/citizens_perception_on_chinese_engagement_in_tanzania_development (accessed April 30, 2016).

3 My friend's open question to her "Tanzanian friends" might be compared to the trend in the literature on Africa–China to hold the "agency" of African states responsible for the outcomes of Chinese investment or influence. A corrective to tropes of African passivity in narratives of Chinese agency (cf Brown and Harman 2013), it is nonetheless important to closely watch how such arguments are deployed in political practice; particularly when the phenomena in question are co-constituted by a variety of actors.

4 Nonetheless, it would be presumptuous to assume the responsibility for this argument is entirely my own. The possibility of making this connection was learning of the existence of an abandoned Chinese cemetery containing the remains of the small community of Chinese who lived in Dar es Salaam prior to the construction of TAZARA. I learned of the cemetery from a Chinese woman working for a Chinese conglomerate in Dar es Salaam. She herself had discovered the cemetery after it appeared to her in a dream. It was this dream that led her to search for it. From one perspective, it could be said that it was these spirits who reached out to her, contributing indirectly to me deciding to make such an argument.

References

Bodomo, Adams B. 2012. *Africans in China: A Sociocultural Study and Its Implications on Africa-China Relations*. Amherst, NY: Cambria Press.

Boyer, Dominic. 2006. "Conspiracy, History, and Therapy at a Berlin 'Stammtisch'." *American Ethnologist* 33(3): 327–339.

Bräutigam, Deborah. 2009. *The Dragon's Gift: The Real Story of China in Africa.* Oxford: Oxford University Press.

Bräutigam, Deborah. 2015. *Will Africa Feed China?* Vancouver: Oxford University Press.

Brown, William and Sophie Harman. 2013. *African Agency in International Politics.* New York: Routledge.

Clifford, James and George E. Marcus. eds. 1986. *Writing Culture: The Poetics and Politics of Ethnography: A School of American Research Advanced Seminar.* Berkeley, CA: University of California Press.

Corkin, Lucy. 2013. *Uncovering African Agency: Angola's Management of China's Credit Lines.* Farnham, MA: Ashgate Publishing, Ltd.

Geertz, Clifford. 1973. *The Interpretation of Cultures: Selected Essays.* New York: Basic Books.

Goffman, Erving. 1959. *The Presentation of Self in Everyday Life.* New York: Anchor Books.

Hirono, Miwa and Shogo Suzuki. 2014. "Why Do We Need 'Myth-Busting' in the Study of Sino–African Relations?" *Journal of Contemporary China* 23(87): 443–461.

Keane, Webb. 2014. "Affordances and Reflexivity in Ethical Life: An Ethnographic Stance." *Anthropological Theory* 14(1): 3–26.

Kockelman, Paul. 2007. "Agency." *Current Anthropology* 48(3): 375–401.

Matthews, Gordon. 2011. *Ghetto at the Center of the World: Chungking Mansions.* Chicago, IL: The University of Chicago Press.

Pedersen, Morten Axel and Morten Nielsen. 2013. "Trans-Temporal Hinges: Reflections on an Ethnographic Study of Chinese Infrastructural Projects in Mozambique and Mongolia." *Social Analysis* 57(1): 122–142.

Sautman, Barry and Yan Hairong. 2012. "Chasing Ghosts: Rumours and Representations of the Export of Chinese Convict Labour to Developing Countries." *The China Quarterly* 210: 398–418.

Scott, James C. 1985. *Weapons of the Weak: Everyday Forms of Peasant Resistance.* New Haven, CT: Yale University Press.

Strathern, Marilyn. 2004. *Partial Connections.* Walnut Creek, CA: Rowman and Littlefield.

Strauss, Julia C. 2013. "China and Africa Rebooted: Globalization(s), Simplification(s), and Cross-cutting Dynamics in 'South–South' Relations." *African Studies Review* 56(01): 155–170.

Sun, Irene Yuan, Jayaram, Kartik and Omid Kassiri. 2017. *Dance of the Lions and Dragons.* Retrieved from McKinsey and Company.

Wekesa, Bob. 2017. "New Directions in The Study of Africa–China Media and Communications Engagements." *Journal of African Cultural Studies* 29(1): 11–24.

10

REFLECTIONS ON THE ROLE OF RACE IN CHINA–AFRICA RELATIONS

T. Tu Huynh and Yoon Jung Park

In late May 2016, a Chinese laundry detergent commercial went viral on social media. The blogosphere went wild with comments about "the Chinese" being racist, suggesting that there is something exceptional about the Chinese people's racism toward the people from Africa. The commercial shows a Chinese woman luring a Black man close enough to shove a detergent gel-capsule into his mouth and heaving him into a washing machine. The man pops out (white-)washed and transformed into an Asian man. The commercial, produced for and targeted at a national audience, was globally circulated on Facebook as a YouTube video by an American musician and French photographer, both living in China, with the caption: "Chinese ad for a local brand of clothes (*sic*) detergent. Raw racism … ."[1] It has since been reposted on various social media platforms, including Twitter, Weibo, and WeChat. Social media followers' comments were accompanied by glut of rapid fire analyses by journalists and a few China–Africa scholars on "the Chinese" perception of race and "Chinese racism." Many scholars cautioned against the (mostly) Western media's exploitation of Chinese exceptionalism, which presented both Chinese people and the Chinese state as implicitly "more" racist than White Euro-Americans and Western countries.

The jostling between scholars and journalists on this issue parallels some of the contestation over broader discourses on China's engagement with African countries, wherein scholars often struggle to be heard. More importantly, however, the racial assumptions embedded in existing "China–Africa" discourses have yet to be unpacked. The abovementioned commercial, and reactions to it, can be seen as one form of "bottom-up" backlash against this silence in a new context in which China is not just rising, but has gained relative power vis-à-vis Europe and the United States. More than ever, the "yellow peril" evoked by Euro-American journalists reporting on "China in Africa" since the early 2000s requires attention. The fear-based writing raises concerns about whether the core of Western-dominated capitalism is shifting East and further implications for hegemonic whiteness. An alternate notion, that Chinese people are somehow "ignorant" of race, has been intimated by some in Western countries, made explicit by Chinese authorities, and supported by a few scholars. Claims from these various camps, that Chinese people are more racist, practice racism without race, or are racially unaware, ignore the argument that race and racialization are (and have long been) key elements that structure the lives of all who are racialised Black, Brown, Red, White, and Yellow, including Chinese nationals and ethnic Chinese in the diaspora. Works such as Robinson's (2000) on racial capitalism and DuBois's (1903) articulation of the colour line as *the* key problem of the twentieth century have already pointed to this.

In our view, whether we are focused on nineteenth-century indentured labourers or twenty-first-century investors, China and the Chinese people are a constituent part of a global racial hierarchy. That said, the fact that China and diasporic Chinese communities have historically been racialised by others does not mean that China does not have its own racial imaginings; nor does it mean that such race thinking is static. As Wilensky (2002) and Wyatt (2012) reveal, although perceptions have changed over time, Chinese people have possessed views on dark-skinned people, or *kunlun* (崑崙), since before the 1500s. The term first referred to dark-skinned Chinese people and extended to those in Southeast Asia and India before it was applied to the African slaves of Arab merchants in China. The challenge at this juncture, when China wields increasing power to shape global political–economic processes,[2] is to better understand and theorise Chinese identity formation in China and comprehend how these notions interface with global notions of Chineseness and, more broadly speaking, race and racial identities. With regard to China–Africa engagement, the challenge is to understand how these forces of identity formation inter-relate and play out in contemporary China's relationship with African countries.

We hope to point the way by examining how race has been treated in the existing body of work. The core of China–Africa literature is comprised of studies by economists, political scientists, scholars in development studies, and, to a lesser extent, migration studies; these have tended to ignore how race shapes trade, investment, and migration patterns that are said to be linked with China's economic rise and have presumed a sort of colour-blindness (Bonacich et al. 2008). If race comes up at all, it is often mentioned only as one of few social issues that may impact on relations between Chinese and Africans. Our focus here is the subset of literature that does, whether implicitly or explicitly, deal with race. While ours is not an exhaustive review, we attempt to highlight moments when the concepts of race, racism, or racialization were utilised to discuss Chinese, Africans, relations between them, or perceptions they have of one another.[3] Importantly, our aim to disrupt restrictive binary constructions and ahistorical conceptions of race, which appear in many of the available China–Africa studies, builds on the works of Ethnic Studies scholars in the US, who have already been triangulating race in their studies (see Bow 2010; Espiritu 2008; Joshi and Desai 2013; Robles 2013; Wu 2002). Monson (2015), Africanist historian of China's role in constructing the TAZARA railway, was one of the first to point toward this direction, as she called for scholars to jettison a polarised view of "China–Africa" relations by re-inserting and making explicit Whiteness and Euro-America (or, simply, race) in future analyses. Our survey of the scholarship reinforces Monson's call by attempting to chart a way forward for a deeper excavation of race that takes into account an often unconscious adoption of Western frames. We hope to shift the focus to foreground China's new role in the racial capitalist world-economy, while simultaneously taking cognizance of Chinese exceptionalism. Race and racism, as all analytical categories, are ultimately socially constructed; part of the problem, we argue, is that a focus on Chinese and African bodies and even (nation)states sometimes obscures our view of larger global structures and processes of power which have been, are, and continue to be racialised and racializing.

Chinese in Africa

Chinese people in African countries have attracted a great deal of attention, particularly since the mid-2000s when journalists began writing about "masses" of Chinese "invading" the continent. The re-introduction of the "yellow peril" discourse from voices outside of the continent, followed by increasing reports of anti-Chinese sentiment and criticisms of the behaviours of "the Chinese" at local and state levels, motivated academic researchers to pay closer attention to migrations of Chinese people to the continent in the twenty-first century. While local, state, and international (mainly, some European countries and the US) views homogenised all Chinese actors, conflated the Chinese state and people (creating narratives of "China, Inc."), and framed this new phenomenon in terms of China's rise to economic power, few explicitly used the term race.[4]

Against that backdrop, Park and Huynh (2010) organised and published a special journal issue that started to expose the heterogeneous composition of the Chinese people on the African continent, with specific studies of Chinese in Equatorial Guinea, Mali, Senegal, Zambia, and South Africa. Esteban (2010), Kernan (2010), Yan and Sautman (2010), Huynh, Park, and Chen (2010), and others challenged the notion that the Chinese on the continent were part of one homogenous mass directed by Beijing. Esteban argued that while the Equatoguinean political and economic elite, who stood to gain from the long-standing friendship between Malabo and Beijing, held positive views of China, local entrepreneurs and members of the opposition voiced frustrations and complaints about the rapidly rising Chinese population and their questionable practices. Kernan pointed out the Chinese migrants in Dakar and Bamako work across various sections ranging from small retail shops to large multinationals. Yan and Sautman identified differences between Chinese farmers from the 1960s and 1970s from those of the present in Zambia, and Huynh et al. identified at least three distinct cohorts of ethnic Chinese in South Africa. Going back to an earlier study, Hsu (2007) found time of arrival (and length of time in Africa) impacted upon class and sense of allegiance amongst Chinese Zanzibaris. These ethnographic studies challenge a prominent neo-colonial representation of China that suggests that Chinese people are only drawn to oil or mineral rich African countries and, also, tacitly address the problem of imagining the Chinese "race" as a homogeneous population.

Subsequent studies like Zi (2015) locate the Chinese merchants in Botswana in a "middleman minority" frame to discuss their role as foreigners in a situation of hostility and opportunity. Arsene (2014) also picks up on the theme of conflicts and tensions between Chinese shop-owners and local employees in Uganda, emphasizing the relevance of *enjawulo* or local socio-cultural expectations and practices around work that supplement formal wages. Arsene's work forms part of Giese's (2014) special journal issue that comprises studies aimed at delineating everyday interactions and moving beyond the usual perceptions of competition, tension, and conflict. In that issue, Lampert and Mohan's (2014) contribution on Chinese businesspeople in Ghana and Nigeria points out that tensions and conflicts between Africans and Chinese differ according to the Africans' class position (i.e., African traders versus distributors or consumers). Indeed, similar to Esteban's (2010) above-mentioned study, they shift the focus from Chinese migrants to African actors. Lampert and Mohan also mention culture and race, pointing out that African labourers sometimes engage in negative racialised stereotyping; however, they find that conviviality that is premised on interpersonal interactions between Africans and Chinese can serve as a tool for Africans to leverage benefits from China–Africa relations.

In a more recent co-authored manuscript, Mohan, Lampert, Tan-Mullins, and Chang (2014) introduce the concept of "intersectionality." They argue that their research revealed too many "factors shaping these tensions [in Sino-African encounters] ... [such] that we cannot use the lens of race or nationality to explain what we found" (p. 101). Race is defined as a form of social difference and they seem to equate "expressions of racial difference" ("racial prejudice" or, what we would call, racism) with "outright xenophobia" (p. 102). Sautman's (1994) earlier work on the presence and experiences of African students in Nanjing in the 1980s, which we shall return to later, rejects this sort of conflation of racism and xenophobia. In his study, African students were specifically targeted for violence because of their appearance – their Blackness; as evidence he points out that other foreign students, namely, White men, were exempted from the same type of physical (and psychological) violence. Had Mohan et al. considered the works of African scholars like Flockemann, Ngara, Roberts, and Castle (2010) who invoke Fanon's notion of "Negrophobia," and Koenane's (2013) references to the 2008 xenophobic violence as "racism and/or Black on Black hatred" (p. 107), they might have agreed that xenophobia is another way to avoid acknowledging the embeddedness of race. That is, the use of xenophobia to point out that antagonisms toward the Chinese people as "outsiders" are unexceptional disregards the relevance of racial structures as well as racialization that historically include China/Chinese people (see Bright 2013; Harris 1998; Huynh 2013; Park 2008a; Yap and Man 1996). As in many studies privileging political economy, socio-cultural factors are reduced to a mere series of discrete identities that can be added or removed, to analyze at will. In other words, the "social dimensions" (Mohan et al. 2014) are conceptually separated from economic and political processes, as evidenced by their three examples of tension and conflict between Africans and Chinese. They conclude, "[a]s yet the tensions have not become overtly politicised or racialised, though such potential is there" (p. 124). While they point to some of the potential downfalls of intersectionality (the very additive problem referred to above), they seem to fall into the same trap. More useful is their argument that intersectionality can potentially be used to examine relative privilege and empowerment (p. 103). The notion that social dimensions are relational is particularly valuable; understanding this helps us to delineate the complex positionalities of a Chinese person in Africa, which might result in privilege as regards their class position but also vulnerability in terms of minority and migrant status vis-à-vis the general population.

Sautman and Yan (2014 and 2016), whose work spans many years of laborious field research in Zambia, have gradually constructed an argument that anti-China/Chinese views, including claims of a neo-colonial China and amoral practices of "the Chinese," are embedded in a racial hierarchy that privileges the West/Western counterparts. For example, while Chinese have been subjected to figurative and literal "bashing" during protests, Whites (or "expatriates," as the term is only reserved for them) have never experienced strike violence in Zambia. However, Indians, Sautman and Yan (2014) point out, have also experienced strike violence. Accordingly, racial and ethnic differentiations, which require a consciousness of such differences on all sides, are simultaneously present in the local context, affecting labour relations, and creating different strike outcomes. Sautman and Yan point out that after the 2010 strike that resulted in the death of 13 Zambian miners Chinese were exposed to "much online racist abuse and calls for harsh retaliation…" (p. 12). Zambian racist views of Chinese were expressed through words and imageries that derided linguistic, phenotypical, and civilizational/developmental differences like "'choncholis,' 'ching chongs,' 'squinty eyes,' 'yellow savages,' 'Chinese piglets' and 'chinks'" (ibid.).

From shedding light on racial views of Chinese held by Zambians, Sautman and Yan's (2016) recent work turns toward analyzing the discourse of labour racialization associated with Chinese enterprises in Africa. They suggest that this discourse in the global South diverges from that of the global North and that the racializing that takes place at the Africa–China interface should be analyzed in a South–South setting. Echoing Monson's (2015) concern, Sautman and Yan propose a triangular framework to better comprehend how Africans, Chinese, and Western actors "co-constitute racialization." They point out that Western discourse already predetermined Chinese employers as racialisers before actual labour relations were even formed between Chinese and Africans; however, they argue, racialization occurs in multiple directions, as inferred also by Mohan et al. At one level, Chinese migrants might hold racist views, as aptly documented by Cheng (2011); however, Sautman and Yan argue that the Chinese "lack political power, determinative influence, or cultural hegemony" in the African context "to create a public discourse to inferiorise their African hosts" (Sautman and Yan 2016, 3). They argue that the African host, on the other hand, "can regulate, racialise or even expel Chinese." That is, the racialiser is the one with the power to influence social structures, including attributing racial meaning to social groups and earmarking them for unequal treatment. Chinese migrants, they claim, do not possess such power in Africa to make statements about an entire population. As such, "the discourse of race among Chinese at enterprises in Africa is generally different from the one found by others for Global North enterprises in the Global South" (p. 2). With several examples, they argue that Chinese entrepreneurs might appear to be racist, but, in practice, they do not discriminate or essentialise African differences.

Their argument brings to mind works like Weitz's (2002), exposing the Soviet Union's racial politics while never mentioning race. According to Weitz, the ideology of race was explicitly denounced in the development of Soviet nationalities, but still found its way into national policies (p. 3). We believe that racial thinking, which is both global (Monson, 2015) and integral to capitalist expansion (Robinson, 2000), influences both beliefs/perceptions and practices at multiple levels. At the individual and organizational level, policies of employing Black managers at a Chinese enterprise, a lack of explicit discrimination, or good behaviour in the workplace do not necessarily indicate the absence of racism or racialization. Sautman and Yan, concerned with "who" has the power to define difference and structure hierarchies, argue that the Chinese in Africa do not hold the race card. We would caution about making such assertions, as racism and racialization can and do take place at multiple levels (interpersonal, organizational, societal, national, and global) and involve a multiplicity of actors at all these levels; minority or migrant status does not necessarily translate into a lack of power to discriminate or engage in racist behaviours. That said, we believe that Sautman and Yan's latest publication is, thus far, the most substantial contribution toward thinking about the role of race in twenty-first-century China–Africa relations.

While all of the studies mentioned above focus on the more recent migrations of Chinese people to Africa, one body of work focused on groups of Chinese present on the continent in earlier times. What sets these works apart, however, is not just the focus on a different population of Chinese in Africa in an earlier timeframe, but also the spotlight aimed on (race-making) structures and processes. It is also no accident that all of these studies were set in South Africa. South Africa's unique history in the African continent with regard to Chinese – including a long history of Chinese presence, some of the highest numbers of Chinese people, and record of laws specific to Chinese – in addition to its racist history, sets it apart from other African countries; as such, the research coming out of South Africa has been more explicitly focused on race, race-making, and racism. These earlier works on Chinese in South Africa include literature focused on the long history of anti-Sincism (Harris 2006, 2010) and explicitly anti-Chinese legislation to keep more Chinese from entering (Harris 2014; Yap and Man 2006; Park 2008a), the importance of Chinese miners in national racialization projects (Huynh 2008a and 2008b; and Bright 2013), and the ambiguous position of Chinese people during apartheid and in the post-apartheid era (Park 2008a and 2008b; Erasmus and Park 2008) – focus specifically on race and race-making.

Africans in China

The Africans-in-China literature shares a similar pattern with Sautman and Yan's work, focusing on Africans with the aim to reveal something about race. There has been a long presence of other non-Black foreigners in China and yet neither their presence nor China, in general, has been subjected to extensive racial analysis. Dikötter's (1990, 1992, 1994, 1997, and 2015) body of work on racial discourse in China has stood virtually on its own for many years, with the exception of a crop of writing in the aftermath of the Nanjing riots, which we shall get to below. This has changed only in the last few years partly due to increased "sightings" of African–Chinese couples with their bi-racial children, several highly publicised claims of racist treatment reported by Africans in China, and a number of public protests carried out by Africans in Guangzhou.

As Wilensky's (2002) and Wyatt's (2012) abovementioned studies indicate, Chinese people in China have possessed views of dark-skinned people since before the 1500s (Ming Dynasty). The term *kunlun* only gradually came to refer to the African slaves that Arab merchants brought with them to China. *Kunlun*, Wyatt insists, did not mark human difference or foreignness by skin colour, but by geography or place of origin (pp. 16–23). And, in Wilensky's study, the dark-skinned characters found in popular literature always possessed magical/mythical powers that were simultaneously feared and admired. That is, the notion of inferiority was not originally part of the imaginary of dark-skinned foreigners. The nineteenth-century brought about a tremendous shift in race thinking that many contemporary scholars, especially those equipped with Western knowledge of race, have not yet come to terms with in their analyses of race in twenty-first-century China.

Historical studies show that the nineteenth century was a period when China was occupied by multiple European powers who viewed the Chinese people as an inferior race. Chinese intellectual elites also went into exile overseas (many to Japan) and actively engaged in thinking about race, as they experienced being "Other" first-hand and attempted to grapple with China's crisis under Qing rule. The influence of Darwin and Spencer amongst the reformers and revolutionaries near the end of the Qing Empire resulted in Chinese versions of racial hierarchies with Whites and Yellows at the top – superior and intelligent – and "degenerate" Browns and Blacks at the bottom (Liang and Tan in Dikötter 1994, 407). With the stage set, Chinese began constructing an imaginary "Yellow lineage," with the (mythical) Yellow Emperor as the common ancestor and source of social cohesion and advancement, competing with the "White lineage" (Dikötter 1992, 71). Revolutionaries advocated overthrowing the Qing Empire as a way to rescue the country from the Manchus and modernise it. Manchus were excluded on the grounds of a narrowly defined idea of the "Chinese race," based on Han ancestry (rather than on territory) (Dikötter 1992, 97).

By the early twentieth century, this (racialised) notion of nation was invoked when the term was used conterminously with the state (Jakimów and Barabantseva 2016, 168). Sun Yat-sen (1866–1925), who became the first president and founder of the Republic of China, used the term *minzu* (*min* meaning people and *zu* descent) to politically circumscribe the Han people in a shared imaginary genealogy (Dikötter 1994, 406). Significantly, in serving to unify Han Chinese, the term politicised both race and nation, putting the people in what the revolutionaries perceived as a battlefield of contending races and nations (Dikötter 1992, 112). Notwithstanding the myths underpinning it, *minzu*, Jakimów and Barabantseva observe, "also marked a start to the development of the theory of the Chinese nation, granting China and its people a rooted history and culture" (2016, 168).

Essentialist thinking about the Chinese race continues into the twenty-first century, being renewed with other myth-making projects like the annual staging and televising of various people that make up China, with ethnic minorities made to wear their colourful traditional costumes and the Han in business suits (Anderson 2001, 39). Such celebrations normalise the Han as the quintessential Chinese. Another such project is state-sponsored archeology. Chinese paleoanthropologists have upended the widely accepted monogenic theory that the human species originated in Africa, with the discoveries of "Peking Man" and "Yuanmou Man" (Sautman 2001). Referring to this state-led project as a form of racial nationalism, Sautman explains, it "holds that each of us can trace our identities to a discrete community of biology and culture whose 'essence' has been maintained through time" (p. 95). Put differently, the Chinese cannot "come out of Africa," an issue that has not yet been further discussed.

The Nanjing protests against African students and instances preceding these have received considerable scholarly attention (Crane 1994; Dikötter 1994; Sullivan 1994; Sautman 1994; Johnson 2007; Cheng 2011) and should be understood in the context of these Chinese nation-building projects and very specific racialised notions of others. According to Sautman (1994), these protests were specifically anti-African and far from isolated, having cropped up earlier in Tianjin and Nanjing Universities in 1986 and 1988. He argues that they had precursors and reflected evolving views of Africans that were not extended to other foreigners in China; in other words, the protests were not expressions of a generalised xenophobia. A small number of African students arrived in China in the early 1960s supported by education scholarships from a Chinese government program that was part of Mao's alliance with Asian, African, and Latin American countries after splitting with the Soviet Union. In aligning with these countries, Mao articulated a powerful anti-imperialist and anti-racist stance, supporting national liberation movements in African countries as well as taking measures to suppress biological theories of race and racist thinking within China (Dikötter 2015, 123; Mark 2013, 53). On the ground, Sautman writes, "Red Guards held rallies to support oppressed peoples, including blacks, and no one would have openly expressed hostility to students from developing countries" (1994, 414). Official views and state-led support of African liberation movements and development that were part of global South–South solidarity priorities, however, were not always shared by the people, especially in the 1970s and 1980s.

Though there were indications of social isolation and dissatisfaction with living conditions among the African students, anti-African/Black hostility in the 1960s only manifested itself once, in the beating of a Zanzibari at a Beijing hotel (ibid.). Horrified Africans organised sit-ins and hunger strikes; repatriation ensued. For a brief period, the African scholarship program was discontinued. In the 1970s and 1980s, the scholarship program resumed; African students returned to China. During these decades, their expressions of dissatisfaction about living conditions were met with antagonism, racial epithets, and assaults. In Shanghai, Beijing, She-nyang, Guangzhou, and Wuhan, Chinese and African students clashed, sometimes violently. Chinese students hurled bricks and used the term "black devils" (*hei gui*). They petitioned school administrators to discontinue admission of Africans on the grounds that they were "'polluting Chinese society with their relations with Chinese women'" (Sautman 1994, 422). Another view that surfaced was Africans were uncultured peasants or poor recipients of aid from China who have caused Chinese to become second-class citizens in their own country. Mao's internationalism became the backdrop for a reactionary anti-Africanism (Cheng 2011, 564). The racial ideology that underpinned actions of the Chinese students was dismissed by authorities and observers as "a general xenophobia with roots in China's past humiliation at the hands of foreign powers and resentment at the higher living standards of foreigners" (Sautman 1994, 424). No similar claims were ever made against other (White) foreign men, who also had relations with Chinese women.

Cheng (2011) extends Sautman's above history to the articulation of racial dis-course and Chinese nationalism at a time when China is rising as a global power and there is mutual migration from China and Africa. He argues that race and nationalism are being merged without the public's awareness. He observes these processes, especially a re-emergence of anti-African racism in cyberspace, promoted by a group of educated Chinese netizens who have worked in Africa. Unlike Sautman's earlier descriptive study, Cheng attempts to explain anti-African racism as under-pinned by "a racial discourse that was constructed in the early stage of modern Chinese history to define China's place in the world order" (p. 562). Building on Dikötter's work, Cheng proceeds to show that the development of a Chinese nationalist ideology is inseparable from race discourse. As Monson puts it, "Cheng's work (on Chinese nationalism) show[s] us … that race history is in essence a 'history of meanings' that are produced in dynamic contexts of political, economic, and social relations" (2015, 4). Cheng argues that race is not imposed by others, but comes from within China where "racial language and concepts have emerged to help articulate nationalistic claims" and the deepening of the race/nation discourse has received little resistance from the media or intelligentsia (Cheng 2011, 562).

Hood's (2013) work takes the argument a step further exposing the role of the state-owned media in racializing and making inferior the African body. She con-veys how health, race, and gender are constitutive of the nationalist state formation since China's opening up policy in the late 1970s. She argues specifically that the media's role in health communication is largely responsible for how city-dwellers, who are mostly Han Chinese, understand HIV/AIDS transmission and their sense of distance from it. At the start of its spread in the mid-1980s, media constructions of Africa as the original site of the virus and representations of the hyper-sexuality of Africans as the primary vehicle of transmission of HIV/AIDS guided urban Han Chinese understandings about susceptibility and transmission as both "other" and distant. Citing examples from Chinese sources, Hood insists that the use of images of foreign subjects (including gay White men in early campaigns) was deliberate: they deflected memory of a semi-colonial past laden with sex and opium as well as more recent blood scandals that involved capitalist exploitation and local authorities' negligence that left millions of marginalised persons in the rural areas dead (p. 287). Images of unclothed African women were always pictured as HIV positive, while always clothed Han Chinese were located in modern health environments. In doing this, the authorities of HIV media connected, for their Chinese audience, the proliferation of the disease with "'scientifically proven' origin myth heavily reliant on understandings of *yuanshi*" or primitive/ancient, proximity to nature, lack of scientific knowledge, and lasciviousness (pp. 294–295). Such racial framing of the disease not only reinforced extant race ideologies that treated Black Africa (*hei feizhou*) and Black people (*heiren*) as inferior, but also rendered it "a disease with 'dark' consequences that affects cultures unevenly – that is, it has worse effects on the less developed" (pp. 284 and 289).

Zhou, Shenasi, and Xu's (2016) study, which sets out to investigate "Chinese attitudes toward African migrants in Guangzhou, China," seems to overlook Sautman (1994), Cheng (2011), and Hood (2013) and their painstaking reconstructions of a race history in the context of China. Zhou et al. are concerned with racial attitudes manifesting around inter-racial contact between local Chinese residents and African migrants. They want to uncover sources of prejudice and discrimination. Their study, in our view, builds on two implicit essentializing premises: that (1) there is something unique about Chinese perceptions and reception of the rapidly increasing number of Africans and (2) there is something exceptional about African migrants, that would elicit different inter-racial dynamics than between Chinese and other groups of foreigners. They report that Chinese entrepreneurs, rental agents, and residents are more accepting and have positive views of Africans because they have greater opportunities to interact and have become economically interdependent on this group of foreigners. At the same time, miscommunication and enforcement of visa regulations make for hostile race relations and negative stereotyping of Africans by Chinese taxi drivers and the police (including their assistant enforcers). Zhou et al. conclude: "These findings indicate that the Chinese have not yet formed a collective xenophobic consciousness, nor have their attitudes been racialised in ways that are explained by theories of threat or ethnic economy developed from the Global North" (2016, 157). Not only do they conflate xenophobia and race consciousness, but they have privileged concepts from the global North and applied them in a situation of South–South relations (criticised by Sautman and Yan 2016). Relying on theories relevant to ethnic relations and enclaves, Zhou et al. propose that extensive social interactions could mitigate possibilities of anti-Black racism. Restricted by these theoretical frameworks they neglect an entire field of race studies that is critical of the forces that enable and underpin racialization processes, and they ignore Chinese participation in the global circulation of racial ideology (see Kowner and Demel 2013).

Conclusion

Returning to the Qiaobi laundry detergent ad, what some scholars have noted was that while no one was interested in the national audience's reception of the commercial prior to its global circulation, the social media of various countries and their consumers came to dominate the discussions that followed. In subsequent online news sources, journalists seem to implicitly accept the validity of the two White male views, reiterating and assimilating their evidence of "Chinese racism" without question. By ignoring the positionality of these actors – White, male, and privileged – they miss out on an opportunity to critique a number of issues regarding power, perspective, and generalised claims about China.

The focus on "Chinese racism" as a thing in itself, rather than a process of giving meaning to identities (difference) that is interwoven with complex political, economic, and social relations, suggests that racial attitudes and ideas among the people in China are exceptional and possibly more racist than people in other parts of the world. This objectification and moral judgment of "Chinese racism" is in fact part of a longer history of "intellectual imperialism" in the way Said conceived of it in *Orientalism* (see Dirlik 1996, 98). Our argument is that "Chinese racism" is part of a broader discourse about the asymmetry of power – specifically, who gets to make statements about whom and authorise views of the "other," and as global power is rebalanced, discussions about racism in China and elsewhere should also reflect these changes. From this standpoint, drawing the conclusion, based on one Chinese company's racist ad, that all Chinese must be racist is in itself an articulation of Western racism (and Orientalism). As Sautman and Yan ("One Bad Advert Doesn't Make 1.4 billion Chinese Racist." *South China Morning Post*, June 1, 2016) point out, "When racist ads or statements appear in the Western media – and there have been plenty – no one claims they show 'the Americans', 'the French', and so on, are racist."

Race is not just a discourse, but is also a continual process of meaning/myth making by those with the power to do so. Racialization is a process of differentiation and hierarchization. We deliberately take the position that social identities are formed in active, simultaneous (and sometimes state-led) processes of constructing both the group in question and "others" or "outsider" populations. Moreover, we treat race as an historical phenomenon that is intimately interwoven with the expansion of the capitalist system – racial capitalism, as Robinson calls it (2000). Because China has been re-incorporated into this system, while simultaneously claiming to be non-capitalist, it is inadequate to apply extant theories of race that are built upon experiences of White settler colonies. Race thinking entails ideology or what is now referred to as (shared) imaginary. This imaginary, with increasing and accelerating engagements between China and Africa and China's rise on the global economic stage, is shifting; however, existing hierarchies of race and Western conceptions of "Africa" and "China," "Black" and "Yellow" have not yet been completely displaced.

As recently as 1995, villagers in a rural part of the Limpopo Province in northern South Africa referred to one of the authors (both Asian American) as "White" or *lekgoa*, the SiPedi term for foreigner or White person. "Whites" were, at that time, the primary and predominant "other"; as such, anyone who was not a Black Pedi person was identified as such. Fast forward 20 years, and we encounter more and more accounts of White (American and European) researchers and development workers in various, typically remote and rural, parts of Africa being (mis) identified as "Chinese," indicating that in large swathes of the continent Chinese have become the most prominent "other." Across the continent where relatively few can read or speak Chinese, signage, billboards, and even local and national political campaign documents are translated into Chinese. The importance of Chinese development aid and investment, combined with the ongoing shifts in China's global position, has rapidly and drastically altered power dynamics at the national level; these, in turn, affect how Africans and Chinese interact and engage. At the same time, despite the hype over "a million Chinese in Africa," Chinese migrants remain highly visible, numerical minorities, vulnerable to locally changing tides. How these global and national shifts in power dynamics continue to affect (race) policies and (racialised) interpersonal dynamics on a larger scale remains to be seen.

With more and deeper face-to-face encounters between Africans and Chinese, some stereotypes are being replaced with more experience-based, educated views of the "other." In both China and Africa, we find greater instances of African–Chinese relationships (Lan 2015) and more mixed-race children (Frazier and Zhang 2014). Studies of these and the responses they elicit from various sides are only just starting to emerge. At the same time, numerous studies that highlight the diversity within groups, including the Africana migrants in Shanghai and Guangzhou (Adams 2015) and various groups of Chinese in Africa and the tensions between them (Lam 2015, Huynh et al. 2010) point out that generalization even within a single migration cohort is impossible. As the body of China–Africa/Africa–China literature expands, the micro studies illustrate an increasingly complex and dynamic story. As scholars, we would be mindful that race making, othering, and hierarchies of power are entwined and ongoing processes require rigorous analysis that disrupt and displace simplistic models, binary thinking, and application of existing theories based on experiences in the global North.

Notes

1 Video available at www.youtube.com/watch?v=Xq-I0JRhvt4.
2 China's current political and economic position in the world-system surpasses that of Japan in the late nineteenth and early twentieth century: Japan aspired to become a "White man's country," but clearly failed.
3 A glaring gap in our work, however, is that we have been limited by language; Chinese and French works on the topic have been excluded from this study and we hope that others will pick up where we start.
4 The one prominent exception was the American journalist, Howard French, who alluded to the Chinese subjects' racist perception of Black Africans in his monograph (French 2014).

References

Adams, C. Jama. 2015. "Structure and Agency: Africana Immigrants in China." *The Journal of Pan African Studies* 7(10): 85–107.

Anderson, Benedict. 2001. "Western Nationalism and Eastern Nationalism: Is There a Difference that Matters?" *New Left Review* 9: 31–42.

Arsene, Codrin. 2014. "Chinese Employers and Their Ugandan Workers: Tensions, Frictions and Cooperation in an African City." *Journal of Current Chinese Affairs* 43(1): 139–176.

Bonacich, Edna, Sabrina Alimahomed and Jake B. Wilson. 2008. "The Racialization of Global Labor." *American Behavioral Scientist* 52(3): 342–355.

Bow, Leslie. 2010. *Partly Colored: Asian Americans and Racial Anomaly in the Segregated South.* New York: NYU Press.

Bright, Rachel K. 2013. *Chinese Labour in South Africa, 1902–10: Race, Violence, and Global Spectacle.* New York: Palgrave Macmillan.

Cheng, Yinghong. 2011. "From Campus Racism to Cyber Racism: Discourse of Race and Chinese Nationalism." *The China Quarterly* 207: 561–579.

Crane, George T. 1994. "Collective Identity, Symbolic Mobilization, and Student Protest in Nanjing, China, 1988–1989." *Comparative Politics* 26(4): 395–413.

Dikötter, Frank. 1990. "Group Definition and the Idea of 'Race' in Modern China (1793–1949)." *Ethnic and Racial Studies* 13(3): 421–432.

Dikötter, Frank. 1992. *The Discourse of Race in Modern China.* Stanford, CA: Stanford University Press.

Dikötter, Frank. 1994. "Racial Identities in China: Context and Meaning." *The China Quarterly* 138: 404–412.

Dikötter, Frank. 1997. "Racial Discourse in China: Permutations and Continuities." In *The Construction of Racial Identities in China and Japan: Historical and Contemporary Perspectives*, edited by Frank Dikötter, 12–33. Honolulu, HI: University of Hawaii Press.

Dikötter, Frank. 2015. *The Discourse of Race in Modern China.* Second edition. Oxford & New York: Oxford University Press.

Dirlik, Arif. 1996. "Chinese History and the Question of Orientalism." *History and Theory* 35 (4): 96–118.

DuBois, W.E.B. 1903. *The Souls of Black Folk: Essays and Sketches.* Chicago, IL: A.C. McClurg & Co.; Cambridge: University Press John Wilson and Son.

Erasmus, Yvonne and Yoon Jung Park. 2008. "Racial Classification, Redress, and Citizenship: The Case of the Chinese South Africans." *Transformation: Critical Perspectives on Southern Africa* 68: 99–109.

Espiritu, Yen Le. 2008. *Asian American Women and Men: Labor, Laws, and Love.* Lanham, MD: Rowman & Littlefield.

Esteban, Mario. 2010. "A Silent Invasion? African Views on the Growing Chinese Presence in Africa: The Case of Equatorial Guinea." *African and Asian Studies* 9: 232–251.

Flockemann, Miki, Kudzayi Ngara, Wahseema Roberts and Andrea Castle. 2010. "The Everyday Experience of Xenophobia: Performing The Crossing from Zimbabwe to South Africa." *Critical Arts* 24(2): 245–259.

Frazier, Robeson Taj and Lin Zhang. 2014. "Ethnic Identity and Racial Contestation in Cyberspace: Deconstructing the Chineseness of Lou Jing." *China Information* 28(2): 237–255.

French, Howard. 2014. *China's Second Continent: How a Million Migrants Are Building a New Empire in Africa.* New York: Alfred A. Knopf.

Giese, Karsten. 2014. "Perceptions, Practices and Adaptations: Understanding Chinese-African Interactions in Africa." *Journal of Current Chinese Affairs* 43(1): 3–8.

Harris, Karen L. 1998. "The Chinese 'South Africans': An Interstitial Community." In *The Chinese Diaspora. Selected Essays. Volume II*, edited by L. Wang and G. Wang, 277–278. Singapore: Times Academic Press.

Harris, Karen L. 2006. "'Not a Chinaman's Chance': Chinese Labour in South Africa and the United States of America." *Historia* 52(2): 177–197.

Harris, Karen L. 2010. "Anti-Sincism: Roots in Pre-industrial Colonial Southern Africa." *African and Asian Studies* 9(2): 13–231.

Harris, Karen L. 2014. "Paper Trail: Chasing the Chinese in the Cape (1904–1933)." *Kronos* 40: 133–153.

Hood, Johanna. 2013. "Distancing Disease in the Un-black Han Chinese Politic: Othering Difference in China's HIV/AIDS Media." *Modern China* 39(3): 280–318.

Hsu, Elisabeth. 2007. "Zanzibar and its Chinese Communities." *Population, Space and Place* 13: 113–124.

Huynh, Tu T. 2008a. "Loathing and Love: Postcard Representations of Indentured Chinese Laborers in South Africa's Reconstruction, 1904–1910." *Safundi* 9(4): 395–425.

Huynh, Tu T. 2008b. "From Demand for Asiatic Labor to Importation of Indentured Chinese Labor: Race Identity in the Recruitment of Unskilled Labor for South Africa's Gold Mining Industry, 1903–1911." *Journal of Chinese Overseas* 4(1): 51–68.

Huynh, Tu. 2013. "Black Marxism: An Incorporated Analytical Framework for Rethinking Chinese Labor in South African Historiography." *African Identities* 11(2): 185–199.

Huynh, Tu, Yoon Jung Park and Anna Ying Chen. 2010. "Faces of China: New Chinese Migrants in South Africa, 1980s to Present." *African and Asian Studies* 9: 286–306.

Jakimów, Malgorzata and Elena Barabantseva. 2016. "'Othering' in the Construction of Chinese Citizenship." In *Politics of the 'Other' in India and China: Western Concepts in Non-Western Contexts*, edited by Lion König and Bidisha Chaudhuri, 167–178. New York: Routledge.

Johnson, M. Dujon. 2007. *Race and Racism in the Chinas*. Bloomington, IN: Author's House.

Joshi, Khyati Y. and Jigna Desai. 2013. *Asian Americans in Dixie: Race and Migration in the South*. Urbana, Chicago, and Springfield: University of Illinois Press.

Kernan, Antoine. 2010. "Small and Medium-sized Chinese Businesses in Mali and Senegal." *African and Asian Studies* 9: 252–268.

Koenane, Mojalefa LehlohonoloJohannes. 2013. "Xenophobic Attacks in South Africa: An Ethical Response – Have We Lost the Underlying Spirit of Ubuntu?" *International Journal of Science Commerce and Humanities* 1(6): 106–111.

Kowner, Rotem and Walter Demel. 2013. *Race and Racism in Modern East Asia. Vol. I, Western and Eastern Constructions*. Leiden: Brill.

Lam, Katy. 2015. "Chinese Adaptations: African Agency, Fragmented Community and Social Capital Creation in Ghana." *Journal of Current Chinese Affairs* 1: 9–41.

Lampert, Ben and Giles Mohan. 2014. "Sino-African Encounters in Ghana and Nigeria: From Conflict to Conviviality and Mutual Benefit." *Journal of Current Chinese Affairs* 43 (1): 9–39.

Lan, Shanshan. 2015. "Transnational business and family strategies among Chinese/Nigerian couples in Guangzhou and Lagos." *Asian Anthropology*, doi:10.1080/168347X.2015.1051645.

Mark, Chi-kwan. 2013. *China and the World Since 1945: An International History*. London and New York: Routledge.

Mohan, Giles, Ben Lampert, May Tan-Mullins, and Daphne Chang. 2014. *Chinese Migrants and Africa's Development*. London: Zed Books.

Monson, Jamie. 2015. "Historicizing Difference Construction of Race Identity in China-Africa Relations." SSRC Think Pieces: Making Sense of the China-Africa Relationship.

Park, Yoon Jung. 2008a. *A Matter of Honour. Being Chinese in South Africa.* Johannesburg: Jacana Media Pty (Ltd.)

Park, Yoon Jung. 2008b. "White, Honorary White, or Non-White? Apartheid-era Constructions of Chinese." *Afro-Hispanic Review* 27(1): 123–138.

Park, Yoon Jung and Tu T. Huynh. 2010. "Introduction: Chinese in Africa." *African and Asian Studies* 9: 207–212.

Robinson, Cedric J. 2000. *Black Marxism: The Making of the Black Radical Tradition.* Chapel Hill, NC: The University of North Carolina Press.

Robles, Rowena. 2013. *Asian Americans and the Shifting Politics of Race: The Dismantling of Affirmative Action at an Elite Public High School.* New York: Routledge.

Sautman, Barry. 1994. "Anti-Black Racism in Post-Mao China." *The China Quarterly* 138: 413–437.

Sautman, Barry. 2001. "Peking Man and the Politics of Paleoanthropological Nationalism in China." *The Journal of Asian Studies* 60(1): 95–124.

Sautman, Barry and Hairong Yan. 2014. "Bashing 'the Chinese': Contextualizing Zambia's Collum Coal Mine Shooting." *Journal of Contemporary China,* doi:10.1080/10670564.2014.898897.

Sautman, Barry and Hairong Yan. 2016. "The Discourse of Racialization of Labour and Chinese Enterprises in Africa." *Ethnic and Racial Studies,* doi:10.1080/01419870.2016.1139156.

Sullivan, Michael J. 1994. "The 1988–1989 Nanjing Anti-African Protests: Racial Nationalism or National Racism?" *The China Quarterly* 138: 438–457.

Weitz, Eric D. 2002. "Racial Politics Without the Concept of Race: Reevaluating Soviet Ethnic and National Purges." *Slavic Review* 61(1): 1–29.

Wilensky, Julie. 2002. "The Magical Kunlun and 'Devil Slaves': Chinese Perceptions of Dark-skinned People and Africa Before 1500." SINO-PLATONIC PAPERS, Number 122.

Wu, Frank. 2002. *Yellow: Race in America Beyond Black and White.* New York: Basic Books.

Wyatt, Don. 2012. *Blacks of Pre-Modern China.* Philadelphia, PA: University of Pennsylvania Press.

Yan, Hairong and Barry Sautman. 2010. "Chinese Farms in Zambia: From Socialist to 'Agro-Imperialist' Engagement?" *African and Asian Studies* 9: 307–333.

Yap, Melanie and Dianne Leong Man. 1996. *Colour, Confusion and Concessions. The history of the Chinese in South Africa.* Hong Kong: Hong Kong University Press.

Zhou, Min, Shabnam Shenasi, and Tao Xu. 2016. "Chinese Attitudes toward African Migrants in Guangzhou, China." *International Journal of Sociology* 46(2): 141–161.

Zi, Yanyi. 2015. "The Challenges for Chinese Merchants in Botswana: A Middleman Minority Perspective." *Journal of Chinese Overseas* 11(1): 21–42.

11

KENYAN AGENCY IN KENYA–CHINA RELATIONS

Contestation, cooperation and passivity

Maddalena Procopio

China's vastly increased involvement in Africa is one of the most significant developments in the region since the early 2000s.[1] Scholarly focus has, more often than not, been on *China*'s foreign policy's influence, impact, priorities, achievements, and difficulties in Africa. Less attention has been paid to analyzing African sides of the equation: beliefs, interests, resources, and structures that determine the motivations and modalities of action in dealing with the opportunities and challenges that China is presenting. This shortcoming does not come as a surprise if one looks at how Africa has been portrayed in International Relations literature (Clapham 1998; Brown 2006; Lemke 2003, 2011; Williams 2011): Africans' leverage in negotiating with external forces has been chronically neglected, depicting the continent as dependent and powerless (Cornelissen et al. 2012, 2). Similarly, Africa's relations with China have been understood as heavily asymmetrical due to China's overall economic and political strength, leading to the assumption that there is little willingness or ability to negotiate by Africans. While these themes contain elements of truth, it is time that the African side of this equation is analyzed not by studying the "consequences for Africa" of China's arrival, but the *reasons* and *modalities* of Africans' action.

Through an attempt to change the standing point for observing real world dynamics, this chapter aims at contributing to Africa–China, and Africa in IR scholarship more broadly, by exploring Kenyan agency in the context of its relational dynamics with external forces.[2] The aim is thus to explore, though a study of motivations and modalities of action, whether Kenyans have the willingness and ability to safeguard their interests when engaging with external forces (agency). The study does so by looking at how the interactive relation between state and non-state actors is mobilised and negotiated, and how it constitutes a strength in building national capacity in domestic as well as foreign policy.

Scholarship on "African agency" has been increasing in the past few years (see Brown 2012; Brown and Harman 2013; Neubert and Scherer 2014; Shaw 2015;

Whitfield and Fraser 2010; Murray-Evans 2015; Gadzala 2015). However, the conceptualization of agency is often limited to agents' actions, whereby agents are grouped into "elites" or "social groups," thus defining agency as actions either from the top down or the bottom up, rather than as *inter-relational decision-making processes* between the two. To address this issue, what follows unpacks the relation between "agency" and "governance" whereby agency is the control over the content and modalities of governance; it is the protection of "a realm of decision-making [...] from external influence, defending spaces in which African agents can struggle amongst themselves over the nature of appropriate political and economic processes" (Whitfield and Fraser 2010, 343). The processes of governance thus become a crucial site for determining power distribution between Africans and external actors. These sites are identified not merely at the national level, but in sub-national sectoral settings where they most often occur (Sellers 2011, 135). It is important to acknowledge that these sites exist. For a long period, portrayals of Sino-African relations assumed that the engagement was happening in a vacuum.

This chapter first introduces the debate around African agency and clarifies the reasons why treating agency as the negotiation of governance provides more analytical balance to understand actual dynamics. Governance processes are unpacked by examining how Kenyan actors mobilise and negotiate interests and resources, and whether they do so according to existing norms of practice, thus fitting the Chinese into pre-existing frameworks, or whether they deviate from such procedural and normative standards, either for short-term self-interest or to renew obsolete systems of governance.

Agency as the safeguard of governance

Agency is articulated as the domestic definition and determination, through negotiation, of governance processes vis-à-vis the same processes being defined and determined by external actors (Whitfield and Fraser 2010, 343). It is the *willingness* and *ability* to safeguard a space within which domestic forces can negotiate. To fully engage with the concept, it is necessary to transcend the state-defined idea of agency and employ one that understands governance processes as all-encompassing. Agency should thus not be identified with the state's aggregate power, or the sum of state's resources, but rather in the context and the structure of specific negotiations (Habeeb 1988). Acknowledging that a range of actors exercise agency in governance processes means understanding the relational value of power sharing between the governing and the governed (Hagmann and Péclard 2010, 545), and how such relations are constantly re-negotiated through formal and informal decision-making processes.

Governance is thus a process of "social coordination" (Bevir 2011, 1), a "political space in which relations of power and authority are vested" (Hagmann and Péclard 2010, 551). It is dynamic, encompassing actors who have asymmetrical access to resources and power, and whose interests are negotiated across sub-national, national, and international platforms (Ohmae 1995; Turner, 1998). It is historically determined through the definition of formal and informal norms, procedures, conventions of interaction (Hall and Taylor 1996, 936–937; Keohane and Nye 2000; Biermann 2002), and it is subject to change. This multi-level type of governance means that decision-making norms and institutions are forged at specific levels and then negotiated with tiers above and below (Brenner 2004). These interactions occur as competing "economic and political forces seek the most favourable conditions for insertion into a changing international order" (Jessop 2000, 343). It is worth noting that this approach does not view governance as normatively loaded in a way that signifies "good" governance, but rather as the "nature of all patterns of rule" (Bevir 2011, 1), or "real governance" (de Sardan 2008, 1).

Adopting this definition of agency, as the domestic control of the governance process, helps reveal sources of agency neglected in Sino-African relations, such as upper and lower level state bureaucrats, professional associations, trade unions, private sector representatives, or informal sector leaders. Moreover, the asymmetries identified between China and Africa, whereby China's aggregate resources largely outnumber those of African countries, are often the result of cross-country studies focusing on one sector. While comparatively valuable, these studies risk isolating sectoral dynamics from the wider national context.

Instead, combining reasoning about *macro-, meso-,* and *micro-*level governance dynamics – national, sectoral, and issue-specific – and paying simultaneous attention to localised and broader contexts, leverages the strengths and weaknesses that a country-system can deploy vis-à-vis external forces. This approach "emphasises the autonomous political characteristics of distinct policy sectors, hence the multiplicity of political patterns in any one country" (2006, 47), allowing asymmetries to be understood in a more nuanced fashion. As highlighted by Levi-Faur, the assumption behind this design is that "sectoral variations are important" (ibid.), i.e. differences within a country, and these can be as important as national variations, i.e. differences between countries.

Arenas

It is the very confrontational dynamics between domestic players, as they attempt to safeguard their interests that emphasise the suitability of a system of governance to safeguard broader domestic interests. A governance system, or arena, represents the "political space in which relations of power and authority are vested" (Hagmann and Péclard 2010, 551). Domestic arenas vary significantly depending not only on the sector but also the policy area under study: they can be highly regulated, lack established norms of conduct, be defined by domestic rules and procedures, shaped by international structures, bureaucracy-led, or socially orchestrated, and so forth. This makes it difficult to talk about a "Kenyan" approach to safeguarding domestic interests with China.

Those that have been "dominated by longstanding conventions on how and by whom statehood is defined" are called *rooted governance systems*. Those defined as "lack[ing] predefined or commonly recognised procedural modalities for decision making" (Hagmann and Péclard 2010, 550) are defined as *lacking a governance system*. And those that combine elements of the first two, where a governance system either did not exist and is under construction, or did exist but was acknowledged as inappropriate and in need of restructuring, are called *governance systems in-the-making*.

Kenyans' relations with China mostly unfold in contexts where informal and formal governance frameworks are in place, rooted and legitimised by concerned stakeholders. Rooted governance is characterised by highly regulated patterns of rule, but these patterns can vary significantly depending on the specific issue-area. An example of Kenyan rooted governance is that concerning trade unions. Trade unions are entities that have historical significance in Kenya, their presence is recognised by other concerned stakeholders, they are highly regulated, and have become even more so since the 2010 Constitution. However, rooted governance concerns also informal behaviour, such as that of small traders whose activities are not regulated by officially recognised laws, but are nonetheless clearly structured within society. Rooted governance can also be rooted in international structures, whereby internationally recognised norms assume priority over domestic initiatives. The fact that governance is rooted does not equal stating that the actual behaviour conforms to established norms of conduct (further explored below in terms of interaction modalities).

Governance in-the-making, in turn, is characterised by a system of rule that is either being revisited or created. The main triggers for a decision to restructure the system are recognition of gaps in the legal system, the need of institution-building, and lack of implementation capacity. For instance, the fight against counterfeit products came about due to evident deficiencies in the governance system allowing fake goods (mainly from China but also other Asian and African countries) to threaten intellectual property rights, thus hindering innovation and development in Kenya. The creation of the Anti-Counterfeits Act and Agency represented an important moment of democratic governance. This case is particularly interesting as it was generally framed, by interviewees, as a case of acknowledged inadequacy of Kenyan laws, and, later, of implementation capacity, rather than as a case of China-blaming (Procopio 2016, 127–137). Although action taken to address the issue represented a reaction toward Chinese imports, the problem was always framed as domestically more than externally-led. The case highlights how the country's legal, socio-economic, and political environment can leave space for democratic governance to take place. However, understanding the relatively new co-existence of democratic governance and rapid economic development is key to explaining the inconsistency in implementation efforts and results. For governance-in-the-making cases, while the values associated with democratic development are taken seriously during the formulation of new systems of governance, the reality of economic (under)development represents an obstacle, or at least a point of separation, between the official realm of policy-making and the actual possibilities of governance.

Finally, the cases where the system of governance was very weak/absent were characterised by historical neglect of a specific issue in the development agenda of

Kenya or by foreign partners, thus leaving space for the Chinese to operate according to their rules. For instance, the creation of Confucius Institutes occurred at a time when Kenyan stakeholders had not yet started to regulate internationalization of higher education to fulfil increasing demand and market needs. They were thus still lacking a specific system of governance to regulate the presence of a foreign institution within local universities, leaving it open to an anarchic behaviour on both Chinese and Kenyan sides. This lack of strategy or structure to negotiate in the country's best interests is believed to be the most frequent scenario in Africa–China relations. However, while the case of Confucius Institutes demonstrates an inability to domesticate a Chinese initiative, research into Kenya–China relations more generally suggests that weak governance systems are a minority. The question then concerns whether Kenyans are willing and able to maintain preferences at the negotiation table with China, be they in line or not with the Chinese priorities. The answer requires an analysis of motivations and modalities of interaction, and before then a few clarifications on how actors are conceptualized.

Actors

To understand how different groups of actors perceive and engage with the Chinese presence, the analysis relies on Wight's tripartite definition of agents, i.e. agents that have a personality and 'freedom of subjectivity' (agency1), which is shaped by the social context in which they were born and raised (agency2), and which is bound by the position the agent occupies in society (agency3) (Wight 1999, 126-134). Hence, rather than focusing exclusively on a functional distinction (the role occupied in society, or "elites" vs "masses"), it is more useful to look at the "competing repertoires that agents mobilise in their interactions" and the "resources that individuals and organised interest groups have at their disposal" (Hagmann and Péclard 2010, 547).

Repertoires

During decision-making processes, actors rely on ideas and memories (Whitfield and Fraser 2010, 344), based upon history, identity, and discourse (Galbreath et al. 2008, 16), which in turn create individual spheres of decision-making that can impact the broader context within which they act (Alden and Ammon 2012, 65). Hagmann and Péclard call this mix of ideas and memories repertoires. Actors "master symbolic repertoires to further their interests, to mobilise popular support, and to give meaning to their actions" (Hagmann and Péclard 2010, 547). Repertoires are also used to "defend and to challenge existing types of statehood and power relations" (ibid.).

Repertoires emerging in Africa–China relations are mostly twofold: one consists in framing the relations as South–South cooperation through concepts of mutuality, equality, solidarity, non-interference, and economic development. The other frames relations with China as damaging for locals and counter-productive, mainly in the sense of being disrespectful of local social and economic rights. Both types are viewed in opposition to Western practices, norms, influence over the continent. On the one hand, Western political and economic conditionalities and impositions are resented in favour of China's (seemingly) hands-off approach, and emphasis on economic development as a passage to independence. On the other hand, the West's focus on liberal political and socio-economic rights is contrasted to China's perceived disrespect for labour and human rights, associational life, constitutional norms and so forth. These repertoires form the basis of national developmental discourses, as well as personal attitudes toward foreign engagement in national matters.

In Kenya, the government's Look East policy, initiated by President Mwai Kibaki and continued by his successor, President Uhuru Kenyatta and Vice-President William Ruto, was important in creating the basis for challenging existing developmental discourses (mainly Western oriented until then) and the country's relations with external actors. This orientation was complemented by an emphasis Kenyan stakeholders placed on the importance of legal and regulatory frameworks, deemed necessary to safeguard their interests. Such emphasis was not upheld by specific social groups only, i.e. elites, but rather broadly recognised as a fundamental feature of contemporary Kenya. This highlights the significance of the rule of law for Kenyan actors, despite awareness about the challenges of implementing this in practice. Hence, it is possible to find Kenyan small traders who praise Chinese products and Kenyan businessmen that provide them with goods to sell as well as small traders who condemn the Chinese for stealing their jobs or putting them out of business. There are officials in government ministries open to negotiations with China because these are "faster, less prescriptive, more convenient" than with traditional partners, and those that reject Chinese proposals because these are not in line with the ministry's interests, irrespective of potential financial benefits. Sino-African relations have particularly struggled to figure out whether different repertoires such as the "economic development" pushed forward by China and the "democratic governance" one pushed forward by Western countries can co-exist and be blended in African contexts.

Resources

On top of repertoires, it is important to study resources, especially in Sino-African relations where the economic asymmetry between China and most African countries is often perceived as a reason for asymmetrical outcomes. Resources form the major basis of power and, as Olivier de Sardan put it, there are at least two kinds of power: "the power everybody has and the power only some people have" (de Sardan 2008, 186). As a consequence, the type and quantity of resources that actors can rely upon, in the negotiation, varies. Executive and high-level bureaucrats have access to a wider pool of financial resources than lower-level bureaucrats or social groups. However, in Kenya, the mobilization of resources available to executive and high-level bureaucrats is circumscribed by complex systems of governance that reduce power asymmetries between the top and the bottom. This is consequential to the fact that in most cases the expertise resides with lower bureaucratic levels and social groups such as professional organizations, trade unions, or educational institutions. In other words, while politicised high-levels of bureaucracy might have more power to strike direct deals with external actors, and more financial power than lower levels, it is generally lower-level bureaucrats and social groups that, during the implementation of such deals, leverage the decisions taken at the top. It is thus important to look not only at financial, but also human resources available to Kenyans in negotiation processes.

Studies of Africa–China relations tend to focus on structural and financial resources to the neglect of human resources. By looking at negotiating agents in Kenya, the profile that emerges is one of actors with a very good knowledge of the sectors within which they operate. From trade unionists to technical universities' directors, from officers at the Ministry of Education to those in the National Economic and Social Council, from manufacturers associations' leaders to private sector representatives in the pharmaceutical industry, all demonstrate a comprehensive set of skills in negotiating with the Chinese. These include full knowledge of the legal, institutional and regulatory environment within which they operate, good knowledge of the gaps (or lack thereof) that the Chinese aimed at filling, and a relative understanding of Chinese strategies.

For instance, in negotiations between trade unions and Chinese companies, the latter were financially stronger than the former. However, Kenyan trade unionists were significantly more prepared to engage the dispute mechanism. Despite the interference from Kenyan bureaucrats interested in preserving relations with the Chinese, and willing to bypass the established framework of dispute resolution, trade unionists resiliently relied on legally established protections to resolve the dispute with the Chinese and to limit inappropriate interference from Kenyan higher-level bureaucrats (Procopio 2016, 125–127). In other cases, the preparedness and long-term vision of Kenyan human resources were necessary to alter inadequate or obsolete systems of governance to the improvement and benefit of the country. For instance, in the fight against counterfeits, a highly skilled group of bureaucrats and social groups' leaders identified regulatory and legal gaps that did not guarantee the protection of intellectual property rights, hence not only making the country open to the influx of counterfeits, but above all making it unable to prosecute wrongdoers (Procopio 2016, 127–137). Similarly, in the revival of Kenyan traditional medicine, knowledgable Kenyan human resources acted as mediators between the local context (laws, popular perceptions, norms, etc.) and the Chinese wealth of resources to expand the usage of traditional medicine (Procopio 2016, 173–182).

These skills are *necessary* but not always *sufficient* to obtain positive negotiating outcomes. The quantity and distribution of human resources, together with financial resources, are indispensable and often lacking, especially for implementation (vis-à-vis decision making). While the ability to *identify* the "right" set of legal, institutional and regulatory tools for governing specific issues is demonstrated, when it comes to *implementation* there are often gaps that reduce the ability to control outcomes. For instance, in the fight against counterfeits, while legal and institutional measures were adopted (the Anti-Counterfeits Act and the Anti-Counterfeits Agency respectively), inspections at borders remained weak due, mainly, to understaffing. This problem is not unique to the relations with China. In this respect, the development of the Public Private Partnership initiative (legalised by an Act of Parliament and Presidential Assent in 2013) is one way deemed relevant to potentially solve this resource shortage in the long run by merging public and private, financial and human capacities.

Motivations

The *willingness* to exert control in the negotiations with China, in other words, the motivation to do so, is strictly related to the interests that any given actor perceives to be affected in its relations with China. One purpose of the sectoral approach adopted here is acknowledging the existence of competing interests within a national context. Assuming that interests are unitary, and emerge only out of state agents, erases the complexities of country-systems and does not take into account the profoundly different priorities that various issue-areas and policy sectors occupy in the national context and in relation to external presences. In actual governance, macro policy orientations (for instance, Kenya's Look East policy) meet sub-national and issue-specific micro-cosmos, where "national" agendas are re-negotiated according to more specific interests, strategies, priorities. Studying how these *layered interests* are connected and safeguarded renders it possible to determine whether the engagement of African countries with China is part of broader national developmental frameworks or responds instead to more specific, issue-related interests.

Since 2002, Kenya has undergone incredible changes in governance structures. Governance practices have changed due to both domestic and external factors. Domestically, the country experienced an opening to multi-party democracy (since 2002), and constitutional reform in 2010 that envisioned devolution and an expansion of liberal ideals. Externally, political tensions characterised the period when President Kibaki was in power, during which the "Look East" policy was formulated. This was later instrumentalised by President Uhuru Kenyatta and Vice-President William Ruto during their trials at the International Criminal Court (ICC). These dynamics gave rise to shifting patterns of power-sharing between state and society, with bureaucratic apparatuses and societal groups varying significantly in terms of objectives, autonomy, the scope of policy tools, and their capacity to respond to foreign-triggered pressures. Some bodies responded more promptly to constitu-tionalism, while others carried out their day-to-day operations in a more conservative fashion, with no particular capacity or willingness to change.

At the national level, two broad categories of interests can be identified: economic development, and safeguard of constitutional rights and democracy. Both national interests are incorporated in Kenya's development blueprints, and are absorbed by implementing actors, who, at times, favour one over the other, though mostly recognise the value of both. Kenya Vision 2030 can be considered a synthesis of the country's willingness to move toward industrialisation and economic development, while preserving constitutional rights, evident by its focus on collaborative governance to achieve economic and social development. Kenya Vision 2030 was itself the result of a long consultative process with social groups at large, including the private sector (Government of the Republic of Kenya 2008, 3; Okerere and Agupusi 2015, 96–97). As a result, a variety of stakeholders beyond the state identify with it.

While national policy has provided incentives to engage with China, deemed a crucial contributor for implementing Kenya Vision 2030 and achieiving national economic development, research into sectoral and issue-specific levels suggests that engagement with China at these levels is not happening with the same enthusiasm and promptness. Sectoral interests emerge out of complex, long-term dynamics of domestic governance, which cannot be easily and rapidly changed by a shift in national policies. Moreover, even if Kenyan and Chinese interests in developing a sector are broadly complementary, these interests may become void if contextualised within specific policy and implementation environments. Rhetorical interests, which imbue national and sectoral approaches, may be dismissed at the issue-specific level where actors become more concerned with the practicalities of the engagement. Another important element to consider when unpacking motivations is the *initiator* of the relationship. In studying agency it matters whether the decision to interact is based upon a *triggered reaction* or rather a *spontaneous calculation* or no decision is taken at all. Focusing on the initiator tells us who initially identifies the engagement with the Chinese as salient. Murray-Evans suggests that "the conflation of agency (an ontological presupposition) with resistance (empirical claims) potentially obscures the variety of different positions taken up by African agents in international trade negotiations and other interactions with the outside world. If we can only observe African agency when it is expressed as opposition to external forces, we may miss, for example, those instances when African agents agree with or contribute to dominant ideas or cooperate with other international actors" (Murray-Evans 2015, 1847). Three main types of initiators can be observed: re-active (contestation), pro-active (cooperation), and passive. This dismisses the idea that Kenyans only engage in contestation with the Chinese by showing that actions can involve pro-active, cooperation-seeking forms of behaviour as well as passive ones. Focusing on the initiator helps identify the motivations and under what type of circumstances Kenyan actors engage with China, and is fundamental to acknowledging domestic and international power relations.

Contestation is a *re-action* to be expected in cases where Chinese actors/activities challenge existing Kenyan values, norms or frameworks of operation and provoke a re-action to control such change. Re-action appeared all the more intense not only when interests clashed with the Chinese, but also when the interests between different Kenyan groups in charge of "governing" the specific issue diverged. The more a threat was perceived, the more a re-action was expected. This behaviour was particularly evident in Kenya's trade sector, where the arrival of Chinese products and businesses was perceived by local associations, traders, and manufacturers as a threat. The perceived threat is consequential for different situations: in some cases, it is the Chinese occupation of business spaces traditionally belonging to Kenyans and the challenge of existing frameworks of business governance between wholesalers and retailers. In other cases, such as the fight against the import of counterfeit products, the cause of action against the Chinese lies in the perceived weakening of innovation to the detriment of national development. In the case of labour unions' relations with Chinese companies, the threat lay in the lack of abidance to Kenyan laws by the Chinese, exacerbated by Kenyan bureaucrats' mis-conduct, i.e. bypassing existing norms of conduct to achieve quicker returns.

Cooperation is *pro-active* behaviour that emerges in cases where a calculated choice is made by Kenyan actors to include Chinese actors for the implementation of specific policies or projects. It is characterised by the recognition, by Kenyan actors, of specific needs in each sector/area, followed by the identification of Chinese resources as useful for the achievement of desired outcomes. Concerned actors thus initiate a transformation of the system to include Chinese actors for the implementation of specific policies or projects. In contrast to the re-active mode, the situation is non-confrontational and it is interesting to understand how Kenyan actors articulate their motivations for engaging the Chinese vis-à-vis other, often better known and more established partners.

One example is the development of traditional medicine (Procopio 2016, 173–182). The historic demonization of traditional medicine in Kenya created a vacuum in the social and governmental structures of governance, reducing local agents to operate underground while allowing the flourishing of foreign traditional medicine practitioners. Relations between local government organs and social groups, including practitioners and manufacturers, evolved into a highly disjointed, anarchical domestic system of governance, where lack of trust and absence of protection ruled. In this context, the Chinese presence is not important as a trigger for change but rather as a chosen facilitator for such change, an example of success that Kenyan stakeholders choose to learn from to achieve an identified goal, namely the revitalization of Kenyan traditional medicine with the ultimate aim of it contributing to the country's overall healthcare delivery. State organs are leading the re-organization of the sector, in conjunction with a variety of social groups. Full knowledge of domestic needs is complemented by a clear vision as to what external actors can contribute. China is identified as the prime source of support, especially for trade and training-provision. It is also regarded as a partner with which to fight against the imposition of Western medicine, and the limits placed upon the world by international institutions, like the World Health Organization, perceived as Western-led.

While the cases of re-active and pro-active behaviour have questioned the general belief that Africans are unwilling and unable to negotiate to their benefit, it is also necessary to acknowledge that there are cases where a passive attitude is observed in which the governance of a particular issue is left in Chinese hands.[3] Neither the content of relations nor the modality is actively defined to safeguard Kenyan interests. China is left to operate unchallenged. There can be various reasons why passive behaviour occurs: perception of complete domination/inability to act, the issue not being perceived as significant or relevant, and/or the potential threat not yet perceived. In all cases, the system of governance is weak, reliant on external financial resources and not institutionally fit to avoid the fulfilment of narrow self-interests, mostly determined by high-political motivations. The lack of domestic negotiations, mostly due to governance weakness, leads to the national context becoming irrelevant to the outcome of relations with China.

Modalities

The initiator is not the only actor populating a system of governance, and it is thus important to study how it interacts with other actors, i.e. the modalities or mobilization strategies. In order to evaluate whether established norms and procedures in any given sector/sub-sector are maintained or bypassed by various Kenyan actors during negotiations with China, it is valuable to study *tables of negotiation* and their conformity to the broader arena. Tables of negotiation represent "a formalised setting where contending social groups decide upon key aspects of statehood over a given period of time. A wide range of negotiation tables exists, from diplomatic conferences involving heads of states, through donor consultations between international financial institutions and local NGOs, to meetings by customary chiefs under the village tree" (Hagmann and Péclard 2010, 551). "Negotiation tables" vary significantly from case to case so that the range of actors that can potentially sit around them is vast. In other words, they are the settings where actual negotiations take place, characterised by the inclusion or exclusion of stakeholders, at times according to existing systems of governance, at times irrespective of them. While arenas, or systems of governance, are harder to alter, tables of negotiation can be more easily subjected to short-term interests and politicization. Studying "tables" is useful to establish whether mobilization practices in engaging with the Chinese conform to existing procedural norms, and thus whether established norms and procedures are maintained or bypassed.

The most frequent observable behaviour was "light alteration" of existing procedural norms, both in contexts where governance was rooted, and where it was in-the-making. It is deemed "light" alteration because it involved a temporary inability of Kenyan actors to guarantee that negotiation tables conform to the broad system of governance. The temporary nature of these occurrences is resolved by the ability of Kenyan actors to identify the weaknesses that the relations with the Chinese have emphasised and act to resolve them. As Mosley et al. point out, "any negotiating game is bound to change significantly, not least as a result of the fact that after years of experience both sides of the bargaining table figure out ways to neutralise each other's strategies" (Mosley et al. 1991, cited by Whitfield and Fraser 2010, 345).

Light alteration often corresponded to practices such as corruption and improper government interference. In other words, attempts by a minority of Kenyan agents to bypass the system and not adhere to standard negotiation tables/procedures in order to achieve self-interests. However, in most cases, the very existence of a rooted system – previously domestically negotiated and legitimised by the majority of stakeholders – generally functions as a point of leverage over acts of pure self-interest. The case of trade unions is a good example of how the system of governance is safeguarded despite light alteration, or inappropriate actions taking place during negotiations (Procopio 2016, 125–127). The governance system provides for a negotiation table populated by three types of actors: the trade union, the Ministry of Labour, and the company or the employers' association acting on its behalf. On top of these, the negotiation also involves the judiciary (the Industrial Court), for cases where the previous three stakeholders do not manage to settle disputes. The arena is legally and institutionally defined, generally legitimised within the country. The tables of negotiation with the Chinese (the *actual* negotiation) show, however, an alteration of the norms and practices deemed legitimate in the domestic arena. Two deviations bring the relations with the Chinese *temporarily* outside of the domestic system. First, the Chinese are not taking part in the associational life formalised by the arena, providing thus a moment of escape from the locally determined norms of behaviour. Second, Kenyan unofficial, parallel governance dynamics – the bypassing of legitimate negotiation tables through the interference of government officials in negotiations – also play a role, and impact, if partially, the outcome of negotiations. Despite the existence of a strong system of governance for dispute resolution, government interference was based on broader interests in infrastructure development, from which originated the need to maintain good relations with Chinese companies. This alteration is mostly related to the Chinese presence, rather than *foreign* presence. It entails alleged corruption, as well as improper domestic government interference.

The frequent lack of adherence of China to existing governance frameworks has been received with diverse reactions, ranging from exclusion to accommodation. While the Chinese, knowingly or unknowingly, often bypass existing governance frameworks, the instances in which Kenyan agents and structures bring governance back to existing/legitimised patterns are more than the instances in which the opposite occurs. "Full conformity" to existing procedural norms is observed almost as frequently as "light alteration" while non-conformity is rarely observed. The fact that Kenyans are more likely to bring governance back to existing, institutionalised patterns, rather than being dominated by external frameworks, is important not only in the context of IR, but also in a domestic scenario where democratization is not merely constitutional jargon, but a de facto moment occurring through a myriad actions carried forward by bureaucrats and society alike.

Interestingly, it is more often the *modality*, rather than the *content* of the negotiation, that represents the point of fracture between Chinese and Kenyan agents. In other words, most cases studied show that Kenyans are not against relations with China because they believe they would not be beneficial, but because the Chinese do not adhere to existing *modi operandi*. This lack of alignment with norms of conduct is at times exacerbated by Kenyans. At other times it is recognised as a product of weak domestic procedural norms. Eventually what seems to be relevant to the ultimate control/ownership of the negotiation with the Chinese is the conformity of negotiation processes to the general system of norms and practices as understood and legitimised by Kenyans. The Chinese non-abidance to locally accepted frameworks, domestically or internationally defined (i.e. reliance on multi-lateral cooperation), is often reduced by requesting the Chinese to abide by the laws and institutions, if they exist, or by creating more appropriate legal and institutional frameworks.

Conclusion

This chapter critiques "first generation" research on China–Africa relations, which paid disproportionate attention to the Chinese side of the equation. By shifting the focus to the African side, specifically Kenya, a comprehensive approach was used to explore governance as the locus where agency can be found and is exercised. This type of agency is contingent on historical, political, cultural, economic realities and does not easily allow one to speak of "African" agency. However, although Africa cannot be taken as a unitary block, and the choice of Kenya to explain the role of domestic politics within the China–Africa discourse is not representative of the continent, the method used (layered interests), and the relational dynamics that were observed domestically and in relation to China, are likely to facilitate the understanding of China–African dynamics in other countries.

By investigating relations between national and sub-national actors, through interests and modalities of action, it becomes clear that the Chinese role in Kenya is far more bound up in different types, and levels, of domestic politics than was previously believed. The layered interests approach provides a more blurred, but realistic, and less generalised picture of actual dynamics. It helps us understand that while national interests are rhetorically framed around mutual benefit, sectoral and issue-based interests are often revisited to fit existing meanings and patterns of engagement. Although development in Kenya is undeniably linked to clientelistic legitimacy, the picture observed was also one of bureaucrats and common people who believed in the country's development and acted professionally to advance that objective, not motivated by allegiance to a leader, but by their personal and professional ethics. In other words, corruption and clientelism may be part of the story, but so are transparency and competence. This occurs in a country where governance is in flux, where the economy "no longer relies on the state; the multi-party democracy is here to stay and presidential demands provoke as much resistance as obedience" (Hornsby 2012, 7).

Notwithstanding this, Kenya has to rely on external forces to continue on its journey of development. In certain ways, Kenya's pre-existing extraversion strategies continue with the Chinese. However, other elements suggest that dependence on external powers is bound to diminish in the years ahead. This is seen in the sheer dynamism of the country's growing middle class, the nation and sector-wide commitment to industrialisation, the increased awareness that democracy is "here to stay," the purposive detachment from Western aid and the adoption of aid provision declarations that focus on the host country's elaboration of priorities. All these elements suggest that Kenyans are more interested in taking ownership of their development path. After all, relations with China and the 2008 financial crisis have emphasised, once more, that foreign assistance does not last forever, and is vulnerable to changes in donor countries. Moreover, a growing segment of the youthful population is increasingly educated, reliant on technology and dynamic enough to use domestic and international resources to create new economic opportunities within, rather than outside, Kenya.

While dependence implies an asymmetrical relationship, what seems to emerge from this analysis is that *sustainability* in development is becoming an established concept guiding Kenyan approaches to external actors. This does not eliminate dependence on foreign financial and human resources, but suggests that Kenya is gearing up to make use of these resources for long-term development rather than one-off free grants. Part of this picture is the recognition that Kenya does not merely "receive" Chinese trade, investments, assistance, but also "requests" this. For long in the literature, it was believed that African countries were struggling with the forceful Chinese penetration of African spaces. However, this chapter has sought to clarify that focusing on Africa as the *recipient* provides only a *partial* portrayal of actual engagements. Pro-active behaviour can also be observed according to which Kenyan actors *identify and choose* China as a partner in trade, development, and more.

An understanding of how Kenyans engage with China is incomplete without acknowledging that the relative newness of the Chinese in Kenya is a variable influencing perception, ideas, and potentially the outcomes of relations. China is often defined as new and, as a consequence, requires Kenyans to be alert. Even in cases where actors are generally positive about China, a sense of unfamiliarity is emphasised, especially in comparison to traditional external actors. A number of agents claim they lack understanding of the Chinese and thus learn and re-learn mainly by engaging with them: perceived successes and failures in engagement reorients preferences and thus strategies for the next round. In such a push and pull bargaining exercise, the Chinese also become increasingly aware of the context within which they operate, and the weaknesses and strengths of Kenyan governance. In some cases, better knowledge was used to bypass the governance system more easily, while in other cases it was a catalyst for change toward greater adherence to Kenyan rules. The exposure to Kenyan norms of conduct has contributed to the Chinese socialization to a more inclusive system of governance.

Finally, Kenya–China relations are as complex as, and a reflection of, domestic politics. This complexity is to be found, first and foremost, in how the domestic context conceives foreign presences, strategizing them to achieve domestic and foreign aims. Stringent political reasons have remained important in Kenya–China relations, including important elements of interdependence where Kenya needs China but China also needs Kenya (seen with the ICC and the "use" of China to leverage Western influence). Beyond high politics, however, the activities in which Kenyan state and social groups engage with China are generally part of broader development plans, national or sectoral, public or private. This means that in everyday activities, the engagement of Kenyan actors with the Chinese is less exceptional than often believed, and the interests defined at the top levels are re-negotiated at lower sectoral and issue-specific levels in line with existing meanings, patterns of engagement, and according to legitimised systems of governance. It is in these sites that agency should be located and meaning can be found.

Notes

1 Africa is here intended as Sub-Saharan Africa. The chapter's aim is to dismantle an all-encompassing concept of "Africa" to focus on complex sets of specificities that African countries present. Nonetheless, talking of Africa holistically does make sense under certain circumstances, not least in reference to the existing literature within which this study is located.
2 Based on the author's doctoral thesis, "Negotiating Governance: Kenyan Contestation, Cooperation, Passivity toward the Chinese," London School of Economics, 2016. Extensive fieldwork was carried out in Kenya between September 2013 and December 2014.
3 For example, Kenyan interaction with Chinese actors in the set-up of Confucius Institutes was rather limited, adopting a hands-off approach, letting the Chinese carve out their space of management instead of actively engaging in decision-making.

References

Alden, Chris and A. Ammon. 2012. *Foreign Policy Analysis: New Approaches*. Abingdon: Routledge.

Bevir, Mark. 2011. "Governance as Theory, Practice, and Dilemma." In *The SAGE Handbook of Governance*, edited by Mark Bevir, 1–16. London: SAGE Publications.

Biermann, F. 2002. *Earth System Governance: The Challenge for Social Science*. Amsterdam: The Global Governance Project.

Brenner, Neil. 2004. *New State Spaces*. Oxford: Oxford University Press.

Brown, William. 2006. "Africa and International Relations: A Comment on IR Theory, Anarchy and Statehood." *Review of International Studies* 32(1): 119–143.

Brown, William. 2012. "A Question of Agency: Africa in International Politics." *Third World Quarterly* 33(10): 1889–1908.

Brown, William and S. Harman. eds. 2013. *African Agency and International Relations*. Abingdon: Routledge.

Clapham, Christopher. 1998. "Degrees of Statehood." *Review of International Studies* 24: 143–157.

Cornelissen, S., Fantu Cheru and Timothy M. Shaw. 2012. "Introduction: Africa and International Relations in the 21st Century: Still Challenging Theory?" In *Africa and International Relations in the 21st Century*, edited by Scarlett Cornelissen, Fantu Cheru and Timothy M. Shaw, 1–17. Houndsmill: Palgrave Macmillan.

de Sardan, J.P. Olivier. 2008. "Researching the Practical Norms of Real Governance in Africa." *Africa Powers and Politics Programme Discussion Paper*, no. 5, Overseas Development Institute, December.

Flemes, D. and L. Wehner. 2015. "Drivers of Strategic Contestation: The Case of South America." *International Politics* 52(2): 163–177.

Gadzala, Aleksandraa. ed. 2015. *Africa and China: How Africans and Their Governments are Shaping Relations with China*. London: Rowman and Littlefield.

Galbreath, D.J., A. Lasas and J.W. Lamoreaux. 2008. *Continuity and Change in the Baltic Sea Region: Comparing Foreign Policies (On the Boundary of Two Worlds)*. Amsterdam and New York: Rodopi.

Government of the Republic of Kenya. 2008. Kenya Vision 2030, First Medium Term Plan 2008–2012, Government of the Republic of Kenya, Nairobi.

Habeeb, W.M. 1988. *Power and Tactics in International Negotiation: How Weak Nations Bargain with Strong Nations*. Baltimore, MD: Johns Hopkins University Press.

Hagmann, T. and D. Péclard. 2010. "Negotiating Statehood: Dynamics of Power and Domination in Africa." *Development and Change* 41(4): 539–562.

Hall, P.A. and R. Taylor. 1996. "Political Science and the Three New Institutionalisms." *Political Studies* 44(5): 936–957.

Hornsby, C. 2012. *Kenya: A History Since Independence*. London: I.B. Tauris.

Jessop, Bob. 2000. "The Crisis of the National Spatiotemporal Fix and the Ecological Dominance of Globalising Capitalism." *International Journal of Urban and Regional Research* 24(2): 323–360.

Keohane, Robert and Joseph Nye. 2000. "Realism and Complex Interdependence." *International Organisation* 41(4): 725–753.

Lemke, D. 2003. "African Lessons for International Relations Research." *World Politics* 56, (1): 114–138.

Lemke, D. 2011. "Intra-national IR in Africa." *Review of International Studies* 37: 49–70.

Levi-Faur, D. 2006. "A Question of Size? A Heuristics for Stepwise Comparative Research Design." In *Innovative Comparative Methods for Policy Analysis: Beyond the Quantitative-Qualitative Divide*, edited by B. Rihoux and H. Grimm, 43–66. New York: Springer.

Mosley, P., J. Harrigan and J. Toye. 1991. *Aid and Power: The World Bank and Policy-Based Lending*, vol. 1, London: Routledge.

Murray-Evans, P. 2015. "Regionalism and African Agency: Negotiating an Economic Partnership Agreement between the European Union and SADC-Minus." *Third World Quarterly* 36(10): 1845–1865.

Neubert, D. and C. Scherer. eds. 2014. *Agency and Changing World Views in Africa*, Berlin: Lit Verlag.

Ohmae, K. 1995. *The End of the Nation-State: The Rise of Regional Economies*. New York: Simon and Schuster Inc.

Okerere, C. and P. Agupusi. 2015. *Homegrown Development in Africa: Reality or Illusion?* Abingdon: Routledge.

Procopio, M. 2016. *Negotiating Governance: Kenyan Cooperation, Contestation, Passivity Toward the Chinese*. Ph.D. London School of Economics.

Sellers, J.M. 2011. "State-Society relations." In *The SAGE Handbook of Governance*, edited by Mark Bevir, 124–141. London: SAGE Publications.

Shaw, Timothy. 2015. "African Agency? Africa, South Africa and the BRICS." *International Politics* 52: 255–268.

Turner, S. 1998. "Global Civil Society, Anarchy and Governance: Assessing an Emerging Paradigm." *Journal of Peace Research* 35: 25–42.

Whitfield, Lindsay and A. Fraser. 2010. "Negotiating Aid: The Structural Conditions Shaping the Negotiating Strategies of African Governments." *International Negotiation* 15 (3): 341–366.

Wight, Colin. 1999. "They Shoot Dead Horses Don't They? Locating Agency in the Agent-Structure Problematique." *European Journal of International Relations* 5(1): 109–142.

Williams, David. 2011. "Agency, African States and IR Theory." Paper presented to the BISA Annual Conference, Manchester, 27–29 April.

12

BUREAUCRATIC AGENCY AND POWER ASYMMETRY IN BENIN–CHINA RELATIONS

Folashadé Soulé-Kohndou

The study of African agency in international relations has gained much ground in the last decade, with several authors scrutinizing how African governments, civil society organizations and other groups have been dealing with China and influencing the relationship to their advantage (Van Bracht 2012; Gadzala 2015; Fraser and Whitfield 2008; Prizzon and Rogerson 2013). Despite the pioneering character of these studies, few have addressed the particular role bureaucrats play in negotiating with China and how variations of tactics across bureaucracies affect the outcome of such negotiations. Yet, how African states affect and influence negotiations deserves attention, as well as appreciation, and requires Bob Jessop's theory of the need to identify the specific sets of state officials located in specific parts of the state system, i.e. the bureaucrats of specific entities in ministries (Jessop 1990).

This chapter focuses on the negotiation tactics and manoeuvres of bureaucrats of Benin, a small francophone African state, when negotiating infrastructure project contracts with China, and questions the latitude of social action of bureaucrats as substate actors often acting in the shadows of the negotiation process and the agentic dimension of their action. More specifically, it focuses on bureaucrats located in ministerial departments in charge of reviewing calls for tenders, monitoring execution of public works and closing projects. For the purpose of this chapter, civil servants and bureaucrats are used interchangeably. It challenges the assumption that the high dependence on aid and overseas investment of many West African francophone countries necessarily limits their capacity of acting and exercising influence during negotiations with China. Additionally, it challenges the alleged assumption that African governments are passive in their relations with China (e.g. Gadzala 2015, 16–17) and re-evaluates how negotiations vary within a country and across bureaucracies. By connecting theories of bureaucratic institutionalism and social agency, and highlighting the extent to which weaker actors can influence negotiations depending on the structure of their bureaucracies and the role substate actors play during these negotiations, this chapter will analyze how these parameters impact on the issues under negotiation. The connection between the study of bureaucratic politics and agency in the case of Benin–China relations is also a means to more specifically locate African agency in Africa–China relations (see, for example, Mohan and Lampert 2012; Corkin 2013) and, more broadly, African agency in global politics (see, for example, Brown and Harman 2013). It is based on Colin Wight's assumption that "since agents are differentially located in the social structures, their exercise of agency is disparate" (Wight 2004, 269–280). The different case studies show how disparate agencies are exercised by civil servants across ministries, but does not result in the exercise of a "collective agency" of the Beninese government.

Theoretical literature on bureaucratic politics in IR has so far largely focused on case studies of large-size states bureaucracies, like the United States, and more recently on rising powers (see Hill 2003; Allison 2008; Art 1973; on emerging powers see Cason and Power 2009 and Siko 2014). This literature has addressed the organizational process of bureaucratic politics by focusing on the role of executive and high-level ranked bureaucratic branches (Neustadt 1960; Dyson 1986; Kaarbo 1998). It has examined how state bureaucracies impact on foreign policy, highlighting the fragmented and institutionally driven nature of foreign policy making and implementation. This literature has also provided empirical insights into how the administrative structures of government affect foreign relations (Alden and Ammon, 2011).

Through the Benin case, based on empirical field work and access to negotiation materials, draft terms of reference, draft contracts and final contracts in infrastructure projects, this chapter addresses this shortfall by arguing that bureaucrats play a key role during the negotiations, and their coordination largely affects the outcomes and the edition of the infrastructure projects' contracts. It draws on 30 semi-structured interviews with principal actors in the Beninese bureaucracies involved in negotiating various infrastructure projects (government actors, especially low-ranking, middle-ranking and high-ranking civil servants), most of them located in Cotonou, the economic capital of Benin. Civil society actors, Beninese entrepreneurs and diplomats based at the embassies of traditional donors were also interviewed.

The main arguments that this chapter advances are twofold. First, the increasing relationship with China has been largely interconnected with the decreasing involvement of traditional donors in crucial sectors for the Beninese government's general economic and social policy, meaning infrastructure development financing. Second, the intervention of Benin's highly influential executive branch (the presidency) has a major impact on variations in negotiating outcomes: where the executive branch, often promoting the relationship, is less involved, or circumvented, the outcome is more beneficial for Benin. Where the executive branch is more permissive, bureaucrats are more peculiar regarding technical prerequisites and exert agency as agents seeking to preserve national law and regulations. This chapter thus asserts that despite large power asymmetries in Benin–China relations, and favouritism by the executive branch toward Chinese contractors, Beninese bureaucratic agents exert agency; they exercise variable control over the process and outcome of negotiations with China in accordance with their country's national regulations through several structural and tactical means. Despite the constraints of state organization, and the different social positioning of bureaucrats in the government system that more or less limits their "capacity to do," these bureaucrats focus their agency or exercise of power before, during, and after the negotiation process. This confirms the theory that bureaucrats as agents social action can only be captured in its full complexity if analytically situated within the flow of time (past, present, and future) (Emirbayer and Mische 1998; Wight 1999).

Evolution of Benin-China relations

Disruptive political relations

Official relations between China and Benin started in 1964, but a Chinese presence in Benin can be traced back to the late 1950s. Following independence from France in 1960, Benin progressively established diplomatic relations with all countries, including the People's Republic of China. However, the influence of France on Beninese diplomatic relations remained strong. As France supported Taiwan, it imposed this diplomatic line to most of its former colonies, and Benin (then Dahomey) voted in favour of Taiwan's recognition at the United Nations on October 25, 1971. Following this episode, China suspended its relations with Benin. Relations were re-established in December 1972 following the revolutionary coup organised by Mathieu Kerekou in October 1972, who established an authoritarian regime based on Marxist-Leninist ideologies under the name "Popular Republic of Benin" (PRB). Diplomatic relations with Taiwan were suspended, and Kerekou was the first Beninese president to make an official visit to China.

Despite the similarity of ideology between the PRC and the PRB, relations were not mainly based on ideology. The economic and business component of relations increased, albeit slowly. The low level of interactions mainly resulted from high political instability in Benin between 1960–1989, during which Benin experienced six coups, seven national constitutions, and 11 presidents, of which six were military and five civilian (Topanou 2013). Stable democratic transition in February 1990 did not have an impact on bilateral relations with China. The then-president, Nicéphore Soglo, did not restart relations with Taiwan but, being willing to break with past socialist ideology, chose to prioritise relations with France, the European Union, and the US.

Diplomatic offensive under President Yayi Boni

Before running for president, Yayi Boni was governor of the West African Development Bank from 1994 to 2006. According to former and current diplomats at the foreign ministry who participated in negotiations with China led by President Boni, he was personally impressed by the "Chinese economic model and rise," and the main promoter of China's entry into the Bank and turn away from Taiwanese capital. According to interviewees, Boni's friendly relations with the former head of the China Development Bank also played a key role in strengthening these relations when he was elected president of Benin in 2006.

Following Boni's election, which he won after a campaign mainly focused on economic growth, China became Benin's key partner. The focus of relations was largely economic, Benin having turned the page of its socialist era under the presidency of Mathieu Kerekou. Under Boni, Benin–China relations intensified. The president made five official state visits to China between 2006 and 2014 (in comparison with two such visits to France during the same period). A legal framework officially recognizing the PRC as the only representative of the Chinese nation and population was adopted.[1] There was an increase in China-funded projects, largely in infrastructure and agriculture, resulting in the number of Chinese workers in Benin rising from 230 to some 700 workers in 2013.[2] The president is often considered by ministers and the cabinet as "the first diplomat" in Benin–China relations. Moreover, he personally receives Chinese SOE delegations at the presidential palace and officially makes promises of assigning infrastructure projects to these corporations.[3] Under Boni, Benin's debt was progressively erased in 2005 and 2009, amounting to FCFA 20,000,000 (US\$ 32,8000).[4] Moreover, besides the EU, China is the only country whose products are used for bids won by Chinese contractors and are tax-exempt (Houndeffo 2013, 231).

Comparative advantages of China

The Chinese credit loans system is considered by the executive as the best option for addressing Benin's infrastructure deficit, especially in a context where traditional donors decided to decrease or abandon financing infrastructure projects. The 2012 Beninese strategy clearly articulates infrastructure development (especially road transportation) as a key element of national development.[5] However, traditional donors like the EU and France (especially the French Development Bank), who were largely present in this sector before 2006, progressively disengaged, citing corruption by Benin's government and the low profitability of the projects. According to interviewees, these sectors are increasingly considered less competitive for these traditional actors.

In this context, China is viewed by Benin's executive as more "demand-driven," in the sense that its cooperation and bilateral relations are largely focused on Beninese priorities (infrastructure), and Chinese cooperation is considered less procedural. Interviewees repeatedly mentioned fewer procedures, seen in the reduced number of feasibility studies, allowing for projects to be executed at a fast pace. In return, Benin offers very preferential tax exemptions and contractor preference to China in road, agriculture, and administrative infrastructure projects. Most of Benin's road infrastructure projects are executed by Chinese contractors, especially SOEs like China Railways, SynoHydro, Anhui Foreign Economic Group, Shanxi Coking Coal Group Co Ltd, or Zhejiang Teams International Economic & Technical Cooperation.

The modus operandi of China's involvement in Benin is similar to its cooperation in other African countries (Alves 2011) and can be summarised in four pillars:

1. Non-reimbursable subventions and technical assistance;
2. Donations (technical materials; infrastructure): China donated Benin its first multi-sports complex following the reestablishment of diplomatic relations in 1972;
3. Zero-interest loans: reimbursement is spread over 25 years (these loans are often subject to debt cancellation);
4. Concessional loans: interests in this case are not higher than 2 percent.

Benin's government also offers flexible policies to China by adopting a parallel track in its relations with China, by not requiring its participation in donor harmonization processes:[6] China's aid in Benin, according to OECD DAC standards, is very low and mostly consists of technical assistance.[7] China is also not required by Benin's government and executive branch to follow DAC donor requirements: it does not participate in DAC coordination meetings (with the exception of high-level meetings among ambassadors, which the Chinese ambassador has participated in since 2012).

From a Chinese perspective, despite its lack of strategic mineral resources, Benin is considered a strategic hub in Africa relations, especially for economic reasons: market consumption for Chinese goods, road transport for its merchandise as China has not developed many airline routes in Africa and uses Cotonou's harbor to expand in West Africa and especially in neighbouring Nigeria. In this regard, Cotonou also acts as a "warehouse-state" for Chinese contractors' infrastructure projects in Nigeria and the subregion (see Igue 1992). Moreover, by focusing on cooperation on development issues including infrastructure, transportation networks, manufacturing, agricultural industries, and debt relief in Benin,[8] China has been downplaying the role of natural resources and mining cooperation in its Africa relations (Sun 2014).

China: a key actor in the development of Benin's infrastructure projects

Chinese involvement in infrastructure projects in Benin can be categorised into three types. Type 1 takes the form of infrastructure projects with Chinese funding (concessional/non-concessional loans). These generally take the form of loans for infrastructure projects allocated to Chinese companies, a loan prerequisite. The Beninese state also provides funding. Most are road infrastructure projects and do not require any competition or preliminary open or restricted call for tenders. Examples of these projects are the road interchange of Godomey in Cotonou and the rehabilitation works of the Akassato-Bohicon road. Type 2 involves infrastructure projects executed by Chinese contractors following loans provided by development banks to Benin. In this case, the call for tenders is either open or restricted to several companies beyond those from China. Most involve road and ICT infrastructure projects. In Benin, Chinese companies participate in calls for tenders for projects with big financial stakes (i.e. no less than US$ 50 million). Examples include the reconstruction of the Godomey–Pahou road as part of the inter-state national road in West Africa (funded by the World Bank); the N'Dali–Nikki-Chicandou road (funded by the African Development Bank), and the Fifadji bridge (funded by the West African Development Bank). The third type involves infrastructure projects executed by Chinese contractors with Beninese public funding (from the National Treasury). In this case, calls for tenders are either open or restricted to several companies, including Chinese. Most of these projects also take place in road infrastructure projects, with examples including layout works of the Segbana-Samia, Bodjécali-Madécali-Iloua, and the Kilibo-Nigeria border corridors of merchandise transits from Nigeria to other regional countries.

Tracing the negotiation process

The negotiation process in Benin–China relations can be separated in three progressive phases that extend from preliminary political discussions to bureaucratic interactions. The first involved preliminary discussions among Beninese and Chinese officials: these generally happen after official state visits in Cotonou, Beijing, or at FOCAC meetings. When President Boni occupied the rotating AU presidency from 2012 to 2013, he also opened AU high-level meetings up to Chinese attendance, which allowed for bilateral discussions on the sidelines of AU conferences. During these preliminary discussions, most loan and infrastructure projects are discussed. The role of the executive branch is crucial in this phase. The 1991 Beninese constitution, following the democratic constitution, established a "presidential" regime in which the executive branch has considerable decision-making powers, and influence across all legislative and judicial institutions (Topanou 2013). In this regard, the president exercises important control of foreign affairs and the intensification of Benin–China relations is largely a presidential project, leaving the foreign minister in an advisory and implementation role.

For the second phase, following official pledges made by Chinese officials and plans announced at multilateral or cross-regional meetings like FOCAC, priority projects are selected by the Beninese government and its bureaucracies, especially the ministries of development and planning. At this level, Chinese contractors engage in an internally competitive process by approaching different Beninese ministries, and proposing to fund their priority projects by jointly applying for China Exim Bank or China Development Bank funding. This competitive process largely driven by the national "going-out" strategy makes Chinese SOEs more or less prepared actors of FDI in a context where they are decreasingly competitive within the Chinese market (Wang and Zhao 2015). The process can also be inverted when Beninese ministries, following orders from the executive, approach the Chinese embassy directly to request funding for their priority projects. In doing so, sectoral and technical ministries thus set up their own foreign agenda and circumvent the

Ministry of Foreign Affairs, traditionally the focal actor in Benin–China relations in charge of coordinating relations and projects locally.

Finally, whether the project is executed on Chinese funding (concessional or non-concessional loans), development bank loans (e.g African Development Bank, West African Development Bank), public funding (Beninese government), the terms of the agreement (contracts) are discussed, negotiated, and revised among the bureaucrats from the sectorial ministry under which the project falls, and the Chinese counterparts (SOE contractors representatives, delegations from the Chinese embassy).

Who negotiates within the bureaucracies?

As with many African and Asian developing countries, Benin was subject to the slimming down of public bureaucracy and downsizing of government under IMF structural adjustment programs in the early 1990s as a means to make bureaucratic action more effective.[9] These programs did not achieve the expected success: Benin still has a massive, choked bureaucracy and faces various dysfunctionalities: quality of expertise varies across ministries and at the intra-ministerial level; corruption is also endemic and affects the efficiency of delivery (Topanou 2013). Inter-ministerial coordination is a traditional prerequisite to signing contracts with China and requires the presence of the Ministry of Foreign Affairs (Benin's MOFA). However, as speed is a key issue in Benin–China relations, especially under Boni,[10] and overall Africa–China relations, this results in the circumvention of MOFA in order to accelerate project execution. This can create tensions among the various sectoral ministries and the MOFA, as the latter are often asked to intervene when conflicts arise between these ministries and Chinese corporations during the executing phase.

The bureaucrats in charge of negotiating with China consist of MOFA diplomats and technicians from sectoral ministries. For instance, in the case of administrative infrastructure, the bureaucrats from the Beninese Ministry of Urbanism and Habitat consist of civil engineers and architects, whereas bureaucrats from Benin's Ministry of Public Works consist of public works engineers. Negotiations hence include both diplomats and technical staff from sectoral ministries. This configuration is not without issues, as the two categories of bureaucrats do not necessarily adjust on their priorities: technicians tend to be more peculiar and demanding in terms of technical prerequisites and national regulations, whereas diplomats can be more permissive and request more tolerance from the sectoral ministries toward the Chinese partners upon pressing demands from the executive branch.

It is important to note the academic path of the MOFA negotiators: most were trained in national elite universities or graduated from French universities, which makes them less knowledgeable about Chinese culture and more dependent on Mandarin–French interpreters. During the revolutionary period under Kerekou, several Beninese students received full-scholarships to study in China in several fields, including politics, engineering, and management. Most returned to Benin and now act as vectors of China–Benin business development. However, the post-revolutionary governments have not used their knowledge of Chinese culture and languages by involving them in the negotiation, and prefer to rely on diplomats educated in France since their degrees are considered as having higher status.

According to interviewees, through its economic mission in Benin attached to the Chinese embassy in Cotonou, China also engages in direct relations with civil servants within the bureaucracies by providing their ministries with visiting tours and business road shows (e.g. for the Ministry of Trade and the National Chamber of Commerce). These tours include training Beninese agents on private sector development, and how trade and business chambers are structured in China. They provide an opportunity for Chinese officials to socialise Beninese civil servants to Chinese business practices, and engage Chinese public and private enterprise about prospective business opportunities in Benin. Strategic sectors such as agriculture or energy are chosen as topics for the seminars in partnership with China Exim Bank. The follow up process allows China to use the Beninese civil servants that take part in these road shows to serve as focal points and leverage for information, for example about new calls for tenders for various projects in the country.

Comparative case studies and multi-dimensional bureaucratic agency

Although fast-track processes and low-cost spending of Chinese corporations in infrastructure projects was frequently mentioned by interviewees as China's main comparative advantage, China is also under scrutiny for several reasons. The following two case studies highlight the role bureaucrats play during the negotiation process where they intervene (phase three of the negotiation process) and the stratagems and tactics they use in order to influence outcomes to their advantage.

Recurrent issues of content

Recurrent issues of content concerning Chinese contractors in infrastructure projects in Benin revolve around three elements: (1) Lack of employment of local workers, according to interviewees, even the levelling staff are said to come from China; the use of low quality materials, and the lack of respect of local Beninese standards and legal requirements; (2) Non-use of French as a working language, the refusal of Chinese workers to monitor the construction works; and (3) Some episodes of violence against local workers.

Civil society actors, among which labour unions such as the Syndicat des Travailleurs de l'Administration des Transports et des Travaux Publics the Ministry of Public Works (SYNTRA-TTP), the official labour union connected to the Ministry of Public Works, also recurrently accuse the government of favouritism, unfair competition, connivance with dumping strategies from China in Benin and contribution to national unemployment and closing of local businesses in the road infrastructure sector, as companies like China Railways Suisiju Group Corporation very often win the contract attribution.[11] Moreover, Chinese contractors sometimes get the contracts attributed without taking part in the official Call for Tenders.[12]

Bureaucratic agency in asymmetrical negotiations

In order to shed light on how bureaucracies exert agency in negotiation processes and outcomes in Benin–China relations, this section will briefly examine two specific cases in road infrastructure, administrative infrastructure, and ICT infrastructure.

Case Study 1: Akassato–Bohicon road infrastructure project

Project characteristics

The Akassato–Bohicon project is a Type 1 road project with an estimated cost of CFA 107 bn (US\$175,480,000) whose aim is to connect two departments in Benin (Atlantique and Zou) funded upon subvention and zero-interest loans provided by the Chinese government (Lot 1 and 2) and by the China Exim Bank, and led by the Ministry of Public Works. Selection of corporation and execution of the project was undertaken in China and assigned to the China Geo-Engineering Corporation and Xinxing Group. The Beninese government rejected a grant provided by the EU for the construction of the road. The project's objective is to provide a liaison between Allada, Sehoué, Zogbodomey, and Bohicon, all localities whose activities are based on agriculture, manufactured goods business, and livestock farming. Circulation on this road is particularly difficult due to high traffic and road deterioration, partly due to soil quality. However, this road section is of particular strategic importance: it is one of the most "economic" roads of Benin, allowing for circulation of goods in other regional countries. ECOWAS has also integrated this as part of West Africa's regional integration program. Overall, road infrastructure projects are strategic for the Beninese economy, being a source of economic revenue, and facilitating mobility of goods and individuals. Moreover, roads also provide potential political gains for the regime in terms of electoral wins.

Negotiation process and controversial clauses

The road project was first discussed at a diplomatic level by Benin's executive branch and its Chinese counterpart on the sidelines of the 2009 FOCAC IV meeting in Sharm-el-Sheikh, Egypt. Both sides agreed on a zero-interest loan affected to finance road infrastructure projects agreed to by the Beninese and Chinese sides in November 2010. Tensions arose around several controversial clauses around the third batch (lot 3) of the project: the refusal by the Chinese counterpart to appoint an independent controls bureau and the threatening of withdrawing the loan if a Chinese control office was not appointed. The Chinese counterpart also charged the Beninese government with a loan of CFA 6,09 bn (US\$9 987 699) in order for the appointed China control office to provide a feasability study whereas a

Beninese control office (ETRICO Ingenieurs Conseils) had already been appointed and had estimated the total costs of the construction at CFA 52 bn (US$85,280,000) (in comparison to CFA 107 bn (US$175,480,000) of the Xinxing corporation). Despite national legal requirements, very few local materials and workers were used for the project. It has also been affected by delays, strikes by local workers regarding their working conditions, and complaints about the quality of the materials used.

Bureaucratic minorities' agency

Within the Ministry of Public Works, the direction of roads and the general direction of Public Works are composed of bureaucrats, most of them civil engineers, who work under the minister's cabinet. In this regard, they are subordinates, and form a minority within the government whose power is concentrated around the cabinet. However, they are the most active civil servants within the SYNTRA-TTP union, and their action confirms Putnam's assessment that a minority within a government can influence the two-level games played by leaders in international negotiations (Putnam 1988). In order to understand the conditions under which these minorities have influenced the resulting outcomes in bureaucratic organizations, it is necessary to analyze their strategies. In the case of SYNTRA-TTP, one of its leaders, Jacques Ayadji, occupies the No. 2 position within the union,[13] is a public figure and charismatic civil engineer who is also in charge of roads within the Ministry of Public Works. The process by which these bureaucrats inside the government have bargained with other actors reveals bureaucratic politics at play.

These stratagems and tactics revolve around three elements: rewards-and-costs stratagems, informational maneuvers, and procedural maneuvers. By using SYNTRA-TTP, a union whose legal status is independent from the ministry, as a platform for their contestation, these bureaucrats have become "reactive bureaucratic agents," a phenomenon previously unseen in Beninese national politics. First, rewards-and-costs stratagems consisted of coalition-building inside and outside the bureaucracy (Ministry of Public Works). These stratagems have revolved around alliances with other departments and unions within the government, and mobilizing the support of influential groups outside the executive branch through the use of media in order to mobilise outside supporters and influence constituency opinion (Halperin 1972). Coalition building was set up through alliances with other unions (CSA-Bénin, CSTB, CGTB, Fensetraf-Bénin) from other ministries, namely the Ministry of Economy and Finances, a key player in the negotiations. Following successive denunciations by Syntra-TTP[14] through public press conferences, TV and radio participation by the unions, including on highly popular radio broadcasts like "Caravane du Matin" on Radio Tokpa where Jacques Ayadji often participates, the union gained the support of the minister, Lambert Koty, who decided to set up an independent standards control bureau for all lots, a decision highly contested by the Chinese counterpart who finally agreed to an independent standards control on Lot 3 of the project. The refusal of the Chinese contractors to allow access to the ministries' staff members to the construction site was also heavily covered by the local press.

Second, procedural stratagems consist in manipulation of the legal environment by inclusion of discriminatory criteria in the call for tenders diffused by the Ministry of Public Works in order to exclude Chinese companies. This stratagem allows for bureaucrats to determine the procedure in order to affect the choice outcome without having to induce individual preference change (Maoz 1990). This action had led bureaucrats to impose several clauses when drafting the call for tenders making independent standards control compulsory on projects under Type 2 and Type 3. This action has led several Chinese companies, although not all, to withdraw from international bids which participates in "drawing the circle" (i.e. intentional exclusion) of (some) contractors bidding to international call for tenders, by reducing the group of participants (Halperin 1972). By first succeeding in making the government appoint an independent standards control bureau, made possible through EU funding, with the support of the minister, the bureaucrats also succeeded in persuading the executive branch to act in their favour.

Third, informational stratagems adopted by these bureaucrats consist of presenting arguments in order to support their cause and enhance the likelihood of persuasion (Kaarbo 1998). These tactics are exerted through association with supposedly independent actors but also by justifying their position by making normative and moral arguments through denunciation. In this case, the bureaucrats use highly moral rhetoric (for example, Chinese corporations "plunge Benin in continued under-development"; the "dignity" of Benin is being flouted). In so doing, the bureaucrats use an "issue escalation strategy" by converting a practical question into a matter of principle (Lindell and Persson 1986) and gain popular support.

These denunciations are also made through association with popular civil society groups like the Association for the Fight against Communitarianism and Racism through joint press conferences, and through the organization and support of riots of Beninese local workers on the construction sites against their work conditions. Other informational stratagems have consisted in sending an open letter to the National Assembly asking for the members of parliaments not to sign the loan agreement for the benefit of the Chinese standards control bureau. The combined stratagems used by these bureaucrats succeeded in converting neutrals to supporters, and opponents to neutrals, and exert bureaucratic agency (Allison and Halperin 1972).

Effects on the outcomes

Despite these different actions, the Beninese executive branch has unilaterally decided to take away the standard controls investigation of Lot 3 of the Akassato–Bohicon, despite support from successive Ministers of Public Works Lambert Koty and Ake Natondé, and give it to a Chinese bureau. According to interviewees, this unilateral decision was due to pressure by the Chinese ambassador about cancelling the loan and pressure from the executive branch on members of parliament (whose majority originates from the party of the president). However, bureaucratic action and organization has led to popular support and has allowed for pressure on the executive branch. Salaries and health assurance of the Beninese workers were reviewed and upgraded. The combination of stratagems has also led the Chinese corporations, either not to bid on several international call for tenders, where intentional discriminatory clauses were integrated (e.g. compulsory independent standards control), or for the executive branch to take into account multiple bids by Chinese SOEs on Type 3 projects, as was the case on the Pahou–Ouidah–Hillacondji road project whose bid was won by a Tunisian consortium.

Case study 2: The administrative tower of Cotonou

Project characteristics

The conception and execution of the administrative tower of Cotonou, which comprises three ministries, is a project executed by the China Anhui Foreign Economic Construction Group (AFECC) under the authority of the Ministry of Urbanism and Habitat. The project total cost is of US$95 million and is funded through a loan provided by China Exim Bank. It was initially a Beninese project led by the National Agency of Works of Public Interest (AGETIP). AFECC took the first step by approaching the Ministry of Urbanism and Habitat in order to consult them on priority projects of their portfolio. The initial decision was made to ask for financing from China and execution by AGETIP. However, AFECC requested to entirely execute the project through funding (loans) provided by China Exim Bank.

Bureaucratic agency

According to civil servants from the Ministry of Habitat, working with the Chinese is not difficult as long as all clauses are stipulated in the contract. This makes the negotiation process crucial, although time consuming. Most bureaucratic agency was exercised during the negotiations. The civil servants at the Ministry of Urbanism and Habitat are mostly architects, and civil engineers. Hence, they succeeded in making amendments and observations on the architectural plans. Several rounds of negotiation between AFECC and the civil servants led to the imposition of several clauses, which were not present, nor mentioned in the Akassato–Bohicon contract. These include:

1. Providing assistance to the Chinese counterpart to acquire local material construction[15] and fuel, and assisting the Chinese counterpart in hiring local workers;
2. Submitting on a regular basis, and two weeks after the start of the projects, the roadmap of the works;
3. Monthly communication of the state of the project by the Chinese counterpart to the ministry;

4.   Free access for the ministry officials and engineers and free access to the construction site by the latter, upon respect of security rules;
5.   Respect for the national laws of Benin in regard to salaries, work accidents, social security, and employment;
6.   Daily diary of progress of execution phase and the use of French and Chinese as working language;
7.   Use of local Beninese materials such as wood, sand, gravel, and cement;
8.   Training of Beninese technicians in the use of machinery and equipment provided by the Chinese counterpart;
9.   Specification of tax-exemption on specific materials used for the project.

Effect on the outcomes

Despite several attempts to reduce control of the bureaucrats and pressure from the executive branch, the clauses agreed upon by both parties were negotiated before the signature of the contract. The Ministry of Urbanism and Habitat used several techniques for delaying the negotiation process in order to get the requested clauses into the contract before its signature. Comparison between contracts of works led by Chinese corporations from the Ministry of Public Works and the Ministry of Habitat show that the latter exerts more bureaucratic agency during the negotiation, as they circumvent or resist pressure from the executive branch or the Chinese counterpart, whereas the Ministry of Public Works is more affected by executive pressure during the negotiation phase.

Most interviewees highlighted the difficult relations with their ministries, the executive branch and the Ministry of Foreign Affairs. The latter two are said to affect the negotiation phase as they ask for conciliating measures from the civil servants, in order not to affect diplomatic relations between Benin and China. Most of the demands that arise from the Chinese counterpart are subject to negotiation by the foreign ministry with the respective ministries, which creates a lack of coherence, synergy and similar position from Beninese diplomats, and technicians, and other civil servants involved in the negotiation process.

Conclusion

Using Benin, this chapter has asserted that bureaucrats in small African states, despite negotiating in starkly asymmetrical relations, are not passive and conforming agents during negotiations; they use influence strategies in order for civil servants minorities' views to prevail. Often acting in the shadows, they act as agents preserving national law in opposition to the executive branch, often preserving its narrow, political interests. They thus engage in domestic internal power politics with the executive branch in their foreign relations. This case study demonstrating the agency exerted by Beninese bureaucrats in asymmetrical relations and negotiations with China has shown that this agency lies in the influence tactics and moves that these minority actors employ and combine, whether bypassing or exploiting their status in order to influence policy (Kaarbo 1998). Although not a strategy, in the sense of a calculated plan to achieve a goal over a long period of time, this chapter showed that these bureaucrats set up stratagems, tactics and maneuvers in order to influence negotiation outcomes. Beninese bureaucrats combine rewards-and-costs, procedural and informational strategems as modes of resistance and action. Although the outcome is more or less successful, due to the politically motivated interference of the executive branch, agency exerted by these actors influences future executive branch decision-making. This case study also shows the lack of collective agency within the government of Benin. The lack of coordination and ministerial competition do not allow for a coherent "China strategy" in the framework of infrastructure projects negotiation.

In and beyond Benin, this chapter also shows how a more specific location of agency in Africa–China relations allows for a better, more critical understanding of the underpinnings and manifestations of African agency in global politics. Further research on the exercise of agency in Africa–China negotiation could include comparative analysis among other small African states such as Togo, Burkina Faso, or Niger, by questioning the effect their different political regimes may have on the stratagems and tactics of different bureaucrats across ministries from democratic and authoritarian regimes.

Notes

1 Decret n°2011–599 September 12, 2011; "40 years of cooperation – achievements of China in Benin," Diplomat, n°2, special edition, December 2012 (French version – local access).
2 "40 years of cooperation – achievements of China in Benin," op. cit.
3 "Boni Yayi décidé à réaliser les barrages d'Adjaralla et de Kétou Dogo-Bis," La Nation, June 19, 2015, http://lanationbenin.info/ (accessed June 21, 2015).
4 "40 years of cooperation – achievements of China in Benin," op. cit.
5 Benin Poverty Reduction Strategy Paper (PRSP/SCRP 2011); "Orientations Stratégiques de Développement du Bénin (2006–2011) – le Bénin emergent." Both documents are accessible on http://eeas.europa.eu/delegations/benin/documents/ (accessed July 17, 2015).
6 Benin Poverty Reduction Strategy Paper (PRSP/SCRP 2011).
7 "40 years of cooperation – achievements of China in Benin," op. cit.
8 Information Office of the State Council, "China's Foreign Aid," Beijing, July 2014.
9 African Development Bank (2003), Benin Structural Adjustment Programs I, II, III – Programme Performance Evaluation Report (PPER).
10 In comparison, under former president Mathieu Kerekou, negotiations were conducted through inter-bureaucratic coordination among several ministries. This inter-coordination, which requires a longer process, is less feasible nowadays, according to interviewees, because of the "speed" requirements by President Boni.
11 China Railways No. 5 Engineering Group Co Ltd (CREG 5) successively won the road projects N'Dali-Nikki-Chicandou; Bodjécali – Madécali-Illoua, and Godomey-Pahou.
12 Press Conference of SYNTRA-TTP on scheming procedures in market attribution of the Ministry of Public Works, SYNTRA TTP, Cotonou, August 13, 2014.
13 In September 2015, Jacques Ayadji was elected secretary general of SYNTRA-TTP.
14 Press Conference of SYNTRA-TTP at the National Assembly, October 8, 2013.
15 The Beninese presidency adopted a legal framework in 2005 for using local materials in public construction works. The Ministry of Habitat succeeded in having 25 percent of materials used.

References

Alden, Christopher and Aran Ammon. 2011. *Foreign policy analysis: new approaches.* Routledge.

Allison, Graham. 2008. "The Cuban missile crisis." In *Foreign policy: theories, actors, cases,* edited by Steve Smith, Amela Hadfield, and Tim Dunne, 256–283. Oxford: Oxford University Press.

Allison, Graham and Mark Halperin. 1972. "Bureaucratic politics: a paradigm and some policy implications." *World Politics* 24: 40–79.

Alves, Ana Cristina. 2011. "Chinese economic and trade cooperation zones in Africa: the case of Mauritius." Johannesburg: SAIIA Occasional Paper 74: South African Institute of International Affairs.

Art, Robert. 1973. "Bureaucratic politics and American foreign policy: a critique." *Policy Sciences* 4: 468–469.

Brown, William and Sophie Harman. eds. 2013. *African agency in international politics.* London: Routledge.

Cason, Jeffrey and Timothy Power. 2009. "Presidentialization, pluralization and the roll back of Itamaraty: explaining change in Brazilian foreign policy in the Cardoso-Lula Era." *International Political Science Review* 30(2): 117–140.

Corkin, Lucy. 2013. *Uncovering African agency: Angola's management of China's credit lines.* Farnham, VA: Ashgate.

Dyson, Stephen Benedict. 1986. "Personality and foreign policy: Tony Blair's Iraq decisions." *Foreign Policy Analysis* 2(3): 289–306.

Emirbayer, Mustafa and Ann Mische. 1998. "What is agency?" *American Journal of Sociology* 103(4): 962–1023.

Fraser, Alastair and Lindsay Whitfield. 2008. "The politics of aid: African strategies for dealing with donors." *GEG Working Paper* 2008/42, Oxford University.

Gadzala, Aleksandraa. ed. 2015. *Africa and China: how Africans and their governments are shaping relations with China.* London: Rowman and Littlefield.

Halperin, Mark. 1972. "The decision to deploy the ABM: bureaucratic and domestic politics in the Johnson administration." *World Politics* 25: 62–95.

Hill, Christopher. 2003. *The changing politics of foreign policy.* Basingstoke: Palgrave.

Houndeffo, Sèdèhou Carlos G. 2013. *Fiscalité des entreprises et des particuliers.* Cotonou: Editions Jurifiscom.

Igue, John. 1992. *L'Etat entrepôt au Bénin, commerce informel ou solution à la crise?* Paris: Editions Karthala.

Jessop, Bob. 1990. "Putting states in their place." In *State theory: putting capitalist states in their place,* edited by Bob Jessop, 338–369. Cambridge: Polity.

Kaarbo, Juliet. 1998. "Power politics in foreign policy: the influence of bureaucratic minorities." *European Journal of International Relations* 4(1): 67–97.

Lindell, Ulf and Stefan Persson. 1986. "The paradox of weak state power: a research and literature overview." *Cooperation and Conflict* 21: 79–97.

Maoz, Zeev. 1990. "Framing the national interest." *World Politics* 42: 77–110.

Mohan, Giles and Ben Lampert. 2012. "Negotiating China: reinserting African agency into China-Africa relations." *African Affairs* 112(446): 92–110.

Neustadt, Richard. 1960. *Presidential power: the politics of leadership.* New York: Wiley.

Prizzon, Annalisa and Andrew Rogerson. 2013. "The age of choice: Ethiopia in the new aid landscape." Research report. London: Overseas Development Institute.

Putnam, Robert D. 1988. "Diplomacy and domestic politics: the logics of two-level games." *International Organization* 42(3): 427–460.

Siko, John. 2014. *Inside South Africa's foreign policy: diplomacy in Africa from Smuts to Mbeki.* London: IB Tauris.

Sun, Yun. 2014. "Africa in China's new foreign aid white paper." Brookings Africa in Focus blog entry, July 16, 2014 www.brookings.edu/blog/africa-in-focus/2014/07/16/africa-in -chinas-new-foreign-aid-white-paper/ (accessed on January 5, 2016).

Topanou, Victor. 2013. *Introduction à la sociologie politique du Bénin.* Paris: Editions l'Harmattan.

Van Bracht, Gérard. 2012. "A survey of Zambian views on Chinese people and their involvement in Zambia." *African East-Asian Affairs* 1: 54–97.

Wang, Duanyong and Zhao, Pei. 2015. "Mismatching structures: a new explanation for the "unsatisfactory" labour conditions in Chinese mining companies in the Democratic Republic of Congo", paper presented at the South-South Labor Conference, Cornell University, October 9–11, 2015.

Wight, Colin. 2004. "State agency: social action without human activity." *Review of International Studies* 30(2): 269–280.

Wight, Colin. 1999. "They shoot dead horses don't they? Locating agency in the agent-structure problematique." *European Journal of International Relations* 5(1): 109–142.

13

DEPENDENCY AND UNDERDEVELOPMENT

The case of the Special Economic Zone in Mauritius

Honita Cowaloosur and Ian Taylor

At the Forum on China–Africa Cooperation (FOCAC) in 2006, President Hu Jintao committed to "[e]stablish three to five trade and economic cooperation zones in Africa in the next three years" (Hu, 2006) as a means to further cooperation between China and Africa.[1] Seven Special Economic Zones were launched across five African countries,[2] one being the Mauritius JinFei Economic Trade and Cooperation Zone Co. Ltd (JFET).[3] The pilot dissemination of these zones by China is a pertinent subject of study as the latter adapts a practice deployed by developed countries in their quest for devolution of their traditional industries to lesser-developed countries, that of Export Processing Zones (EPZs). Through this, developed countries had aimed to forge on their national advancement agenda by prioritising innovation and production of goods of higher value at home. As China adopts this model and establishes its own spatial extensions in Africa, it reverberates dimensions of Dependency Theory.

Founded on the basic understanding that the development of one means the automatic underdevelopment of a hierarchically inferior 'Other', Dependency Theory is being revamped and revived by the Chinese Special Economic Zones in Africa (CSEZA) project. This Chinese initiative introduces new modalities, actors and instruments to the processing and delivery of the same conventional end envisaged by Andre Gunder Frank in 1967; that is, the realisation of a dialectic distribution of development and underdevelopment. Frank argued that underdevelopment is a result of the relationship of colonised territories with the colonisers, that development and underdevelopment are the two aspects of the same system, and that historical accounts of the relationship of dependency of the less developed countries on the rich European ones explains the formers' relative poverty.

Frank was a Dependency Theorist whose line of thought was inspired by the 1959 Cuban Revolution and Lenin's theory of imperialism. During his teaching job in Latin American universities in the 1960s, Frank witnessed expressions of nationalism such as Fidel Castro's and Che Guevara's Committees for the Defence of the Revolution and the Ministry for the Recovering of Misappropriated Assets, which worked towards the nationalisation of land, businesses and properties owned by the Cuban upper classes and foreigners. At the time, Cuba had decided to eliminate all parasitic connections between the developed and underdeveloped countries. Guevara (1964) warned LDCs about this destructive relationship:

> [T]here is the inherent contradiction between the various developed capitalist countries, which struggle unceasingly among themselves to divide up the world and to gain a firm hold on its markets so that they may enjoy an extensive development based, unfortunately, on the hunger and exploitation of the dependent world [...] If at this egalitarian conference, where all nations can express, through their votes the hopes of their peoples, a solution satisfactory to the majority can be reached, a unique step will have been taken in the history of the world. However, there are many forces at work to prevent this from happening. The responsibility for the decisions to be taken devolves upon the representatives of the underdeveloped peoples. If all the peoples who live under precarious economic conditions, and who depend on foreign powers for some vital aspects of their economy and for their economic and social structure, are capable of resisting the temptations, offered coldly although in the heat of the moment, and impose a new type of relationship here, mankind will have taken a step forward.

Lenin's theory of imperialism reinforced Frank's ideas (Lenin, 1961). Lenin identified a relationship of exploitation between the few capitalist oligarchies made up of European states, corporations, peripheral elites, and the poorer states and their general population. He postulated that capitalists manipulate the state into acting as an instrument of exploitation of the periphery and ascertain the domination of peripheral elites, European capitalist states and corporations. The state encourages producers of raw materials to produce and sell their goods to the core, at a cheap rate. The state also hinders the development of indigenous industries in order to keep the periphery dependent on finance from the core. The core ascertains that the peripheral elite class is sustained so that it can continue gathering export income from the luxury goods sold to the peripheral elites. As a result, the wealth of the periphery flows to enrich the core and generates imperialism. Armed with a concrete Latin American illustration of the divergent development dichotomy inherent within capitalism, Frank developed an ideological framework to translate his experiences into a series of causal relationships.

Through the prism of CSEZAs, this chapter assesses the new modalities, actors and instruments of dependency that dispense the development and underdevelopment confirmed by Frank. Categorised broadly under geographic and economic practices, today these new components are often taken at face-value and for granted. This chapter deconstructs these instruments, showing that they are in fact highly value-laden and stimulate continued dependency. In so doing, it negotiates the distance between the rhetoric of diplomacy exercised through geographic and economic instruments, and the ground realities of the resultant development, or underdevelopment.

This chapter starts with a discussion of traditional Dependency Theory and underlines how its core concepts of development and underdevelopment are still relevant. It uses the new CSEZAs to evaluate the roles of China and its host African countries in metropolis and satellite terms. As it establishes the state-level relationships of expropriation and subordination between the two, this chapter introduces the CSEZAs, the touted messiah of development, in this interaction and observes the changes, or reinforcement, that it brings to the hierarchical connection between China and Africa. The instrumentality of assumed *de facto* carriers of development used under the CSEZs namely, Foreign Direct Investment (FDI), fiscal methods such as Special Purpose Vehicles (SPVs) and risk-mitigating and facilitating spaces such as CSEZAs, are deliberated. A visualisation of the above evaluation is provided through reference to the case of the Chinese SEZ in Mauritius. The chapter ends by reflecting on the merits of the basic guiding logic of Dependency Theory and, moreover, on the need to update its modalities for the purpose of capturing the nuances of normalised instruments and practices of development that, in fact, do not necessarily deliver what they promise. CSEZA stakeholders are therefore warned of the practices that need to be clarified and revised if these are to deliver development to African hosts.

Traditional dependency

Frank's version of Dependency Theory (see Figure 13.1) sees capitalism as having created a singular network of development consisting of two levels: the metropolis and satellites. The relationship between the two was characterised by exploitation and subordination. The singular network in which these two levels repeatedly occur allows the metropolis to connect to the furthest satellite in a single stretch. In this liaison of dependency, its immediate metropolitan superior claims the economic surplus generated by the subordinate satellite. Such interaction between the metropolis and satellite thus carries out development and underdevelopment simultaneously, with the former developing at the expense of the latter. Frank does not only grant the status of metropolis and satellite to nations. This categorisation is valid within nations as well, where similar patterns of domination, subordination and surplus appropriation prevail among the different classes and communities (Frank 1968).

Contemporary relevance of dependency theory

Many argue that in a global world, defined by technological, political, economic, social and environmental connectedness in which instruments of power like capital are fluid, Dependency Theory is out-dated in dividing the world between metropolises and satellites. However, even if Dependency Theory is unfashionable,

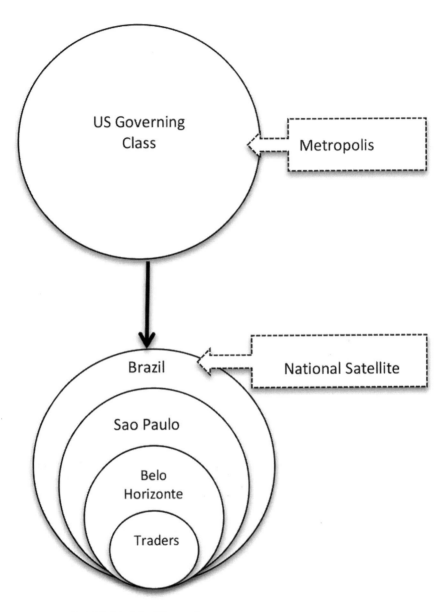

FIGURE 13.1 Frank's Model of Dependency
Source: Based on Frank 1967

conditions of dependency on which it rested still pervade the capitalist global economy. Today, neoliberalism has assumed the position of a function of dependency; dependency is the reality that is being acted upon by neoliberalism. Neoliberalism has certain prerequisites in order to enable growth and development, namely privatisation of state enterprises, tax reforms, trade liberalisation, subsidies cut, and deregulation. Though implemented at the domestic level, these arrangements are mere contextual conditioning in order to fully integrate the nation and push it to participate actively in global trade transactions. The belief in the universality of this approach to development is evident through its endorsement and enforcement by the regulatory pillars of contemporary world affairs, the Bretton Woods institutions. Nonetheless, in this process, countries and entities get further hidebound by externally occurring political and economic activities. It would, therefore, not be wrong to surmise that dependency is now institutionalised in the form of the World Trade Organisation (WTO), International Monetary Fund (IMF), Structural Adjustment Programmes (SAPs), the debt allocation/repayment system and the entire idea of free trade (Ahiakpor 1985, 537; Strange 1998). Ferraro and Rosser's (1994) discussion about how the IMF assumed a new position as a guarantor of debt repayment from developing countries, represented through its adoption of the Brady Plan,[4] rather than prioritising its original responsibility of aiding development in the periphery, reveals how the IMF has entrenched developing countries' dependence on international organisations and developed nations.

Nevertheless, as underdevelopment persists in Least Developed Countries (LDCs), neoliberals argue that the obstacles to development emerge from the home governments' incorrect implementation of the advised pre-conditions. They project underdeveloped states as obstinate and reluctant to let go of protectionism. In the neoliberal revision of its development strategies in 1989, the *African Alternative Framework to Structural Adjustment Programmes for Socio-Economic Recovery and Transformation* of the United Nations Economic Commission (UNEC) emphasised the need for a transformation of the "economic, social and political structures in Africa that hamper development [...] the various forms of inequalities inherent in African societies and [...] the lack of democratic political structures ..." (Owusu 2003, 1659). Therefore, by advocating for greater openness to private, often external forces, in order to further development, and by curbing the autonomous role of the state, neoliberalism reaffirms domination by external forces. Haynes (2008, 72) asserts that Dependency Theory still reigns through neoliberalism mainly also because the peripheral elites accept the logic of neoliberalism. Strange argued that there is no withdrawal route from the capitalist liberal market system, especially for developing states, which need the connection to the core, in terms of technology and markets, in order to facilitate development (1998, 23). As Leander puts it: "although openness may not bring development, foregoing openness amounts to foregoing development" (2001, 116). It appears that Chinese SEZs are being assimilated in Africa in a similar spirit.

China: metropolis or satellite?

According to Frank (1968, 70), world metropolises are those nations (or sectors) that were historically never underdeveloped, although they might have been undeveloped at a certain point. These are countries and entities of the world system that experienced independent development in pre-colonial times. As per this criterion, China is not a world metropolis. Starting in 1842, China was subject to impositions from the British Empire, Japan, and other European actors. There are many historical examples confirming China's non–world metropolis status: having to forcibly open its ports to foreign settlers and trade; the loss of territories such as Hong Kong, Korea and Taiwan; and eventually, the adoption of a complete Open Door policy imposed by the US. These are clear instances of the exploitation and extraction of surplus that China had to undergo as a satellite to the already developed world metropolises. Frank (1978) dismissed the position of China as a world metropolis in his response to Lippit's *The Development of Under-development in China* (1978). Lippit ascribed the responsibility of China's under-development to China itself and discards the possibility that the imperialist behaviour of colonial nations had influenced this situation. In response, Frank stated:

> After 1636 [...] and until our days [1978], China has not been at the center of world economic, political, and cultural development and has occupied a sin-gular intermediary role between the Western powers on the one side and their Asian colonies and junior partners on the other [...].

Even in the classical imperialist period of the nineteenth and twentieth centuries China's semicolonial economic position in the world was peculiar and almost unique among those then suffering the development of underdevelopment. Unlike other such "Third World" countries, which had a consistent merchandise export surplus, China had a merchandise import surplus with the developed countries of Europe, derived from its entreport trade position between the colonial powers and their colonies elsewhere in Asia (Frank 1978, 348–349).

It is important to note here that the allusion to an intermediary role does not mean that China is a semi-peripheral country as per Wallerstein's definition (1976). Recent conflicting discourses about the rise or limitations of China, and whether the role of Western powers will therefore be modified, suggest that the Chinese global situation is definitely more elusive than being fixed at the intermediary role. Hence, China's present reality rallies more with Frank's suppositions as it fits into a shifting middle role, whereby it adapts the twin role of metropolis and satellite depending on whom it is interacting with, through which medium, and possibly for which end. China is a metropolis as it outsources its manufacturing; expro-priates surplus economic value from those host countries; invests in foreign desti-nations (especially in countries with which it has Double Tax Avoidance Agreement [DTAA] treaties);[5] provides technical developmental assistance in the form of infrastructure; and eventually establishes Chinese SEZs abroad.

Nevertheless, it remains a satellite in relation to the world metropolises and the institutions endorsed by them, notably the Bretton Woods organisations, and it is unlikely that new innovations such as the Asian Infrastructure Investment Bank will radically alter this situation. China's realisation in 1979 that to fund its Four Modernisations program, it needed to implement an Open Door policy and attract FDI, established that given its current level of development, it could only be a satellite to the world metropolises. The second indication towards China's position as a satellite came in the form of its WTO membership in 2001. Wang (2000) discussed the many ways in which China was set to reinforce its subordination and underdevelopment by joining the WTO. After reviewing the immediate effect of China's membership into the WTO, Shaun Breslin (2013: 629) stated: 'A strong argument can … be made that China has been one of the main national bene-ficiaries of the spread of a post-Fordist form of neoliberal globalisation that is typically associated with Western and capitalist power and interests. And this is not just a "passive" acceptance of existing norms and processes – China was not drag-ged into WTO membership against its will.' Those who argue that China offers an alternative economic model may wish to explain then how China differs from the West, other than possessing a somewhat stronger residual role for the state:

> In contrast to the pre-reform period, almost all economic activity is now market determined. And, while the state continues to dominate in many stra-tegic sectors, such as finance, energy, and transportation, the great majority of value added in the all-important manufacturing sector is now produced by profit-seeking, private firms.
>
> *(Hart-Landsberg, 2010: 17)*

As China continues to adjust to its WTO membership, it constantly feels the pressure of the world metropoles and is pushed to compromise upon its ideological principle of non-interference, for example, in order to play by the rules of the established world order.

Africa: the ultimate satellite

There is a widespread assertion that irrespective of their richness in natural resources, African countries are all essentially satellites to most of the world's structures. Starting from histories of colonisation to more recent discourses of neo-colonialism, African countries have been exploited by world metropolises and its surplus eco-nomic value extracted, be it by European powers, the US, India, or China. In an adaptation of Frank's Dependency Theory and the concept of underdevelopment to the African context, Rodney (1973) wrote that:

> African economies are integrated into the very structure of the developed capitalist economies; and they are integrated in a manner that is unfavourable to Africa and ensures that Africa is dependent on the big capitalist countries.
>
> *(1973, 43)*

A general consensus among development theorists makes it undeniable that African countries are national satellites. As a Sub-Saharan African country, Mauritius is also a satellite within the global capitalist system, more so since it is entirely economically dependent on externally sourced capital and investment. However, as per Frank's ideas, although an African country is a satellite, in comparison to its internal subordinate sections whose surplus value it extracts, the African state is a domestic metropolis.

The Chinese SEZ in Mauritius

Mauritius has been adjudged the most equitably governed and constitutionally democratic of the five CSEZA host states; the one scoring highest in Africa on the World Bank Doing Business Index 2017 and the Heritage Foundation Index on Economic Freedom 2017; and the most politically stable and democratic countries in Africa, according to the Mo Ibrahim Index of Governance 2017 and the Democratic Index 2016. Though equipped with the most favourable business environment of the Chinese SEZs across Africa, the Mauritian Chinese SEZ has failed to a greater degree. Since its introduction, the Mauritian Chinese SEZ has been marked by a variety of difficulties, each of a different nature and whose respective responsibility is attributable to a distinct actor or practice.

Initially called the Mauritius Tianli Economic and Trade Cooperation Zone Co. Ltd, the company was incorporated in Mauritius on 4 May 2007. Shanxi Tianli Enterprise Group was the sole investor. Mauritius Tianli Economic and Trade Cooperation Zone Co. Ltd leased 51 hectares of land at Riche Terre for a 99-year term. To this end, 120 farmers were evacuated from the plot. However, the company closed down in 2009 and the lease agreement was nullified. Shortly afterwards, on 13 August 2009, the company reopened under a new name, the Mauritius JinFei Economic and Trade Cooperation Zone Co. Ltd. (JFET) as a private, domestic company, limited by shares. At a parliamentary session on 20 October 2009, the Mauritian government confirmed that JFET shareholders were Taiyuan Iron and Steel (Group) Co. Ltd (TISCO) with 50 per cent of the shares; Shanxi Coking Coal Group Co. Ltd holding 30.2 per cent; and Shanxi Tianli Enterprise Group with 19.8 per cent of the shares. These companies' activities range from the import of coke, to coal operations, hotel management, construction, agriculture and financial services (Shanxi Tianli Enterprise Group Corporation, 2010). These three shareholding companies formed a China-based consortium called the Shanxi JinFei Investment Co. Ltd through which they directed capital into JFET.

The zone is in Riche-Terre village, 5 km from the harbour and the capital city, Port-Louis. The arrangement between Mauritius and China concerning the zone is covered by a Framework Agreement and a Land Lease Contract. The latter contract's exit clauses state that if the lessee fails to abide by the construction schedule specified in the lease agreement, the lessor may redeem the land. The construction schedule comprised phase 1 which extended from September 2009 to September 2012, and phase 2, which began in September 2010.

By the end of 2014, the SEZ remained largely undeveloped with the exception of the following infrastructure: the Noah Wealth Centre, whose activity is unknown; two blocks of flats accommodating Chinese resident workers, and partly available for rental to outside parties; one building housing Goldox Construction Ltd and Goldox Travel and Tours Ltd; and one building acting as the JinFei

Warehouse. Therefore, phase 1 of JFET, which was supposed to end in 2012 and make way for phase 2, was never completed.

After a change of government at the end of 2014, the new Mauritian government reviewed the JFET project in 2015. In April 2015, it took back the 211 hectares initially leased to Shanxi Jinfei Investment Co. Ltd and JFET and granted them a new lease for only 30 hectares covering the land under which the Chinese developers had already put up buildings. With this, the Chinese SEZ chapter in Mauritius closed only to be reopened later under another concept, the Smart City. The new lease of 30 hectares for JFET was followed by another agreement between the same Chinese investing parties and the government of Mauritius. The second lease agreement gave an additional 40.4 hectares to the Shanxi JinFei Investment Co. Ltd for the development of a Smart City project. The Smart City is yet another form of an integrated zone project that focuses on the use of renewable energy and thrives on ambitions of sustainability. According to officers at the Board of Investment, Mauritius (interviewed in November 2015), the administration and management of each Smart City would be the sole and exclusive responsibility of the developers. The management board of these units would not include any representative of the national or local government, or from civil society.

Evaluation of JFET's development elements

As an African island with no natural resources, and a skewed reliance on its human capital and geographic position, the national objectives attributable to Mauritius are made up of standard elements generally endorsed by economies that are strongly dependent on external relations. Some of these national objectives are the creation of employment, dissemination of skills, technology, know-how, research and development, backward linkages to local manufacturers, increase in foreign exchange income, among others. Yet, it is not clear at all how the JFET adds to the developmental project of Mauritius.

First, the initial stated areas of investment of JFET and its existing activities overlap with the high income generating industries of Mauritius (see Table 13.1).

The presence of Goldox Travel and Tours Ltd and Goldox Construction also adds to competition in two particular local sectors. Thus, all that JFET does is introduce strenuous competition over markets in Mauritius. With similar activities taking place in both the Chinese SEZ and the domestic trade area, there will be no exchanges of know-how, skills and technology, allegedly a key aim of the SEZ.

Another key issue is the number of jobs the zone is supposed to create for the local community. Since its inception, the number of prospective jobs announced has been incongruent with projections. In 2008, Mauritian authorities estimated the creation of 7,500 jobs, and in 2010, the number rose to 35,000, only to be drastically lowered by a figure of 5,000 jobs provided by the Chief Executive Officer of the project (Mauritius Parliament 2010, 22 June; Barbé, 2010). Due to the controversial nature of the project, JFET stakeholders refuse to provide information on the status of activities. However, judging by existing activities on the

TABLE 13.1 Overlapping Activities of Mauritian Domestic Trade Area and JFET

Main industries contributing to GDP growth in Mauritius	Areas of investment in JFET
Manufacturing (garment, processed fish, beverages, watches, clocks, toys, optical goods, jewellery, travel goods, handbags, textile yarns, fabrics and made up articles, pearls, semi/precious stones, wood)	Garment manufacture/Souvenir manufacturing[6]/Food processing
Real estate and business	International conference centre/Staff dormitories
Hotels and restaurants	Hotels
Wholesale and retail trade: repair of motor vehicles, motorcycles, personal and household goods	Electric home appliances/Light engineering/Wholesale and retail/Shopping centres
Financial intermediation	Financial services
Other services	
Transport and communications	Information and communication technology
Construction	
Health and social work	State-of-art medical centre/Pharmaceuticals
Education	Boarding school
Public Administration	
Electricity, gas and water supply	

site on which there are only three functional buildings, the number of locals working on the site is unlikely to have been more than 30. During a visit in May 2016, the only Mauritians encountered on the site were security guards. While at various times, the Chinese government and zone promoters have denied their intention to bring Chinese labour to JFET, the inclusion of a staff dormitory in the zone indicates that the zone in fact houses Chinese labour.[7]

The awareness that the Chinese SEZ in Mauritius hardly has any positive contribution to the development of the island pushes us to question the larger repercussions of this malfunctioning model of development. How do Mauritius and China, through the JFET, entrench dependency relations and what general applicability does this involve if replicated elsewhere – either as a whole or in parts?

A traditional dependency reading of JFET

A true representation of the equation between China and Mauritius in respect of the Chinese SEZ established at Riche-Terre, adapting Frank's principles to the context, would assume the following form, on two levels. Level 1: China assumes the role of the metropolis vis-à-vis Mauritius, the satellite (see Figure 13.2).[8] Level 2: the Mauritian state, together with China (represented by the Shanxi SOE stakeholders) combine to become the metropolis of JFET, and through JFET, also of

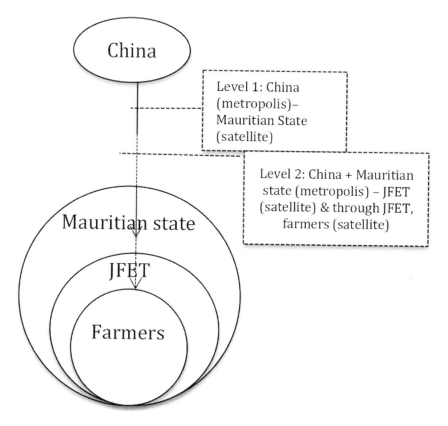

FIGURE 13.2 Expected Structure of Relation of China–Mauritius–JFET as per traditional Dependency Theory

the farmers who were evacuated to make place for the zone. They will derive developmental value and profit out of the activities of the zone, at the expense of the national periphery (the Mauritian people).

Thus, China and the Mauritian state, through JFET, exploit the farmers as its metropolises. JFET, then, symbolically stands as a metropolis to the farmers. The evacuated farmers form the ultimate satellite of this equation. Under this traditional Dependency Theory interpretation, the schema of economic exploitation and subordination is direct. But in this case, the actors involved in the Chinese SEZ in Mauritius do not adhere to straightforward traditional dependency tendencies. Instead, their deployment of economic transactions and the geographical structure of the zone modify the percolation of hierarchy across the different levels involved.

The actual dependency exercise in JFET

The deployment of JFET, however, demarcates itself from traditional dependency readings. As per traditional Dependency Theory structures, the schema of economic exploitation and subordination is usually simple and direct. But in this case, JFET actors form two sets of overlapping dependency relations: one structurally indirect actor cuts across intermediate levels to reach below and intervene directly to exploit a subordinate level, which is already being exploited by its own immediate superior metropolis. Eventually, there are strata that face double or even triple-sourced dependencies. This process is, however, marked by changing roles assumed by each actor involved:

1. *China*: Represented by its SOEs from Shanxi, China is the patron and developer of the zone. The Chinese companies are 100 per cent shareholders of JFET, and are supposed to fund all on-site infrastructure. The Chinese developers are responsible for recruiting investors for the zone, and handling the administration and development of the zone affairs. Conferred Mauritian status, the JFET company functions as a private entity; it is not accountable to the host Mauritian government. As part of the DTAA agreement Mauritius has with China, the stakeholders managing JFET are allowed to repatriate their capital, dividends and profit to China, tax-free. Some Chinese developers have even been given Mauritian citizenship, allowing them to carry on with their activities independently of Mauritian state monitoring of foreign investors. As soon as the Chinese SEZ venture transitioned from being a project plan into JFET the company, China no longer needed to operate through the local authority of the Mauritian government. It could connect to the zone by bypassing the Mauritian state altogether.

2. *Mauritius*: The Mauritian government has no equity stake in JFET. The Mauritian state holds no role in the administration of zone affairs. Its responsibilities are to provide land, off-site infrastructure, passports to the investors (hence, nationality), and rights of local incorporation to investing parties as companies. Investors and companies naturalised as Mauritian have the right to function unhindered. In short, the Mauritian government has completely removed itself from any potential direct involvement in JFET. Its act of exploitation of its own internal sub-national entity, the Chinese SEZ, prevailed only up until the site was being cleared and the farmers were displaced, after which the role of the Mauritian state as a metropolis to that space, thereafter JFET, the company, ended. It would be fair to describe the intervention of the Mauritian state in JFET, in the former's capacity as a metropolis, as being limited only until the planning and site preparation period.

3. *JFET*: Before becoming JFET, the zone area was a plot of land in Mauritian territory occupied by local farmers. At the point when it was a mere plot of land chosen as the site to build the Chinese SEZ, the Riche-Terre plot was a space of the Mauritian government, which exercised its authority to terminate cultivation on the land and evacuated the occupant farmers. As soon as the land was written off to the Chinese developers in the form of an asset of the JFET company, however, the equation changed. As it gained value in its prospective ability to generate income, the plot switched from being a satellite to a

metropolis of the Mauritian state. This is because Mauritius was depending upon high returns and FDI inflow in that zone in order to fund domestic development, and also to garner domestic political goodwill. Nevertheless, for China, JFET is a satellite, mostly due to its overseas location, through which China will be able to extract surplus value at the expense of the Mauritian state.

4. *Farmers*: Regardless of the phase through which the Chinese SEZ in Mauritius was developed, the farmers of Riche-Terre maintained a constant position as the ultimate satellite. As they were evacuated from their land in order to accommodate the Chinese SEZ developers, they were at the mercy of the Mauritian state, which assumed the position of their metropolis. The fact that the Chinese developers themselves chose the site and replaced the farmers also makes JFET the metropolis of the displaced community of farmers. However, contrary to anticipation, China is able to efface itself from displaying direct exploitative links with the farmers. China's direct positioning as a metropolis of the farmers is avoided by the fact that China contractually delegated the responsibility of all off-site infrastructure provision to the Mauritian state, and chooses to exercise its role in relation to JFET only in terms of its status as JFET developer, and not as the bilateral government partner of the Chinese SEZ project in Mauritius.

In terms of the roles defined above concerning JFET, the relationship of dependency between China and Mauritius went through two distinct phases. Phase One covered the period of preparation for the Chinese SEZ project, and the furnishing of all the support amenities that would aid the zone. During this phase, the zone was only a plot of land. Phase Two started when all off-site infrastructure was laid down for the zone and it remained for the Chinese developers to begin on-site construction. At this point, the zone gained the status of an asset transferred to the name of JFET, the private company. Figures 13.3 and 13.4 show how the relationship amongst the relevant entities changed as the Chinese SEZ project unfolded over these two phases. The divergences in the way these dependent relationships are exercised in JFET are shaped by the involvement of particular components that escaped Frank's consideration. These are the intervention of a geographic structure, the presence of FDI, and use of SPVs.

Frank lists a number of global dynamics through which metropolis–satellite relations exist. His list comprises economic, political and social structures. He also refers to local rural–urban hierarchies that, nonetheless, remain economically and socially laden. One dynamic through which dependency relations are performed in the case of JFET that Frank does *not* take into consideration is the 'geographic' structure, which also mediates relationships of exploitation and subordination. The geographic structure is a relevant carrier of a metropolis–satellite relationship. This is clearly demonstrated by JFET. As the property of the Mauritian state, and lying within its national boundaries, the logical position of the Chinese zone vis-à-vis the Mauritian state should have been one of subordination. But this is not the case. As it exists in the shape of a bounded zone, JFET becomes an independent geographic entity and

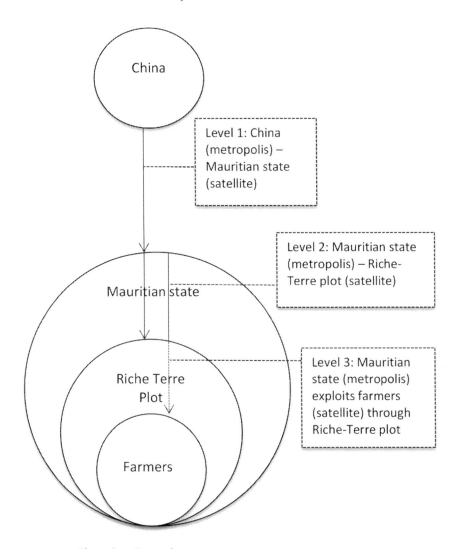

Level 1: China
(metropolis) –
Mauritian state
(satellite)

Level 2: Mauritian state
(metropolis) – Riche-
Terre plot (satellite)

Level 3: Mauritian
state (metropolis)
exploits farmers
(satellite) through
Riche-Terre plot

China

Mauritian state

Riche Terre
Plot

Farmers

FIGURE 13.3 Phase One Dependency

surmounts the only bond of subordination it could have shared with the Mauritian state. Taking leverage of this geographic separatism, JFET is able to compete with the Mauritian state on an economic basis. It even makes the latter its satellite.

Frank did not fully consider what an upgrade of financial instruments within the capitalist system might mean for the alternating, yet overlapping, metropolis/development and satellite/underdevelopment dichotomies. Touted as one of the major sources of capital in the capitalist system, there is a high degree of competition among states to attract FDI. Not only does FDI mean capital inflow, but it is also a source of immediate development through backward linkages such as local job creation, infrastructure, or export income. Hence, in order to secure FDI, countries are forced to provide incentives and preferences. These preferences are

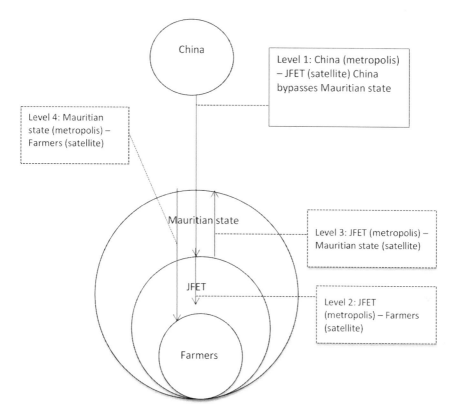

FIGURE 13.4 Phase Two Dependency

usually to the detriment of their own economic objectives, and at the expense of local producers and businesses. Consequently, by welcoming FDI, economies within the capitalist system willingly subscribe themselves to the status of satellites. They even go to the extent of under-developing themselves and their subordinate local entities in order to serve the FDI-exporting metropolis whom they perceive as a messiah of development. The presence of FDI in the equation between China and Mauritius, in this case, falsifies the proposition of Frank that exploitation of a satellite can only end if the satellite distances itself from the metropolis. After the establishment of JFET, the distance between China and Mauritius was set; there is no direct interaction between the two. Contrary to Frank's idea, this does not entail the termination of Mauritius's underdevelopment. Frank does not account for the possibility that though it acts distantly, through appropriately structured FDI, the metropolis can create a local metropolis within the target national satellite.

It is not only JFET's geographical separatism within Mauritian boundaries that allows its patron China to skip the Mauritian state in its dealings with the zone. This elusive transaction is further aided by the fact that JFET, set up as an SPV company under Mauritian jurisdiction, also has the power to remain economically detached from the authority of the Mauritian state. The Chinese SEZ in Mauritius involves double SPVs. Shanxi JinFei Investment Co. Ltd is the first. Created in China, it is composed of Shanxi Tianli Enterprise Group, Shanxi Coking Coal Group Co. Ltd and TISCO and has a fund of USD $80 million for investment in the zone project in Mauritius. JFET is the second SPV, set up under Mauritian jurisdiction. Its shareholding company is Shanxi JinFei Investment Co. Ltd and it has a total fund of USD $10 million. Instead of direct investment, the three ultimate share-holders, Shanxi Tianli Enterprise Group, Shanxi Coking Coal Co. Ltd and TISCO, thus built two protective layers of financial instruments in between themselves and the Chinese SEZ in Mauritius. As it gets bypassed during in-coming and out-going dealings of China with JFET, the status of Mauritius as a national satellite is rein-forced. It witnesses its underdevelopment at the expense of the development of Chinese entities settled on its own land and using its own assets.

Rather than alienating the concept of CSEZAs from Dependency Theory, this analysis only strengthens the links between this new phenomenon and the con-ventional theoretical perspective. Albeit through convoluted routes, this capitalist expression of a nominally socialist state loyally carries out the development and underdevelopment of China and Mauritius respectively. What are even more remarkable in this exercise of dependency are the strategies through which China directs Mauritian entities to carry out their own underdevelopment.

Conclusion

After an acquisition of spatial and economic autonomy in Mauritius in the form of JFET, and through its capacity to act independently as a foreign private company, China bypasses the Mauritian government and enforces JFET in a direct expression of dependency. As a result of being ignored by China on the way in and out of Mauritian jurisdiction, while interacting with JFET and its more profitable adjacent businesses, the Mauritian government, local economy and the public end up as dependencies of, initially 211 hectares and now 70.4 hectares of land, relegated to the administration of China. This reality cannot be hidden by rhetoric of 'mutual benefit', 'win-win situations' or 'cooperation'.

Scholarship on China's relations with Africa tend to treat CSEZAs as empirical events. Since their inception, these zones have been subject to so many con-troversies that most researchers have had to focus on unravelling the substance of these accusations. By collating data indicating exploitative or developmental features of CSEZAs, scholars have attempted to find an answer to the recurrent question about the nature of China's role in Africa. Will these CSEZAs confirm that China exploits Africa, or do, indeed, these propose an alternative route to economic development for the continent? The absence of a theoretical prism in studies of CSEZAs has resulted in a dominant reading of the CSEZA as insular entities. Their relevance to the global arena, and to other actors who do not directly partake in the CSEZA composition, has not been assessed. This deficiency resulted in the inability of scholars to hypothesise about the impact CSEZAs can have on the nature of global interactions. As they are strictly studied in their domestic context, the learning experience that other countries could derive from the innovative policy procedures and strategies employed by the CSEZAs is ignored. Most importantly, by denying the CSEZA a theoretical identity, academia has prevented these ambitious pilot development projects, which are equally endorsed by China and Africa, from rectifying its shortcomings and delivering more equitable zone-based developmental experiences. The association of Dependency Theory to the CSEZA is therefore as much as a suggestion for reform as a critique of it.

The current global capitalist structure is inherently hierarchical and characterised by numerous, different configurations of dependency. There are relations of domination and subordination between countries at the international level, between economic classes and communities within a nation. Over years, the relation of dependency between each of these configurations has been channelled through distinct processes: initially through colonisation, and subsequently through the spread of neoliberal practices. This chapter shows that the advent of the CSEZA model has brought innovation to these conventional dependency practices. It departs from traditional understandings of dependency and refines Frank's perspective to interpret what can be now defined as new instruments of dependency.

The CSEZA transgresses the structural boundaries that dictate relationships of dependency as exercised under neoliberal global capitalist structures. In a dependency situation defined solely by neoliberal structures, each segment is aware of who its metropolis and satellite are and, from that, derives an understanding of the role it should accordingly assume. In light of this knowledge, each metropolis is able to calculate how much it should take away from its subordinate satellite in order to be able to sustain itself after its own superior level has expropriated its surplus value. Thus, each strata subordinate to the world metropolis prepares adequate strategies against the probability of suffering from acute underdevelopment. The CSEZAs challenge this organised devolution of the production process, which characterises the current system through its complex configuration, formed by state, private and community actors. Dependency here is not exercised over alternating levels, as in Frank's drawings.

Surmounting the divergences from Frank's dependency premises, this chapter demonstrates how dependency does not necessarily have to transfer the surplus it has extracted from its own local satellites to the world metropolis. It can fuel the development of the world metropolis by containing it within the CSEZAs. A world metropolis can eventually subdue a national satellite into becoming a comprehensive satellite economy to a specific small space internal to its own locality. The CSEZA-based dependency experience evaluated here also accommodates new FDI and SPVs practices in the modalities of Dependency Theory. In so doing, it leads the dependency perspective away from the organised structural and procedural analysis of dependent relations that it is used to and places the focus more onto the delivery of determinants of dependency, meaning development and underdevelopment.

The unchanged positive interaction between Mauritius and China, despite the stagnating CSEZA, clarifies that CSEZAs are indeed not *the* determinant representative of Chinese engagement with Africa. Neither do the shortcomings of these zones imply that all Chinese cooperative endeavours with Africa are problematic, nor will their positive contribution promise a similar experience across other Chinese projects. What the CSEZAs are representative of, nevertheless, is how knowledge of the roles of each of the actors involved, and of the way in which a project unfolds, can provide a less extremist understanding of the role of China in Africa. The study of CSEZAs encourages Africanists to not only investigate the demeanours and actions of the Chinese in Africa but also get an insight into the responsiveness of African actors to the Chinese. When viewed through the prism of CSEZAs, Africa is as complicit as China in its own exploitation.

If the initial aim of the CSEZA project, which basked under the legitimising rhetoric of cooperation and mutual development, is compared to the dependency-entrenching experience that has been delivered by these zones since FOCAC 2006, the CSEZA's vain conceptual attempt to combine 'cooperation' and dependency can be observed. The CSEZA epitomises China's ambitious plan to harmonise several such dichotomies and create a winning formula for development, one that is more appealing to developing African countries than that offered by the proponents of purist neoliberalism. It thereby seeks to merge cooperation with dependency, socialism with neoliberal capitalism, and Chinese socialist capitalism with African particularism. Though it is a commendable effort on the part of China, and has attracted African interest, the spatial format through which China attempts to implement this project somewhat foils its efforts. As it tries to diffuse cooperation, Chinese socialist capitalism exercised through an enclosed Chinese SEZ space situated within an African country, in turn characterised by subordination, neoliberal capitalism and African particularism, the walls of the CSEZA prove to be too impermeable to allow equitable wins for all stakeholders. Worsening the divisiveness enabled through the geographic separatism of the CSEZA from its host country, China implements strategies which although enhance its ability to extract more profits, simultaneously cut-off Africa from partaking in this profitability and possibilities of development. The stance adopted by the host African countries, especially Mauritius, in the face of this treatment resulting from the geographic and economic separatism exercised through CSEZA, can easily be described as self-deprecating. The examples of Nigeria, which renegotiated the terms of the Lekki SEZ, and Ethiopia, which revised its policy framework to guarantee a better monitoring of the impact of the Eastern SEZ on its domestic economy, are overshadowed by the inactivity of the host countries in the remaining five CSEZAs. Mauritius, as representative of an extreme case, goes to the extent of facilitating its own underdevelopment by employing strategies that incessantly reproduce dependency.

Notes

1   The term 'economic and trade cooperation zones' denotes the function that these zones are meant to carry out. The technical and popular name by which they are otherwise referred to is Chinese Special Economic Zones in Africa.

2 Egypt, Ethiopia, Nigeria, Zambia and Mauritius.
3 This chapter is based on fieldwork conducted in China and Mauritius from 2011 to 2017. The main sources of data in China were the diplomatic offices of the Chinese SEZs host African states, China–Africa Development Fund, research wing of Ministry of Commerce, and other academic institutions. In Mauritius, the points of contact were the Registrar of Companies, Board of Investment and other government bodies.
4 The 1989 Brady Plan was an attempt by the US Treasury towards debt reduction, which called upon the goodwill of the creditor banks to forgive part of the debt credited to the developing countries, in return for guarantees for payment of the rest of the debt.
5 Countries with which China holds DTAAs are particularly conducive to reinforcing China's status as a metropolis as often the fiscal arrangement allows a tax-free or low tax repatriation of profits, capital and dividends to China.
6 Souvenir manufacturing in JinFei will create difficulties for local souvenir manufacturers who consist mostly of women and laid-off workers. Organisations like National Women Entrepreneur Council (NWEC) and Small and Medium Enterprises Development Authority have specialised schemes encouraging unemployed and laid-off workers to opt for such low cost businesses. NWEC currently registers 240 handicraft manufacturers.
7 This is because given the size of Mauritius, all Mauritians commute daily to their work place.
8 China is treated as a metropolis entity in its entirety, without granting Chinese government the role of the metropolis and the investing companies that of satellites. This is because the Shanxi companies developing JFET are mainly SOEs, hence technically equalling the metropolis in status and abilities vis-à-vis foreigners. Moreover, the Mauritian context is the focus.

References

Ahiakpor, James C.W. 1985. 'The Success and Failure of Dependency Theory: The Experience of Ghana'. *International Organization* 39(3): 535–552.

Amal, Munhurrun, Senior Investment Executive. 2015. Board of investment, Mauritius, interviewed in December, Port-Louis, Mauritius.

Barbé, A. 2010. 'Jin Fei sort de terre, Interview of Xie Li, CEO Jin Fei SEZ'. *L'Express*, July 22.

Breslin, Shaun. 2013. 'China and the Global Order: Signalling Threat or Friendship?' *International Affairs*, 89(3): 615–634.

Davies, Martin, Hannah Edinger, Nastasya Tay and Sanusha Naidu. 2008. *How China Delivers Development Assistance to Africa*. Stellenbosch: Center for Chinese Studies, University of Stellenbosch.

Ferraro, Vincent and Rosser, Melissa. 1994. 'Global Debt and Third World Development'. In *World Security: Challenges for a New Century*, edited by Michael Klare and Daniel Thomas, 332–355. New York: St. Martin's Press.

Frank, Gunder, André. 1967. *Capitalism and Underdevelopment in Latin America: Historical Studies of Chile and Brazil*. New York: Monthly Review Press.

Frank, Gunder, André. 1968. 'Le Développement du Sous-Développement'. *Cahiers Vilfredo Pareto* 6(16/17): 69–81.

Frank, Gunder, André. 1976. 'That the Extent of Internal Market Is Limited by International Division of Labour and Relations of Production'. *Economic and Political Weekly* 11 (5/7): 171–190.

Frank, Gunder, André. 1978. 'Development of Underdevelopment or Underdevelopment of Development in China'. *Modern China* 4(3): 341–350.

Guevara, Che. 1964. Speech delivered 25 March 1964 at the plenary session of the United Nations Conference on Trade and Development.

Hart-Landsberg, Martin. 2010. 'The U.S. Economy and China: Capitalism, Class, and Crisis'. *Monthly Review* 61(9): 14–31.

Haynes, Jeffrey. 2008. *Development Studies*. Cambridge: Polity Press.

Heritage Foundation. 2017. *Index of Economic Freedom*. Available at: www.heritage.org/index/ranking (Accessed July 22, 2017).

Hu, Jintao. 2006. 'Address by Hu Jintao President of the People's Republic of China at the opening ceremony of the Beijing Summit of the Forum on China-Africa Cooperation'. *FOCAC.org*, 4 November.

Leander, Anna. 2001. 'Dependency Today-Finance, Firms, Mafias and the State: A Review of Susan Strange's Work from a Developing Country Perspective'. *Third World Quarterly* 22(1): 115–128.

Lease Contract. 2007. Between State of Mauritius and Mauritius JinFei Economic Trade and Cooperation Zone Co. Ltd.

Lease Contract. 2009. Between State of Mauritius and Mauritius JinFei Economic Trade and Cooperation Zone Co. Ltd.

Lenin, V.I. 1961. *Imperialism: The Highest Stage of Capitalism* London: Penguin Books.

Lippit, Victor D. 1978. 'The Development of Underdevelopment in China'. *Modern China* 4 (3): 251–328.

Mauritius, Parliament. 2010. Debate No. (n.d.) of June 22. Available at: www.gov.mu/portal/goc/assemblysite/file/orans22june10.pdf (accessed May 13, 2011).

Mo Ibrahim Country Profile. 2017. Available at: www.moibrahimfoundation.org/ (accessed April 10).

Oatley, Thomas and Jason Yackee. 2004. 'American Interests and IMF Lending'. *International Politics* 41: 415–429.

Owusu, Francis. 2003. 'Pragmatism and the Gradual Shift from Dependency to Neoliberalism: The World Bank, African Leaders and Development Policy in Africa'. *World Development* 31(10): 1655–1672.

Rodney, Walter. 1973. *How Europe Underdeveloped Africa*. Dar-Es-Salaam: Tanzanian Publishing House and Bogle-L'Ouverture Publications.

Sachs, Jeffrey. 1989. 'Making the Brady Plan Work'. *Foreign Affairs*, 68(3): 87–104.

Shanxi Tianli Enterprise Group Corporation. 2010. Group Introduction. Available at: www.shanxitianli.com/English/Detail.aspx?typeid=652 (accessed April 19, 2013).

Strange, Susan. 1998. 'What Theory? The Theory in Mad Money'. *University of Warwick, CSGR and International Political Economy Department*, CSGR Working Paper No. 18/98.

The World Bank, International Finance Corporation. 2017. *Doing Business*. Available at: www.doingbusiness.org/ (accessed April 19, 2017).

Wallerstein, Immanuel. 1976. 'Semi-Peripheral Countries and the Contemporary World Crisis'. *Theory and Society*, 3(4): 461–483.

Wang, Shaoguang. 2000. 'The Social and Political Implications of China's WTO Membership'. *Journal of Contemporary China*, 9(25): 373–405.

14

IVORY TRAILS

Divergent values of ivory and elephants in Africa and Asia

Stephanie Rupp

In both African and Asian contexts, elephants are in crisis. This crisis is propelled by economic desires, on both the African and Asian sides; it is facilitated by political networks, in both Africa and Asia; and is underpinned by paradoxical cultural values of both elephants and ivory, in African and Asian contexts. Elephants are caught up in a complex vortex of economic and political, social, and cultural pressures that impact their ecologies and the stability of their herds. Although elephants were widely distributed throughout both Africa and Asia less than a century ago, today all three sub-species of elephants are reduced to fragmented ecological zones. In Africa, *Loxodonta africana* (savannah elephants) and *Loxodontis cyclotis* (forest elephants) have come under severe demographic pressure in 37 elephant range states; in Asia, populations of *Elephas maximus* have been reduced to restricted habitats in 13 range states (International Union for the Conservation of Nature [IUCN] Red List). Despite international efforts to regulate the flow of ivory over the four decades through quota systems (Elephant Management System), restrictions, and bans (the Convention against Illegal Trade in Endangered Species, or CITES), elephant populations in most regions of Africa and Asia continue to plummet as human interest in ivory is sustained (Barbier et al. 1990; Maisels et al. 2013). Even in remote forest pockets within the Congo River Basin, herds of forest elephants have been decimated in recent years (Poulson et al. 2017).

Across recent decades – indeed, across centuries – the pathways of ivory have been diverse and dynamic; even where international institutions attempted to regulate ivory's flow through policy mechanisms, ivory has consistently found alternative channels through which to flow. In a simplified model, contemporary flows of ivory comprise a three-tiered system: elephant range states, where living elephants are reduced to tusks; manufacturing centers, where tusks are converted into carved ivory objects; commercial centers, where ivory objects are transformed into personal property and individually owned emblems of status (Barbier et al. 1990).

But connecting these three tiers are many interstitial actors and networks, each holding, advocating, and investing in particular values of ivory: hunters in equatorial African forests; the *patrons* or well-placed bosses who provide weapons, ammunition, and incentives to local hunters; middlemen hiding tusks in sacks of cocoa or other commodities for transport to markets; (un)aware government officials, willing to turn a blind eye for financial gain; aggregators, syndicates, or criminal groups, stocking ivory investments for future sales; (un)suspecting customs officials at points of exit, points of transit, and points of arrival; carvers, manufacturers, merchants, vendors, clients, consumers – and perhaps even further networks among luxury consumers, as final owners purchase, transport, and exchange ivory objects among themselves. The links in global ivory networks are many and varied, are geographically widespread, culturally divergent, and players at different links in the chain are largely unaware and uninvolved with any links other than those that are immediately proximal to themselves.

As a result of the dynamic movements and connections that are made possible by such interlocking chains in global ivory networks, throughout the Congo River basin elephant populations have declined by 62 percent between 2002 and 2011; by 2012 the population of forest elephants represented less than 10 percent of its potential size and covered less than 25 percent of its potential range (Maisels et al. 2013). Increasing human population density within the Congo River basin has propelled social and economic changes, resulting in mounting pressure on elephant herds. For example, the accelerating penetration of roads and other human infrastructure has led to ecological fragmentation and increasing overlap between human and elephant spaces. At the same time, pressure on African forest elephants is mounting because of external pressures from other interests: international appetite for ivory products, in particular from Asian consumers but also because of long-standing consumption by European and American purchasers; increasing trade and commercial links between African and Asian nations; and the parallel, mutually compatible lack of transparency in institutions and government offices in elephant range countries and in ivory markets. If we are to shape an effective international policy for the management of ivory and the conservation of elephants, we need to pay attention to interlocking local, regional, and global systems in both equatorial African and Asian contexts. This global analytical perspective enables us to analyze the repercussions of local cultural values of ivory among consumers in Asia on local cultural and economic values of elephants in the Congo River basin, recognizing the layers of interconnected actors, agendas, and actions connecting elephants, ivory, and world systems.

This chapter examines the divergent cultural values of elephants and ivory at distal ends of the global networks that link them: the Congo River basin, an important elephant range area, and Thailand, an important location of ivory processing, marketing, and purchasing. Figure 14.1 illustrates the complexity of actors, exchanges, values, and geopolitical dynamics in global ivory flows; full analysis of the dynamics illustrated in this figure is beyond the scope of this chapter. The analysis presented here focuses on the divergence of symbolic values for ivory and

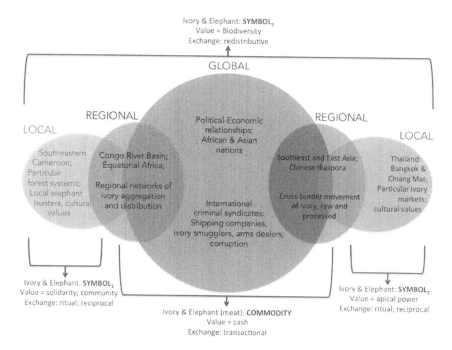

FIGURE 14.1 Overlapping systems and values involved in international ivory trade

elephants in two local spheres, the Congo River Basin and Thailand, where communities and elephants share space, resources, and relations. The values of ivory and elephants in these local contexts are annotated "SYMBOL$_1$" and "SYMBOL$_2$" in this figure. This chapter argues that divergent and even contradictory symbolic values are enshrined in elephants and ivory – social relations of kinship, reciprocity, and distribution on the one hand and individualistic accumulation of status, wealth, and power on the other – in these different positions in the global system of ivory flows. As ivory moves from forest communities in the Congo River basin, where it evokes values of long-term reciprocity among people and between people and elephants, the value of ivory changes substantially. Passing through many layers of middlemen in transit between a forest in Cameroon and a consumer in Asia, the dissociated ivory becomes a commodity with increasing monetary value, having shed its local cultural significance. Once the ivory is processed into a carved object, it reassumes cultural and symbolic values – but fundamentally different values than it embodied when it was held by people in the Congo River basin, for example in the Lobéké Forest of southeastern Cameroon. Instead, among consumers in Thailand, ivory objects symbolise apical power that is simultaneously ritual, political, economic, and social. The current surge in ivory accumulation and elephant decimation signals that values of individual status, wealth, and power constitute the prevailing forces that drive global flows of ivory.

This chapter also makes several more theoretical arguments by way of conclusions that pertain to the broader field of China–Africa studies. First, it highlights the importance of expanding the geographical scope of "China–Africa" studies beyond the expectation that the salient relationships emerge from dyadic pairing of China as a nation with Africa as a continent, or from analyzing relations between China and specific African nations. Second, this chapter highlights the importance of looking at relationships concerning a particular commodity, such as ivory, in networks of flows between Africa and Asia over the *longue durée* (Braudel 1958). Third, this chapter highlights the importance of framing economic flows in terms of overlapping economic spheres, each of which is shaped by distinct values. It highlights the importance of considering the coincidence of reciprocal, market, and redistributive spheres in economic analysis (Polanyi 1957).

The research on which this chapter is based connects two disparate regions of the world where cultural values of elephants and ivory influence the dynamics of the crisis: southeastern Cameroon on the western rim of the Congo River basin, and two locations in Thailand: Bangkok, the capital city; and Chiang Mai, a town in northern Thailand. I am deeply knowledgable about one of these regions, and possess adequate knowledge of the other. This project builds on over two decades of research that I have conducted in the Congo River basin. I conducted my doctoral research in the Lobéké forest region of southeastern Cameroon from 1995 to 2000, and continue to undertake ethnographic fieldwork there on a regular basis. My field methods in southeastern Cameroon are many, varied, and intensive: participant observation, kinship analysis, forest mapping, gathering of oral histories, and interviews in many formats on many topics, over long periods of time, with many interlocutors – conducted in the local language, Bangando. I have gathered materials on hunting patterns and the distribution of meat and other forest products, and have gathered significant, qualitative data concerning elephants and the stories that people tell about them. My friends, neighbours, and informants include elephant hunters, past and present.

My fieldwork in Thailand was brief and thin by comparison, but is anchored in my long-term residency in Southeast Asia. My first academic position was at the National University of Singapore, and as a result of this position and my engagement with Southeast Asian literature and contexts, I came to know the region well. My fieldwork in Thailand was preliminary; I was guided by both the previous research and the personal advice of Daniel Stiles, who has conducted research on ivory sites throughout Thailand over the past fifteen years.[1] With his help, I was able to identify in advance which markets, boutiques, and locations in both Bangkok and Chiang Mai I should visit, in my effort to learn about Thai cultural values of ivory and the current state of ivory markets, including the presence of African ivory in Thai markets. In Thailand, in contrast to my intimate and intensive ethnographic fieldwork in Cameroon, I visually surveyed shops, boutiques, and markets for ivory products, and spoke with those engaging in the ivory trade in informal interviews about their interest in ivory. Thus while my ethnographic data from Cameroon is robust and detailed, my findings from Thailand are impressionistic, suggesting questions for future research and providing a glimmer of insight into the cultural significances of ivory and elephants in Thailand.

A comparison of cultural values of elephants and ivory in these two regions yields insights concerning how such values shape economic markets for ivory, and how these markets impact people's lives in divergent local contexts. Thailand is significant, and often overlooked, in discussions in the literature on "East Asian" or "Chinese" markets for ivory as the driving factor in the surging ivory trade. The role of intermediary nations, including Hong Kong, Macau, Singapore, and the Philippines, is often overlooked in analyses of ivory flows in international ivory networks (Barbier et al. 1990; Stiles 2009; Doak 2014). It is perhaps even more imperative to consider Asian nations such as Thailand, Myanmar, and Laos, countries that have their own herds of elephants producing domestic ivory, thereby generating a pool of legal, domestic ivory into which illegal (African) ivory can be integrated and laundered. I hypothesise that these two regions, the forests of Cameroon and the markets of Thailand, represent significant contexts in the larger elephant-ivory crisis. Local hunters and historically deep hunting networks in the Congo River basin are being engaged in new ways as a result of the international surge in demand for ivory, while consumers throughout the increasingly affluent Chinese diaspora – and especially in Asian elephant range states where ivory has been venerated for many centuries – signify their economic, social, and political status by purchasing ivory objects.

Congo River basin: elephant crisis, elephant values

Throughout the Congo River basin, elephant populations have come under increasing pressure despite strong conservation efforts in forested regions of many nations in the region. Alarmed by this widespread devastation to elephants, a network of ecologists working throughout the basin conducted integrated, regional research from 2002 to 2011, including 80 surveys, to establish the current demographic status of elephants (Maisels et al. 2013). This survey demonstrated that although the Congo River basin region covers 95 percent of the current "known" and "possible" ranges of forest elephants, throughout 75 percent of this region elephant populations have declined to extremely low density (IUCN 2008; Maisels et al. 2013). For example, forests in the Democratic Republic of Congo, which housed the highest density of forest elephants based on colonial-era estimates, are today nearly devoid of elephants. The pockets of forest that still shelter significant populations of elephants are coming under increasing pressure. Even in Gabon, which houses 62 percent of the known forest elephants, only 14 percent of Gabonese forests hold high densities of elephants (Maisels et al. 2013). A more recent study of the elephant populations in northern Gabon demonstrates that herds even in these remote forests have been decimated; over the past decade, 78–81 percent of elephants in northern Gabon have been killed (Poulson et al. 2017).

Pressure from international networks is increasing in southeastern Cameroon, where national parks such as Lobéké, Boumba Bek, Nki, and the Dja Reserve anchor the western region of the Congo River basin. In March 2013, poachers killed 28 elephants in southeastern Cameroon in a coordinated attack using AK-47 assault rifles. Such instances of the large-scale killing of elephants for ivory illustrate significant trends in the decline of forest elephant populations. The rate of elephant killing is accelerating because of the opportunistic synergy between armed militias in several regions of Africa, which turn to wide-spread elephant hunting by use of military weapons and technologies (infrared goggles, helicopters, automatic weapons, grenades), selling the ivory on the international market to fund their ongoing operations. A surge in global demand for ivory provides the impetus for the current crisis, while armed conflict provides the context for widespread killing of elephants (UNEP et al. 2013).

But for communities in southeastern Cameroon, ivory did not – and does not – necessarily embody or express the value of an elephant. In the Lobéké region of southeastern Cameroon, as in the Congo River basin more broadly, elephants are ritually, politically, economically, and socially important. Relations between elephants and people are characterised by the creation and reinforcing of social relations of integration and collectivity. The circulation of resources from elephants, primarily meat, affirms social and economic values of reciprocity, redistribution, and mutual entanglements between people and between people and the elephants with which they share the forest.

Among forest communities on the western rim of the Congo River basin, ritual specialists indicate that they can transform themselves into elephants (and other charismatic animals) in a variety of contexts, and that elephants will protect people in situations of conflict or danger (Rupp 2011). Such contexts of elephant transformation usually take place in social settings that emphasise a larger collective goal or outcome. For example, in the midst of turmoil during the nineteenth century when forest communities were in flux because of regional conflict and migration, elephants protected Bangando families from violent incursions by their adversaries. Bangando recount that during this period of conflict, elephants gathered around families belonging to the elephant clan – the *bo folo*, or people of the elephant – and guided them along narrow paths to safety deep in the forest. The elephants trampled the people's footprints to obscure the direction of their flight. If children cried or were disruptive, elephants blustered and trumpeted to drown out the human noises, lest their enemies follow the children's cries to attack their families. In exchange for and recognition of this special relationship between these people and elephants, the people took elephants as the totem of their lineage and pledged not to eat elephant meat, following the classic formula of clan-based totem and taboo structure (Rupp 2011).

In the Lobéké forest of southeastern Cameroon, forest communities have hunted elephants for centuries using a variety of techniques including pit traps, spears, and axes (for cutting the tendons of the back legs of elephants, to immobilise them), techniques that required collaborative hunting. Bangando hunters traditionally hunted elephants in collective parties, which included men from neighbouring communities, in particular Baka. Hunters followed paths through the forest, either tracking a group of elephants directly or heading to clearings in the forest where elephants were known to shape the landscape, creating *bais* or open patches in the forest. In these openings in the forest, elephants dig up mineral rich earth to access salts, drink the clear, flowing water, and consume vegetation, keeping the canopy of the forest open and providing plentiful sunlight to support a diversity of plants and associated animal life. In these rich clearings, such as the region at the heart of the Lobéké National Park in southeastern Cameroon, forest peoples established villages, taking advantage of the rich natural resources alongside of other creatures of the forest. Elders today recall their youth in these clearings, reveling in the memory of plentiful game and honey, as well as relations of respect and solidarity among the people (Rupp 2011). One elderly man recalled that men gathered all of the products of hunting, including meat, ivory, and skins, at the house of the village head. If an elephant or other large forest animal had been killed, the community organised a party (the *mokopolo*) to go into the forest to assist with the butchering, smoking, and carrying of meat back to the village. Men and women, young and old, took up their baskets and machetes to go into the forest to be part of the collective work of the meat. The meat and other animal parts were carried back to the village head, who was in charge of the distribution of meat, dividing it into baskets called *kata* to send to each household. Through this system of distribution, community members participated collectively in and benefitted collectively from elephant hunting. And because the bulk of the elephant was divided and consumed by the entire community during the celebration of a central ritual, elephants were hunted in moderation.

Finally, the ivory tusks of the elephant were given to the head of the community. If a woman married into the community, she would rest her feet on an ivory tusk as she reclined, covered from head to toe in palm oil reddened by the fine dust of ground wood of a particular hardwood tree, marking her integration into her husband's family, clan, and community. Thus prior to the mid-twentieth century, elephant hunting was a collective activity that induced a powerful sense of social solidarity through the acts of hunting, distributing meat, and welcoming a stranger into the community.

Today, however, the techniques and social relations surrounding elephant hunting have changed significantly because of the pressures from the regional and international trade in elephant parts such as ivory and meat. Whereas in decades past, elephant hunters by necessity hunted in cooperative groups, today elephant hunters tend to work alone or in very small groups. This shift in the organization of the hunt stems from changes in technology: an elephant can be brought down by means of a powerful firearm wielded by a single hunter. Furthermore, because access to such automatic weapons as AK-47s requires extensive networks with political *patrons* and/or arms dealers (usually in Ouesso in the Republic of Congo, or in the Central African Republic), individuals who cultivate such relationships and take on the risk and expense of obtaining weapons and ammunition tend to maintain a degree of autonomy from social networks in their home villages. Finally, because hunting elephants is now illegal, the risk of making the details of a hunting expedition public is significant both for the hunter and the (often political) *patron* who supplied either the weapon, ammunition, or cover for the operation. As a result of these factors, elephant hunting today is usually undertaken by a very small number of participants, and tends to be conducted with great discretion.

The late 1990s witnessed a surge of trenchant village disputes resulting from the secrecy of contemporary elephant hunting that undermined the traditional networks of cooperation, social integration, reciprocity, and sharing. After a successful elephant hunt, two local hunters returned with baskets of meat and two small tusks; the meat was all that the hunters could carry with them out of the forest. When they arrived in the village, rather than being grateful for the distribution of meat, the hunters' neighbours, friends, and even family members were furious that the hunters had not sent for the *mokopolo*, the collective hunting party that traditionally would come to the site of an elephant kill to butcher, smoke, and transport meat back to the village. The hunters refused to divulge the location of the elephant carcass, knowing that if conservation authorities were made aware of the location of the kill, the hunters could be threatened with fines or even arrest. In the resulting controversy, people argued about the relative balance between the risk of illegal elephant hunting and the individualistic benefit from hunting elephants for profit, and bemoaned the loss of collective spirit of sharing and distribution of elephant meat. The impact of international demand for ivory and regional demand for elephant meat has undermined local traditional practices of elephant hunting and the collective social structures that had emerged to ensure that resources from an elephant kill would be widely shared and distributed. These examples of the deep cultural importance of elephants, elephant hunting, and the distribution of elephant products come from the particular region of southeastern Cameroon, yet they illustrate the social, economic, and political centrality of elephants and elephant products to social and ritual relationships the Congo River basin.

Entanglements: values of elephants and ivory in Asian markets

The overall scope of the crisis facing African elephants today can be seen clearly in the Ivory Transaction Index, compiled by a consortium of international organizations – UNEP, CITES, IUCN, and TRAFFIC – that have been actively engaged in monitoring the ivory trade. In this index the year 1998 is held as a baseline data point, representing a low point in the international ivory trade because of the falling off of ivory transactions resulting from the 1989 CITES ban on international trade in ivory and ivory products. From 1998 to 2011, a sharp rise in ivory transactions is apparent, representing a 200 percent increase (UNEP et al. 2013). On the demand-side of this crisis, rapidly increasing income levels throughout Asia have buoyed a new generation of consumers. The competitive and conspicuous consumption of newly wealthy consumers in Asia propel the contemporary ivory trade, and is palpable in ivory markets in major Chinese cities and in other Asian contexts such as Thailand. Markets in Bangkok sell ivory conspicuously; sellers range from small-scale vendors peddling ivory amulets alongside of elephant tail bracelets and other wildlife products, to antique dealers, to jewelers and purveyors of ivory objects, new and old, intended for upscale, wealthy consumers.

Sale of ivory objects in Thailand is largely targeted at Asian consumers. In Cha-tuchak market, a weekend market on the northern side of Bangkok, several large ivory shops are clustered together in one area of the marketplace. Numerous shops overtly target Chinese consumers; at one establishment a sign reads "ivory goods/ commodities shop" in Chinese characters, with no accompanying Thai or European language translations to attract purchasers from non-Chinese speaking backgrounds. Casually browsing the display cases of ivory bracelets, I asked a saleswoman about prices of various objects. When I inquired whether the ivory objects were old or new – implying that I was interested in knowing the age and provenance of the ivory, whether it was new (often illegally imported) stock, or pre-1989 ban (legal) ivory? – the saleswoman grew agitated and nervous. She quickly dug into a drawer under the counter and brought out hand-written, rumpled signs that looked as if they had been posted and removed many times previously, and hurriedly taped them onto the glass display cases. The lettering was scrawled in all capital letters in blue ballpoint ink, leaving deep grooves in the soft paper: "NO IVORY ONLY BONE." These defiant signs, hastily taped to the outside of the cases, contradicted the neatly hand-lettered signs in Chinese, which were arranged carefully inside of the cases near the objects: "100% 象" – 100 percent ivory. The saleswoman marched briskly away, not wanting to have anything to do with me.

In glass cases throughout Bangkok, ivory bracelets and bangles nestled up against each other, the smooth luster of each ring of ivory sliding into its neighbours. Ivory pendants, beaded necklaces, combs, picks, cigarette holders cluttered red velvet-lined cases; merchants displayed their ivory objects and offer prices, gauging the thickness of the ivory with calipers and metal rulers, crunching the sums on calculators. Some glass cases included ivory "blanks," unprocessed cylinders of ivory ready to be cut and carved to meet the wishes of particular customers. In several ivory shops, customers studying pieces of ivory for potential purchase acted as agents for clients in China. Making calls to their clients by cell phone in the midst of an ivory purchase, agents were able to relay information about the ivory objects to the final consumer in China, and ensuring that the ivory object would meet the consumer's wishes. In some cases, the agents carried on conversations with the Chinese client and the Thai salesperson simultaneously; the middle-person trans-lated the consumer's wishes to the ivory salesperson, who then produced from the cases or from storage ivory pieces that fit the stated specifications.

A glimpse inside a typical glass case of ivory objects for sale offers a sense of the demand for ivory by tourists from various backgrounds. A large piece in the front of the case was intricately carved with Chinese motifs: floral decorations on one end, and an impeccably carved woven basket, complete with several carved crabs in the hollow interior space of the ivory lattice. Japanese name-chops – *hanko*, ivory cylinders topped by elephants – were also on display. While Japanese name stamps are cylindrical, Chinese name stamps are rectangular; this ivory dealer clearly anticipates Japanese customers as well as Chinese. Furthermore, Japanese prefer the comparatively hard ivory of African forest elephants for their *hanko* and other objects, such as the plectrum, or *bachi*, of the *shamisen*, a traditional stringed instrument (Nishihara 2012). Mass-produced cigarette holders, phallic amulets, and other amulets and pendants are on display, with a target market of Chinese tourists. Finally, small pendants made from chips of ivory left over from more elaborate carvings beckon tourists of more ordinary means. Contemporary ivory dealers in Thailand seemed to be targeting consumers from a variety of Asian cultural backgrounds and at a variety of socioeconomic levels, with the most impressive works apparently intended for wealthy Chinese customers, or customers who appreciate Chinese-motif objects.

Elephants also hold political significance in mainland Southeast Asia. In the history of state building and interstate conflict in Southeast Asia, elephants were both weapons of war and reasons for conflict between states. Thailand and Cambodia historically fought wars over control of elephants; and elephants were important vehicles of war between the kingdoms. Elephants in Thailand continue to be the central emblem of the monarchy. Elephant statues flank the grand staircase to the Palace; the royal flag is the flag of Thailand with an elephant emblem in the center. Royal places throughout Thailand are decorated with elephants: elephant fountains, elephant figures supporting gates, elephant statues. The Royal Elephant Museum is the former stable of the white elephants belonging to generations of kings in Thailand. The museum charts the complex interweaving of history, spiritual power, and kinship among generations of elephants and generations of Thai royalty.

Elephants are also spiritually important throughout Asia. For example, Ganesh, the Hindu deity notable for his elephant head, is ubiquitous, smoothing the way and removing obstacles for believers. Ganesh represents many positive cultural qualities, but his encompassing traits are to help people achieve their goals through concentration, listening, intelligence, and a variety of tools: a rope (to pull one toward a goal); a single tusk, indicating a willingness to dispense with something of value in the search for a larger goal; and an axe to cut unnecessary bonds. Ganesh appears as a statue fashioned from many materials and in manifold contexts of everyday life: a stone protector of business interests at a street-side stall selling amulets; an elaborate ivory Ganesh that a shopkeeper retains to ensure his own success; and a pink, gaudy, plastic Ganesh in the case at a Thai ivory carver's shop. In Thailand elephants are emblems of individual agency and power, and ivory objects signify social status and economic influence. Elephants and ivory symbolise Thai values of individual accumulation of power and wealth.

If surging interest in ivory objects by Asian consumers is driving demand, how is ivory reaching these Asian markets? Only male Asian elephants produce ivory; and elephant populations throughout Asia have declined precipitously, hovering around 40,000 wild elephants and 13,000 elephants domesticated for labour, with about half of these elephants living in India alone (Pravettoni, 2013). In addition to the paucity of ivory from Asian sources, all shipments of illegal ivory confiscated in Asian ports since 2000 have originated in Africa (Moyle, 2014). This evidence strongly indicates that the bulk of new ivory stock circulating in Asian markets originates in

Africa, including forests in the Congo River basin that supply the coveted hard ivory. On the supply side of these international ivory networks, it is clear that ports along the East African coast of the Indian Ocean, such as Mombasa and Dar es Salaam, have emerged as primary exit points for ivory. From these ports the ivory is shipped by sea. For example, illegal shipments of ivory have been found hidden within sacks of groundnuts in Mombasa, Kenya and in sacks of sea shells in Dar es Salaam, Tanzania. Both of these shipments were bound for Asian ports (Kenya Wildlife Service, 2013; Russo, 2014).

It is important to emphasise, however, that the ivory seized at any particular exit point may not necessarily have originated anywhere near that exit point. Before ivory even reaches African ports of exit, the networks of hunters, patrons, middlemen, aggregators, and transporters criss-cross the continent – sometimes moving from one side of Africa to another before being shipped off to an Asian intermediary, then final, destination. Kisangani is a significant hub of commerce, as is Bangui, with ivory moving out from these central continental locations to ports in both West and East Africa (UNEP et al. 2013).

The volume of ivory that has been seized since 2010 has been increasing steadily, providing evidence that these ivory networks are run by criminal syndicates, which have the wherewithal – the contacts, the organizational abilities, and the armed force – to work in partnership with poachers and to pay off corrupt officials, to aggregate and secure the tusks, to organise and protect the lines of transport both on land and by sea or air, and to coordinate with port officials and dealers at intermediary ports and Asian ports of entry (Stiles, 2009; Gao and Clark, 2014). These networks comprise illegal global syndicates. The important collaborative work by organizations such as UNEP, IUCN, and TRAFFIC has resulted in a comprehensive report on the status of elephants in both Africa and Asia, published online in 2013. This work includes maps of ivory transactions within Africa and between Africa and Asian destinations, revealing entanglements of ivory networks that link local, regional, and global movements of ivory. Seizures of African ivory continue today, despite increasing international pressure on Thailand and other Southeast Asian nations to uphold CITES regulations; in March 2017, 330 kg. of ivory was confiscated at Thailand's Suvarnabhumi Airport, in a shipment of "rough stone" that originated in Malawi (*Bangkok Post*, March 7, 2017).

On the Asian end of these ivory networks, researchers have identified the most frequent ports of arrival, transit, and destination; it is significant that Thailand is one of the leading destinations of ivory, following China. In Thailand, African ivory can essentially be laundered; Thailand has a significant national herd of elephants, and the carving and sale of ivory from these elephants is legal. Thus while it is illegal to bring ivory into or out of Thailand across international borders, the national ivory market within Thailand is very active and is not well monitored (Stiles 2004; Stiles and Martin 2009; Doak 2014). Once African ivory enters the Thai market, and especially if raw African ivory is carved in Thailand, it is virtually indistinguishable from Thai ivory (Wasser et al. 2008). Thus new illegal stocks of raw African ivory can be laundered in Thailand, processed as if it were

Thai ivory, and then sold in national ivory markets and eventually into international, tourist markets. It is also possible that ivory flows from northern Thai towns such as Chiang Mai and Chiang Rai, and perhaps also from Burmese towns such Mong La (Nijman and Shepherd 2014), overland to Kunming, a bustling trading city in southwestern China that is known for its ivory and woodcarving traditions.

Ivory objects themselves also provide evidence of these global networks. For example, ivory carvings of women's busts for sale in a shop in Bangui, Central African Republic, photographed by ivory researcher Daniel Stiles, strongly resemble ivory carvings in an upscale antique shop in Bangkok, Thailand. The route that such carvings would travel would likely be long and convoluted: from the forests of the Congo River basin or the elephants of northern Cameroon, to the Central African Republic, then to an exit point such as Lagos or Douala or Mombasa, to an intermediary Asian destination through any number of middle-men and on to Bangkok, presumably to a tourist who would keep the objects and smuggle them home to some international destination, or to an ivory dealer who would continue to sell the object until it found its final "home." The global reach of the Chinese diaspora, not only into Southeast Asia but also into African nations, and the increasing presence of African traders and companies in Asian cities such as Bangkok, diversifies and extends the reach of consumers and their access to commodities such as ivory.

Paradoxes

The paradoxes of the ivory trade run deep. In both African and Asian contexts, at both ends of the global trade in ivory, *both* ivory *and* elephants are imbued with cultural and political power, even as the consumption of the prior requires the killing of the latter. In southeastern Cameroon, the distribution of elephant meat follows and reinforces lines of social solidarity, exemplified by the integration of a new wife within a family and village marked by resting her feet on ivory tusks belonging to the village head. The penetration of commercial ivory markets, initially during colonial times but accelerated today through participation in elephant poaching networks, has undermined long-standing patterns of reciprocal and distributive social and economic relations. In Thailand, where elephants represent apical ritual and political power, the accumulation of ivory objects has become emblematic of individual status and wealth accumulation. In both equatorial Africa and Southeast Asia, cultural values of elephants and ivory vary over time, and in relation to the values attributed to the animals and their tusks at distal ends of the networks that connect them. We are engaged in a meshwork (Ingold, 2011) of ivory and elephants, a system of fluid global connections across disparate regions of the world, connecting diverse societies, economic processes, and political systems through the exchange of a biophysical substance – ivory – taken from animals and imbued with social meaning. The world is an ever-shifting entanglement of networks bringing elephants and forest peoples, hunters and middlemen, government

officials and criminal syndicates, merchants, carvers, vendors, and consumers, con-
servationists and animal-rights activists together into a multi-directional fabric of
relationships that mutually constitute each other. Indeed, elephants' long-term
viability requires that the divergent values that people attribute to them and to
their tusks shift again, to recapture the sense of enmeshed economic and ecological,
social and political relations shared by us all in our overlapping world of social and
biophysical realities.

Conclusion

In conclusion, three more theoretical arguments pertaining to the broader field of
China–Africa studies can be made. First, this chapter has highlighted the importance of expanding the geographical scope of "China–Africa" studies beyond the
expectation that the salient relationships emerge from dyadic pairing of China as a
nation with Africa as a continent, or from analyzing relations between China and
specific African nations. An in-depth analysis of ivory networks highlights the
importance of situating China in a regional Asian perspective, rather than as an
isolated superpower. In the case of ivory, while Chinese consumers constitute the
majority of individual purchasers of ivory objects today, the Chinese retail market
for ivory objects relies on intermediaries in other Asian nations to serve as ports of
transit, processing and manufacturing centers, and primary marketplaces. Furthermore, consumers of ivory who identify culturally as Chinese may or may not be
citizens of the People's Republic of China, but instead members of the global
Chinese diaspora in a variety of nations, including, for example, the United States
and Thailand.

Second, this chapter has highlighted the importance of looking at relationships
concerning a particular commodity, such as ivory, in networks of flows between
Africa and Asia over the *longue durée* (Braudel 1958). It is in this longer historical
perspective that we are able to appreciate historical dynamics in ivory markets
across time and linking different regions of the world. It is also through broad
historical analysis that we are able to analyze patterns of cultural values attributed to
ivory, and how these patterns index historically contingent values of power, leadership, and individuality. A historical analysis of changing values of ivory reveals
that, although contemporary Asian consumers' desires for ivory emblems of individual wealth and power drive international markets, not so very long ago a different set of consumers, Euroamerican elite and nouveau riche during the Victorian
era, craved and consumed ivory products to represent their wealth, status, and
power in their communities at that time (Walker, 2009). Furthermore, elephant
herds are also historical; the catastrophic decline in elephant herds today has its
roots in the destruction of genetic diversity and robustness of elephant populations
a century ago; as European and American appetites for ivory decimated elephant
herds between 1850 and 1950, the very best tuskers – both male and female, for
African elephants – were removed from the breeding population, weakening future
generations of elephants, including the herds that are under stress from

contemporary consumers. While it is easy for contemporary analysis of the international appetite for ivory to focus on Chinese consumers in particular, and Asian ivory markets in general, the ecological plight of elephants today has its roots in earlier phases of ivory consumption by consumers in Europe and the U.S.

Finally, this chapter highlighted the importance of framing economic flows in terms of overlapping economic spheres, each of which is shaped by distinct values. It highlights the importance of considering the coincidence of reciprocal, market, and redistributive spheres in economic analysis (Polanyi, 1957). Ivory markets function in different ways in local contexts, in long-distance contexts of the transportation of ivory from African locations, through intermediary ports, to Asian processors, and then to commercial markets in particular local contexts. Analysis of ivory networks also demonstrates that distal ends of market systems may be driven by contrasting symbolic values that pit values of elephants, and the reciprocal social ties that emerge from the communal distribution of elephant meat, over and against values of ivory as élite objects of individual economic, social, and political prestige. In regional and global markets, ivory obtains value as a commodity – ivory separated from biophysical relations with elephants and social relations either within the community of origin or the community of final owner of the ivory object. Finally, at the level of the international community, both ivory and elephants assume a fourth valence as ideological and moral statements. At this overarching, symbolic level elephants, regulated under international treaties such as CITES, embody an intrinsic value of biodiversity; on the contrary, ivory is perceived as an emblem of human greed and self-centered consumption.

The contributions of this chapter to China–Africa studies include a call to break away from the insistence to see not only "China" and "Africa" as binary political, economic, or social units of analysis, but also to consider the salient geo-ecological and diasporic contexts of the particular commodity or resource under consideration; to place China–Africa research into historical context of the *longue durée*, which may involve moving beyond "China" and "Africa" as we think of them in contemporary discourse and scholarship; and to consider economic relations between "China" and "Africa" in terms of overlapping and coincidental economic spheres of exchange based on reciprocal, redistributive, and market dynamics of value.

Notes

1 I am grateful to Daniel Stiles for his early and enduring support for this comparative project, and to John and Martha Butt for hosting me in Chiang Mai, Thailand.

References

Alpers, Edward. 1992. "The Ivory Trade in Africa: An Historical Overview." In *Elephant: The Animal and Its Ivory in African Culture*, edited by Doran Ross, 349–363. Los Angeles, CA: University of California Press.

Barbier, Edward B., Joanne C. Burgess, Timothy M. Swanson, and David W. Pearce. 1990. *Elephants, Economics, and Ivory*. London: Earthscan Publications.

Braudel, Fernand. 1958. "Histoire et Sciences Sociales: La Longue Durée." *Annales. Économies, Sociétés, Civilisations* 13(4): 725–753.

Carrington, Richard. 1958. *Elephants: A Short Account of their Natural History, Evolution, and Influence on Mankind*. London: Chatto and Windus.

Cohen, Claudine. 2002. *The Fate of the Mammoth: Fossils, Myth, and History*. Chicago, IL: University of Chicago Press.

Doak, Naomi. 2014. *Polishing off the Ivory Trade: Surveys of Thailand's Ivory Markets*. Cambridge: TRAFFIC International.

Gao, Yufeng and Susan Clark. 2014. "Elephant Ivory Trade in China: Trends and Drivers." *Biological Conservation* 180: 23–30.

Hathaway, Michael. 2013. *Environmental Winds: Making the Global in Southwest China*. Berkeley, CA: University of California Press.

Ingold, Tim. 2011. *Being Alive: Essays on Movement, Knowledge, and Description*. New York: Routledge Press.

IUCN/SSC. 2008. "African Elephant Loxodonta africana: Further Details on Data Used for the Global Assessment." African Elephant Specialist Group. Available: www.iucnredlist.org/documents/attach/12392.pdf (accessed June 14, 2017).

IUCN. Asian Elephant Range States. http://maps.iucnredlist.org/map.html?id=7140 (accessed June 14, 2017).

Kenya Wildlife Service. 2013. "Contraband Ivory Seized at Kenyan Port." www.kws.org/info/news/2013/9julyivorymombasa2013.html.

Maisels, F., Strindberg, S., Blake, S., Wittemyer, G., Hart, J. *et al.* 2013. "Devastating Decline of Forest Elephants in Central Africa." *PLoS ONE* 8(3): e59469. doi:10.1371/journal.pone.0059469.

Martin E. and D. Stiles. 2002. *The South and Southeast Asian Ivory markets*. London and Nairobi: Save the Elephants.

Moyle, Brendan. 2014. "The Raw and the Carved: Shipping Costs and Ivory Smuggling." *Ecological Economics* 107(2014): 259–265.

Nijman, Vincent and Chris Shepherd. 2014. "Emergence of Mong La on the Myanmar-China border as a global hub for the international trade in ivory and elephant parts." *Biological Conservation* 179: 17–22.

Nishihara, Tomoaki. 2012. "Demand for Forest Elephant Ivory in Japan." *Pachyderm* 52: 55–65.

Polanyi, Karl. 1957. *The Great Transformation*. Boston, MA: Beacon.

Poulson, John R., Sally E. Koerner, Sarah Moore, Vincent P. Medjibe, Stephen Blake, Connie J. Clark, Mark Ella Akou, Michael Fay, Amelia Meier, Joseph Okouyi, Cooper Rosin and Lee J.T. White. 2017. "Poaching Empties Critical Central African Wilderness of Forest Elephants." *Current Biology* 27(4): 134–135.

Pravettoni, Richard. 2013. "Asian Elephant Population Estimates" in *Elephants in the Dust – The African Elephant Crisis: A Rapid Response Assessment*. United Nations Environment Programme, GRID-Arendal. www.grida.no/graphicslib/detail/asian-elephant-population-estimates_478d (accessed June 14, 2017).

Reuters News Agency. 2013. "Elephant Conservation Status, Minkebe National Park, Gabon." www.reuters.com/article/2013/02/06/us-gabon-elephants-idUSBRE9150HG2 0130206 (accessed June 14, 2017).

Ross, Doran. ed. 1992. *Elephant: The Animal and Its Ivory in African Culture*. Los Angeles: University of California Press.

Rupp, Stephanie. 2011. *Forests of Belonging: Identities, Ethnicities, and Stereotypes in the Congo River Basin.* Seattle, WA: University of Washington Press.

Russo, Christina. 2014. "To Stem Africa's Illegal Ivory Trade to Asia, Focus on Key Shipping Ports," *National Geographic* online. http://news.nationalgeographic.com/news/2014/08/140827-ivory-elephant-africa-asia-poaching-born-free/ (accessed September 5, 2016).

Stiles, Daniel. 2003. "Ivory Carving in Thailand." www.asianart.com/articles/thai-ivory/.

Stiles, Daniel. 2004. "Update on the Ivory Industry in Thailand, Myanmar, and Vietnam." Bangkok: TRAFFIC. http://www.danstiles.org/publications/ivory/10.TRAFFIC%20Bulletin%202004.pdf (accessed June 14, 2017).

Stiles, Daniel. 2009. "The Elephant and Ivory Trade in Thailand." *Bangkok: TRAFFIC Southeast Asia Report* Bangkok: TRAFFIC Southeast Asia Report (accessed June 14, 2017).

Stiles, Daniel and Esmond Martin. 2009. "The USA's Ivory Markets – How Much a Threat to Elephants?" *Pachyderm* 45: 67–76.

Sukumar, Raman. 2003. *The Living Elephants: Evolutionary Ecology, Behavior, and Conservation.* New York: Oxford University Press.

Sukumar, Raman. 2008. "Elephants in Time and Space: Evolution and Ecology." In *Elephants and Ethics: Toward a Morality of Coexistence*, edited by Christen Wemmer and Catherine Christen, 17–39. Baltimore, MD: Johns Hopkins University Press.

TRAFFIC. 2012. "Cameroon Elephant Poaching Crisis Spreads." www.traffic.org/home/2012/3/29/cameroon-elephant-poaching-crisis-spreads.html.

TRAFFIC. 2012. "Massive African Ivory Seizure in Malaysia." www.traffic.org/home/2012/12/11/massive-african-ivory-seizure-in-malaysia.html.

UNEP, CITES, IUCN, TRAFFIC. 2013. *Elephants in the Dust – The African Elephant Crisis: A Rapid Response Assessment.* United Nations Environment Programme, GRID-Arendal. www.grida.no.

Walker, John Frederick. 2009. *Ivory's Ghosts: The White Gold of History and the Fate of Elephants.* New York: Grove Press.

Wassser, Samuel *et al.* 2008. "Combating the Illegal Trade in Elephant Ivory with DNA Forensics." *Conservation Biology* 22(4): 1068–1071.

Wassser, Samuel *et al.* 2007. "Using DNA Evidence to Track the Origin of the Largest Ivory Seizure since the 1989 Trade Ban." In *Proceedings of the National Academy of Sciences* 104 (10): 4233.

Wemmer, Christen and Catherine A. Christen. 2008. *Elephants and Ethics: Toward a Morality of Coexistence.* Baltimore, MD: Johns Hopkins University Press.

Wildlife Conservation Society. 2013. www.wcs.org/press/press-releases/gabon-elephant-slaughter.aspx (accessed June 14, 2017).

Wipatoyotin, Apinya. 2017. "330 kg. of ivory from Malawi seized at Suvarnabhumi Airport." *Bangkok Post.* March 8, 2017. http://m.bangkokpost.com/news/general/1210489/330kg-of-malawi-ivory-seized-at-suvarnabhumi (accessed June 14, 2017).

Views from upstairs: elites, policy and political economy

15

NEO-PATRIMONIALISM AND EXTRAVERSION IN CHINA'S RELATIONS WITH ANGOLA AND MOZAMBIQUE

Is Beijing making a difference?

Ana Cristina Alves and Sérgio Chichava

Whilst there is little contention that China's foray into Africa largely accounts for the astounding economic revival and geopolitical shift of the continent towards the East in the early twenty-first century, debates around the nature of this relationship have been wrapped in controversy since the early days of China–Africa studies. Some claim China is not concerned with African development, but sees the continent solely as an inexhaustible source of raw materials; others accuse Beijing of replicating hierarchical colonial relationship patterns that reflect underlying economic and power asymmetries; equally, in advancing its policy of non-interference, China is also accused by many of curbing the progress of democracy and governance in Africa. The Chinese government, however, holds on to its normative claims of exceptionalism rooted in a mutually beneficial relationship between equals. Although depictions of China's exceptionalism in Africa are dominant in both African and Chinese leaders rhetoric, the practice seems to remain somewhat contentious as Beijing's ideational aspirations are increasingly strained by myriad challenges on the ground (Alden and Large 2011: 29–32).

Notwithstanding lively debates about China–Africa relations, scholarship appears to have reached a deadlock on whether China's volition and model, or African material structures, are responsible for slow progress in African development since the turn of the century. Although some studies accorded a great deal of African agency in bilateral relations (Corkin 2013), the role of African agents in shaping the rules of engagement with China and its developmental potential remains largely unexplored. This dimension may, we argue, provide a way out of the current predicament. Using the lens of extraversion and neo-patrimonialism, this chapter explores whether China is transforming or conforming to the dominant structural dynamics of Africa's external relations, and how this is affecting Beijing's exceptionalism on the continent. To do so, we scrutinise China's relations with Angola and Mozambique in four parts. We argue that the structural dynamics of China's relationship with both cases largely conforms to extraversion patterns and collusion with neo-patrimonialist ruling elites, and that this complicity with dominant African agents is constraining China's delivery of exceptionalism in both countries.

Extraversion and patrimonialism in Africa

A decade ago an eminent Africanist predicted that China was likely to end up entangled with neo-patrimonialist elites in Africa and constricted by the same engagement pattern that has long shaped the continent's relations with external powers: that of extraversion (Clapham 2008, 361–369). First advanced by Bayart in 1989, the theory of extraversion argues that throughout history African elites have been appropriating resources from their relationships with external powers to concentrate political authority domestically and accumulate wealth at the expense of their societies. This approach has been influential (Jackson 1990; Clapham 1996, Chabal and Daloz, 1999; Cooper 2002). The argument dealt a serious blow to mainstream dependency theories (Africans as passive victims of exploitative external powers) by stating that African elites have by this means actively contributed to the dependency condition of the continent, and that extraversion strategies have structured not only its relations with foreign powers but also internal politics (Bayart 2000, 219–220).

This theory argues that throughout successive historical periods African elites have consolidated their internal authority by capturing rents generated through relations with external powers, starting with the slave trade and colonisation. In the 1990s, structural adjustment programmes and conditional aid from Western countries and institutions, along with NGOs, transnational corporations and private investors, became new sources of rents for African elites who undertook only superficial democratisation reforms to maintain rent flows (Bayart 2000: 226; Peiffer and Englebert 2012, 361). Along with democratisation (including in Mozambique), war came to be the other main instrument of extraversion in this period (Bayart 2000, 228) with many conflict countries receiving support from external powers or selling their natural resources to the West to fund the war effort (including Angola). Importantly, in both situations, external powers have been unable to pursue their agendas. African elites continue to syphon benefits from donors to uphold their authority under the guise of democratisation or peacebuilding, and external powers have been generally unable to control the African agents they armed, financed and trained (Bayart 2000, 229–230). As such, Bayart concludes that "Africans are actors in their own history, always ready to turn external constraints into some new creation", and therefore one should look at African strategies of extraversion as "the means of Africa's integration into the main currents of world history through the medium of dependence" (Bayart 2000, 240–241).

The ability of African elites to simultaneously milk and keep external powers in check is not, however, uniform. One study of the relationship between extraversion and political liberalisation concludes that the vulnerability of African countries to external pressure (on democratisation) is ultimately a function of their extraversion assets. Each country has a "unique extraversion portfolio made up of specific linkages with the international system that its regime can exploit to extract abroad the resources it needs for domestic domination" (Peiffer and Englebert 2012: 361). Extraversion assets can vary from natural resources (fossil fuels amongst the most valuable), strategic location (e.g. for a military/naval base, or trade hub) to geopolitical alignment (e.g. siding with a particular great power). The stronger and more varied assets a regime has, the bigger its autonomy with regards to external agendas; conversely, a weak portfolio makes them more dependent on foreign aid cooperation and thus more vulnerable to external pressure. The end of the Cold War illustrates these dynamics well. The ensuing loss of patronage from the Soviet Union and the US weakened extraversion portfolios across Africa, explaining why conditional aid by Western donors was successful in triggering a democratisation wave that swept the continent in the 1990s, including in Mozambique. However, regimes that controlled vast reserves of natural resources, particularly oil, were able to remain somewhat immune to external pressure (Peiffer and Englebert 2012, 362), especially when donors were the main buyers (or investors) of their commodity exports, such as in Angola, Gabon, or Equatorial Guinea. This means that African elites with stronger extraversion portfolios (i.e. producers of sought-after commodities, a strategic location or strong geopolitical importance) have greater leverage to instrumentalise their relations with external powers in a way that favours their personal interest (consolidation of power domestically) at the expense of the agenda of the foreign partner. Peiffer and Englebert's contribution to this debate thus rests on the concept of 'extraversion assets' and how these explain the diversity across Africa in terms of extraversion capabilities, and also how these vary in each country across time in tandem with the changing value of those assets in the international arena.

Extraversion strategies are not the only challenge that Chinese exceptionalism faces on the continent. The neo-patrimonial structure of the African state also represents a major trial to deliver economic development. According to Médard (1979; 1990; 1992), this novel form of patrimonialism refers to the coexistence of patrimonial and rational-legal dominations, or the absence of distinction between public and private spheres. The concept of neo-patrimonialism is linked to the concept of patrimonialism, which Max Weber developed initially. Weber used the expression to designate a specific style of authority in the so-called traditional societies where the "Big Man" ruled through personal authority (Médard 1992, 167–192). Thus, neo-patrimonialism differs from Weberian patrimonialism in the sense that the latter is purely based on informal rules, while the former is a mix of rational-legal and patrimonial procedures. In neo-patrimonial management, there is both formal and subjective differentiation between the private and the public spheres, characteristic of all legal-rational bureaucratic systems. In this regard, neo-patrimonialism is a modern form of patrimonialism.

The survival of clientelism, and its reconfiguration alongside the emergence of the modern state bureaucracy in post-traditional societies, attracted scholarly attention towards these novel forms of patrimonialism from the mid-1960s (Zolberg 1966, 140–141; Roth 1968, 196; Eisenstadt 1973, 15) followed by an increasing scrutiny of how these manifested in Africa (Médard 1979, 35–84; Callaghy 1984; Clapham 1985; Crook 1989, 205–228; Bratton and van de Walle 1994, 453–489; Hodges 2008, 175–199). Whilst the use of patron–client relations to centralise and cement authority has been a common feature across societies and time, it seems to have achieved unprecedented proportions in post-colonial Africa. Neo-patrimonialist rule in Africa increasingly came to be equated with highly personalistic and autocratic presidential rule, patron–client ties for economic and political advancement, use of state resources to reward supporters and appropriation of state funds for self-enrichment (Pitcher et al. 2009, 132). Many of these studies unpacked how modern institutions and bureaucratic procedures coexist with an informal authority hierarchy based on inter-personal relations, whereby rule is ascribed to a person not the office, and office is often used for personal benefit rather than the public good. The "Big Man" also uses his authority to undermine opposition political parties by co-opting its members in exchange for some privileges and personal favours. Van de Walle (2003) thus considers neo-patrimonialism as one of the major factors undermining the development of stable and competitive politically party systems and democracy in Africa.

Daniel Bach distinguishes between two main forms of neo-patrimonial states: those where patrimonial practice is capped (regulated neo-patrimonialism), and those where patrimonialism is all-encompassing (predatory neo-patrimonialism) (Bach 2011, 277–280). In regulated forms of neo-patrimonialism, leaders rule mostly through co-optation and there is a degree of redistribution and delivery of public goods, generally associated with increased state capacity (for example, in Botswana). In predatory forms of neo-patrimonialism, leaders rule through highly personalistic, sometimes violent means, and there is an absence of public space and lack of capacity to deliver public goods, gradually leading to deinstitutionalisation of the state (such as Uganda under Idi Amin). Owing to the subordination of economic objectives to political survival of the regime, and the draining of resources through patronage, clientelism and corruption, both forms of neo-patrimonialism seem at odds with economic development prospects. In particular, predatory neo-patrimonialism came to be regarded as the quintessential anti-developmental state (Bach 2011, 281). Whilst neo-patrimonial practices in Latin America, Southeast Asia and Central Asia have been more associated with the regulated kind (Bach 2011, 282–289), predatory neo-patrimonialism has been largely depicted as more pervasive in Africa (Bach 2011, 277–280; Evans 1989, 561–582; Kohli 2004), making it a particularly challenging ground for economic development.

Neo-patrimonialist practices and extraversion strategies seem to permeate all African countries in varying degrees, regardless of regime types. Resources seem to play a major role in both the extraversion and neo-patrimonial rule of African elites, meaning that the availability of resources generate more leverage in both domestic politics and foreign relations, and thus greater resistance to external agendas. Whilst growing Chinese patronage and commodities imports from the continent since 2000 have undoubtedly strengthened the extraversion portfolios of many African elites vis-à-vis Western partners, and granted Beijing some appreciation on the continent, China is by no means exempted from the same fate. So far this type of African elites' agency has proven, in most cases, to be self-serving and anti-developmental.

Angola and Mozambique are no exceptions to this pattern. Like many other African countries, extraversion and patrimonialist practices in both cut across different historical periods, starting with the slave trade (Newitt 1994, 19–40; Lindsay 2014, 142–144; Birmingham 2015, 27–40; Vail and White 1981). This chapter, however, will focus on the post-colonial period so as to unpack how these traits have shaped both countries' relations with external powers, and in particular China, and in what ways this has impacted domestic politics and economics.

Historical patterns of extraversion and patrimonialism in Angola and Mozambique

In Angola, two main extraversion assets used by local elites to access foreign resources in order to assert their domestic authority can be distinguished: ideological allegiance/geopolitical location during the Cold War, and oil/diamonds in the post–Cold War setting.

Following independence internal actors actively sought to maintain the international dimension of the conflict inherited from the colonial war to ensure their domestic survival in the context of the long civil war that followed. China did not figure high in Angola's extraversion strategies in this phase, but played a role as it courted MPLA and actively supported UNITA and the Frente Nacional de Libertação de Angola (FNLA) in the 1960s and 1970s. This justifies in part China's absence in the ruling elite's extraversion strategies until the late 1990s despite diplomatic ties being established in 1983. In the 1980s, the MPLA consolidated its power as de facto ruler in Luanda, relying on Moscow and Cuban military forces. UNITA spread its control over much of the hinterland backed by Washington and South African troops. The FNLA vanished in the early 1980s after losing external support, mainly from the US and Zaire, while China's inward-looking policies removed it from the equation. As the Cold War drew to an end, ideological allegiance lost value as an extraversion asset, leaving both the MPLA and UNITA orphans of external patronage. As a result, an armistice was signed and multiparty elections were held in 1992 under the United Nations. Jonas Savimbi, leader of UNITA, accused the MPLA of rigging the elections, and civil war resumed soon after. Even though increasingly isolated as a result of sanctions, UNITA sustained the war effort for another decade by trading diamonds, mostly with private players in the West (Hodges 2004, 170–198; Alves 2009, 160–162).

The MPLA resorted to oil dividends to foot the war bill and consolidate dos Santos' personal grip over the state bureaucracy (Hodges 2004, 159–169; Alves 2009, 158–160; Soares de Oliveira 2015, 29–49). The longevity of dos Santos' rule (1979–2017), and his undisputed grip over the party, the state and the military, stems primarily from his tight control over Sonangol, the national oil company, which is renowned for its opaque accounts and billions of dollars that have gone missing over the last decades. Sonangol dividends enabled dos Santos' personal wealth to grow (using family members and close friends as proxies), and extend his patrimonial rule well beyond Luanda and the MPLA through clientelism and co-optation (Alves 2009, 158–159; Soares de Oliveira 2007, 595–619). In addition to oil export revenues (mostly sold to Western countries, especially the US at this stage), the use of oil as collateral to access loans from Western commercial banks emerged as another important instrument of extraversion in the 1990s. The Angolan population at large has seen few benefits; the oil industry did not create jobs and oil dividends were not reinvested in productive activities or public goods (Soares de Oliveira 2015, 25–49). By the end of the 1990s, the dos Santos regime neatly fitted the description of predatory neo-patrimonialist rule, and thus the archetype of an anti-developmental state.

Like Angola's liberation movements, the Frente para a Libertação de Moçambique (FRELIMO) received important political, military and financial support from Mao's China. These influenced FRELIMO as a political-military movement and its guerrilla tactics during the anti-colonial war (Henriksen 1978, 443; Brito 1987, 18). However, FRELIMO was not locked in to Chinese influence and interests and, throughout the period of armed struggle, also received support from the Soviet Union, despite the open rivalry between Moscow and Beijing in Africa. China and Mozambique established diplomatic relations on the day of Mozambique's independence. Even though cooperation agreements were signed in various areas, state-to-state relations did not develop further because FRELIMO became increasingly close to Moscow. In 1977, FRELIMO representatives visited China to establish formal relations with the CPC, but the request was declined (Zheng, 2016). FRELIMO's support to Vietnam during the Sino-Vietnamese war and the invasion of Cambodia, and to the Soviet Union's invasion of Afghanistan, further contributed to this. Also, Beijing's support to UNITA in Angola was not well regarded by the Maputo regime (Jackson, 1995).

Even though FRELIMO was considered as a paragon of integrity in the early days of independence (Hanlon 2004), things quickly went downhill as civil war ravaged the country between 1976 and 1992. The onset of civil war in 1976, concurrent with the collapse of the socialist state and adoption of neo-liberal politics, combined with the arrival of massive aid from Western donors and NGOs once the conflict ended, further complicated the matter, placing Mozambique among the most corrupt countries in Africa.

Many scholars have pointed out how local political elites have used the state for their own benefit, to realise primitive accumulation and perpetuate their power (Pitcher 1996; Harrison, 1999; Hanlon 2004). Hanlon (2004, 749–751) points out two aspects that contributed to rising corruption in Mozambique in the 1990s: murky privatisations during the transition from central to liberal market, loans made to FRELIMO military and party officials with no expectation of repayment to persuade them to accept the end of civil war, and the imposition of savage cuts on government spending, the salaries of public servants being the most affected. State resources have also been used to weaken opposition political parties, as well as co-opt independent and critical civil society (Forquilha 2009, 26–28). By the late 1990s Mozambique had taken on the characteristics of an uncapped neo-patrimonial state. Lacking the natural resources Angola possessed in abundance, and with assistance from the Soviet Union drying out from 1986 onwards, Mozambique had little choice but to turn to Western donors for financial aid. Its extreme poverty rapidly turned into a valuable extraversion asset for the ruling elite in the post–Cold War context. By the time the civil war ended, external aid had reached US$1 billion, representing almost 70 per cent of GDP and turning Mozambique into one of the largest aid recipients. The ruling elite, however, was cautious, and avoided openly embezzling donors' funds. It became "highly skilled at ensuring that management of donor money is transparent and clear ...; instead they rob banks, skim public works contracts, demand shares in investments, and smuggle drugs and other goods-and they ensure that the justice system does not work so they cannot be caught" (Hanlon 2004, 758).

Overall, China's relations with Angola and Mozambique remained limited and unimportant throughout their respective civil wars, thus exhibiting little extraversion during this period. This pattern, however, started changing in the context of peace as bilateral political exchanges multiplied and China's economic footprint in both countries became increasingly conspicuous.

China-Angola: fuelling extraversion and neo-patrimonialism through oil-backed credit lines?

The end of Angola's civil war in 2002 coincided with increasing oil demand from the East. This inflated the value of oil as an extraversion asset with major Asian importers. By 2007, China had replaced the US as the major destination of Angolan oil exports. Concurrently, from 2004, oil-backed loans started flowing from China to rebuild Angola's infrastructure. The scale of China's loans is unprecedented in the history of the country making it the main feature in bilateral relations. Between 2000 and 2015, Chinese policy banks extended an estimated total of $19.2 billion to Angola (CARI 2017), nearly one quarter of its combined loans to Africa in that period (Hwang et al. 2016, 1), funding the lion's share of the country's infrastructure development.

To understand the extent to which China's engagement fits the extraversion pattern, the economic and political benefits to the ruling elite in power must be assessed. In Angola the loans extended by Chinese policy banks must be distinguished from those provided by the China International Fund (CIF). As documented by many (Weimer and Vines 2012, 93–95; Mailey 2015: 24–44; Burgis 2015) the CIF fits extraversion patterns and neo-patrimonial practices to a large extent. This $3bn loan originated from a private fund in Hong Kong involving some shady former Chinese officials, Lusophone businessmen and members of the Angolan political elite connected to Sonangol. This fund run by a presidential protege was from the start fraught with corruption allegations. After a few scandals with flagship projects that had seen no progress for years, CIF projects were taken over by Chinese state companies (Croese 2012, 133) in 2008. CIF, however, continued to exist; other investments in collusion with government officials came to fruition, including a brewery, a cement factory and, more recently, real estate projects allegedly benefitting from major tax breaks (Macauhub 2017). CIF is also active in Mozambique, where in partnership with SPI – Gestão e Investimentos, SARL, the Frelimo Party holding company – it established a joint-venture to build a cement factory in Matutuíne, Maputo province.

Chinese official loans by Exim Bank, the China Development Bank (CDB) and the Industrial and Commercial Bank of China (ICBC) appear somewhat different from CIF. The capital stays in Beijing, to be directly paid to Chinese companies upon completion of the projects (Alves 2013a, 107–108). Whilst this procedure leaves little room for embezzlement by the local elite, there is nonetheless some traces of extraversion behaviour here too. Sonangol plays a central role in all these loan agreements, either as guarantor or as the borrower, suggesting a close nexus between Chinese authorities and the core of the president's patrimonial rule. Some form of collusion is further evidenced by Chinese oil companies benefitting from favoured access to oil equity in Angola on the sidelines of some of these loans (Alves 2013a, 108–110). Moreover, these loans are by no means immune to rent-seeking practices by the ruling elite. Chinese companies have complained about corruption/extortion practices by Angolan government officials during the bidding process, with fees allegedly varying from 10 to 50 per cent of the value of the contract.[1]

In addition to this type of economic dividends, dos Santos and his entourage stand to gain political benefits. The timely offer of Chinese oil-backed loans in 2004 enabled the MPLA to continue sidelining IMF and World Bank conditionalities (Alves 2013b, 214), allowing dos Santos to introduce political reforms at his own pace and will (Chabal 2007, 10). By the time the MPLA organised Angola's first multiparty elections in 2008, dos Santos had plenty of time to use Sonangol's fast growing oil revenue to co-opt and divide the opposition while, at the same time, improve the MPLA's image by leading a massive reconstruction campaign across the country, mostly funded with Chinese loans. The MPLA won 82 per cent of the votes in 2008. Despite being marred by many irregularities, the ballot largely legitimised dos Santos' rule in the eyes of domestic and international audiences. Collusion between dos Santos' patrimonial network and Chinese companies and funding was also apparent in a number of business deals. His daughter, Isabel dos Santos, who according to *Forbes* magazine is the richest woman in Africa, was revealed to be a hidden shareholder (40 per cent) of a $4.5 billion contract awarded in 2015 to one of the largest Chinese construction SOEs (Gezhouba Group Corporation, or CGGC) to build a massive dam in the Kwanza river, the Caculo Cabaca Dam, with funding from Industrial and Commercial Bank of China. The first stone was laid down by dos Santos as one of his last acts as president (Pereira 2017).

Whilst the link between Chinese infrastructure loans/businesses and the persistence of MPLA rule may seem tenuous, Chinese construction companies' complaints about tremendous pressure from local authorities to undertake unrealistic deadlines to complete flagship projects before the August 2017 elections corroborates at least some degree of extraversion behaviour in bilateral relations.[2] Moreover, since 2011, most Chinese loans to Angola were commercial loans to Sonangol by the CDB and ICBC, estimated at $7.5bn (CARI 2017). Backed by oil supply contracts, these allegedly were to fund the company's oil exploration and production activities, debt obligations and other projects. Considering that these credit lines came in at a time of low oil prices (meaning companies are less likely to undertake new exploration projects), a severe foreign currency liquidity crisis in Angola and growing opposition to the dos Santos regime, one cannot but wonder what part this capital played in reinforcing dos Santos' patrimonial rule and the MPLA electoral agenda. The bulk of these loans to Sonangol originated from a US$15bn credit line opened by the CDB in 2014, which was suspended in December 2016 due to a break of contractual obligations on the Angolan side (failure to repay) and disagreement over the readjustment of oil shipments servicing the loans (Rede Angola 16 October 2016).

The above suggests that despite some nuances, China's oil-backed credit lines to some extent fit the extraversion pattern as they seem to have provided thus far economic and political dividends for the Angolan elite and its domestic agenda. Moreover, despite the massive scale of the infrastructure delivered over the past decade and a half, these loans seem to have had little spillover effects in the national economy. Only a few jobs were created in the construction sector as most labour was brought in from China, whilst the oil industry is capital, not labour, intensive. Economic diversification has seen very modest progress. Angola remains highly dependent on the oil industry, meaning 15 years of massive Chinese financial assistance have not made much of a difference in this regard. Oil still accounts for nearly half of Angola's GDP and 98 per cent of exports. Whilst the context of lower oil prices seem to have given China the upper-hand in relations, whether this will enable a breakaway from extraversion patterns under the new president, João Lourenço, remains to be seen.

China-Mozambique: fuelling neo-patrimonialism and extraversion through business collusion?

At the dawn of the twenty-first century, Mozambique had successfully developed extreme poverty into its main extraversion asset, becoming one of the largest recipients of foreign aid. Its relationship with China was revived following President Jiang Zemin's 1996 African tour with a US$20 million loan from the China Exim Bank to support Chinese companies in Mozambique (Alden et al. 2014, 4). As aid and investment started flowing in from China in the years that followed, signs of extraversion behaviour by the Mozambican ruling elite became increasingly evident in bilateral relations as can be seen in the two cases below.

Timber, construction, banking, forestry and telecommunications are some of the sectors where economic alliances between Chinese companies and the

Mozambican elite were established. The most controversial case concerns one of the major Mozambican sources of export revenue: timber. By 2006, timber already accounted for over 90 per cent of Maputo's exports to China, making it China's sixth largest supplier (Canby et al. 2008) and seven years on Mozambique was China's main African supplier of timber in terms of economic value. Timber exploitation is largely in the hands of Chinese companies linked to Mozambican businesses, a significant part of which are connected to the country's political elite, including the opposition.[3] The predatory exploitation of timber has given rise to lively opposition within Mozambican society against the collusion between Chinese businesses and the country's political elite (particularly linked to FRELIMO). In 2006, a Zambézia Forum of Non-Governmental Organisations (FONGZA) report accused important FRELIMO leaders, including former President Joaquim Chissano, of deforesting Zambézia in collusion with the Chinese (Mackenzie 2006).

Another report, inked in 2008 by three Mozambican civil society organisations, the Friends of the Forest Association, the Rural Mutual Aid Association (ORAM) and Environmental Justice (JA), stated that there was no longer any timber to exploit in Zambézia. It also stated that people linked to FRELIMO were selling their licences to foreigners. This situation is not specific to Zambézia. Cases of violations of the law are reported across the country (exploitation of timber beyond the established limits, export of unprocessed logs), as is corruption, trafficking and contraband in timber. When the Chinese president visited Maputo in February 2007, some civil society voices and academics publicly denounced the situation. For example, Marcelo Mosse, then head of the Centre for Public Integrity (CIP), and the prominent sociologist Carlos Serra, wrote open letters to former president Armando Guebuza, himself cited as having interests in timber. The former criticised the obscure process of illicit enrichment of Mozambique's political elite in partnership with some Chinese companies, emphasising cooperation with China was welcome if done transparently, to the benefit of the people of both countries and not merely replaying colonial patterns (Mosse 2007). Serra demanded the immediate appointment of a commission of inquiry to investigate what was really happening in Mozambican forests (Serra 2007).

Despite much criticism and public denunciations, however, the situation is yet to improve. In 2013, an Environmental Investigation Agency (EIA) report accused former Minister of Agriculture, Tomás Mandlate, and his successor, José Pacheco (the current Minister of Agriculture) of being the main figures in timber trafficking in collusion with Chinese companies (EIA 2013). The Mozambican civil society organisation CIP disclosed that in 2014 Chinese companies involved in illegal logging had financed FRELIMO's electoral campaign (Mabunda 2014). The illegal logging has also led Afonso Dhlakama, leader of RENAMO, the main political opposition party, which has re-engaged in military conflict with the Mozambican government, to threaten to cancel the licenses of Chinese companies and forbid them from exploiting timber in central and northern Mozambique, where his party has major popular support (Miramar 2015).

This case clearly illustrates persistent collusion between the ruling elite and Chinese private businesses to generate rents and appropriate dividends from one of Mozambique's major exports for self-enrichment and political gain. Whilst the fragility of the Mozambican state is partly to blame for the current state of affairs, since it lacks the resources to control the illegal exploitation of timber, the absence of political will allows this situation to endure, mostly due to the business interests of high-ranking state and party officials. In addition to collusion with the country's political elite, the illegal timber trade implicates lower levels of the public administration, the police and the customs service, further weakening the country's governance.

The second case concerns the capture of an agricultural cooperation project by the FRELIMO elite. In April 2007, the Mozambican and the Chinese governments, through the provinces of Gaza and Hubei respectively, set up an agreement for the establishment of a Chinese 'friendship' rice farm at the Lower Limpopo (also known as Xai-Xai irrigation scheme), located in the capital of Gaza, Xai-Xai. The project was initially started by Hubei Lianfeng Mozambique Co Lda, a provincial state-owned company, and was later turned over to Wanbao Africa Agriculture Development Limited in 2011. Among the main objectives of this partnership was agricultural technology transfer from Chinese to the Mozambican farmers in order to increase productivity, and help Mozambique become a self-sufficient rice producer. In order to benefit from this technology transfer, the Mozambican government asked local farmers to set up an association, called *Associação dos Agricultores e Regantes do Bloco de Ponela para o Desenvolvimento Agro-Pecuário e Mecanização Agrícola de Xai-Xai* (ARPONE). However, the majority of the people who created the association and started to work alongside the Chinese company were FRELIMO members. In the same manner, some senior employees of *Regadio do Baixo Limpopo* (RBL), the public company in charge of the irrigation scheme, joined ARPONE and also began to produce rice. Senior state officials in Mozambique are usually linked to FRELIMO. These officials are the only ones able to pay the fees required to access Chinese technical assistance, because they can easily get access to bank credit, can use state resources to cultivate their plots, or can use their political and state influence to persuade the Chinese company to give them agricultural assistance.

This has led those ARPONE farmers without connections to FRELIMO or the state to distrust local authorities and accuse them of using the project for their own benefit and of doing little to help politically unconnected farmers. Equally, according to the Chinese company, this is one of the reasons behind the failure of technology transfer, as FRELIMO members and state officials' farmers are not fully committed to this. For one politically connected ARPONE farmer, a Chinese manager claimed that instead of using the money received from the bank for agricultural purposes, his patron was spending it on international trips.

This case illustrates well the kind of challenges Chinese technical cooperation initiatives face on the ground when becoming part of extraversion strategies by their counterparts in Africa, leading to frustrated attempts at triggering development. Yet the involvement of state politicians in the agricultural projects is not the only reason for ARPONE farmers' failure to benefit from these collaborative efforts. For instance, accessing bank loans, which have high interest rates, and recurrent floods remain serious handicaps.

Unlike Angola, Chinese participation in infrastructure construction in Mozambique is quite small and thus difficult to assess the extent to which local elites are using it to accumulate wealth and strengthen their power. Corruption scandals are, however, frequent in the construction sector and have plagued major external players in the sector, namely Brazilian construction companies spiralling all the way up to FRELIMO and Brazil's Federal government officials (Alden et al. 2014; Alden et al. 2017).

Between 2000 and 2015, Mozambique received $1.87bn loans from China (CARI 2017), less than one-tenth of the loans extended to Angola. The bulk of the loans were disbursed by the China Exim Bank to build transport infrastructure. Whilst China's share in Mozambique's infrastructure and mining sectors remains marginal, the recent discovery of offshore natural gas and other natural resources (large coal reserves in the north) might change this scenario in the near future as production comes online and becomes available to use as collateral for larger loans. Whilst it enhances prospects of alleviating Mozambique's deep dependency on Western donors and international institutions, at the same time it presents a new set of opportunities to the local elite for economic accumulation in collusion with external partners. Just to illustrate this, in 2014 a group of FRELIMO high military officers set up a joint-venture with the Chinese company Kingho Investment to explore coal in Tete province. It is worth noting that in September 2017, FRELIMO received a "gift" of 500 motorcycles from the Chinese Africa Great Wall Mining Development Company to support its congress. This wasn't the first time: during the 2009 electoral campaign FRELIMO also received 30 motorcycles from the Chinese TENWIN INT LDA (Notícias 2017; O País 2009). This put in question the Chinese notion of non-interference in internal political affairs.

As with Angola, the two business collusion cases analysed above show how little Chinese engagement has contributed to Mozambican development so far, if anything they have been largely detrimental owing to the negative environmental and socio-economic impact of these business ventures. Given the recent history of relations, the prospects for changing this type of elitist collusion and extraversion in the new context of abundant natural resources seem remote.

Conclusion

Although much emphasis has been put on the significance of Africa's turn to the East in terms of liberating Africa from the Western grip, not much has been said regarding the extent to which this new reality has (or has not) changed long-standing extraversion and neo-patrimonialist practices that have arguably hindered the continent's development for centuries. This review of China's relations with Mozambique and Angola reveals that China seems to be conforming to African extraversion strategies much in the same way other external powers did. Whilst there are some signs of China trying to engage differently, the two cases demonstrate how easily these distinct traits get folded into established extraversion strategies. The cases also show that China's economic cooperation has been thus far unable to escape neo-patrimonialist practices that have proven largely incompatible with development. In Angola, oil has been the major extraversion asset in its relationship with China. The Angolan elite has successfully instrumentalised oil exports to China and oil-backed loans from Beijing for economic and political purposes, thus arguably playing a significant role in consolidating and strengthening dos Santos' patrimonial rule since the end of the civil war in 2002. In Mozambique, due to a lack of natural resources throughout most of its post-war period, extraversion strategies with China have been built mostly around joint businesses and cooperation initiatives, with ample evidence of economic and political dividends for the ruling elite. Furthermore, there are few signs of Chinese engagement triggering economic development in either country; in fact, in Mozambique, the impact has been mostly negative.

Despite African praise of China's rhetoric of exceptionalism, the structural dynamics of bilateral relations do not seem to differ much from the way they relate to Western powers. Angola and Mozambique are looking East but ruling elite agency remains largely subservient to the logic of extraversion. The much alluded and celebrated agency of Angola in the context of its relations with China (Corkin 2013) can be thus largely explained in light of extraversion strategies. Rather than ascertaining African standing in bilateral relations, this type of agency may in reality only be deepening African dependency on China as ruling elites perfect their ability to extract resources from yet another external power. Collusion with ruling elites seems to be inescapable, even when unintended. This indicates that China is finding it increasingly costly and difficult to disengage from this pattern, much like other external powers before it. What is problematic here is not the collusion with African elites in itself, but the way it relegates to a secondary position the interests of the majority of the populations in African countries. It is uncertain at this juncture whether the solutions from China in dealing with this predicament will be any different and more effective from those of the West (pressure to improve governance), and whether it will be able to in fact deliver a distinct type of engagement that can trigger development in a neo-patrimonial context.

Further studies have to be carried out across Africa to better reflect the diversity of the continent before these preliminary findings can be generalisable. Future research avenues include assessing whether Chinese is better able to structure the terms of the relationship and deliver some measure of exceptionalism in regulated/capped neo-patrimonial types (like Botswana) as defined by Bach. Also of interest would be to assess whether China contributes to the consolidation or erosion of any of these types (capped and predatory neo-patrimonialism). Although this chapter suggest that extraversion happened in both cases regardless the difference in portfolios strengths, a broader number of cases need to be examined building on Peiffer and Englebert's (2015) methodology to assess the extent to which extraversion assets strengths impact differently on African countries' vulnerability to China's developmental agenda. As China settles in the continent for the long run, China–Africa scholarship is poised to become increasingly intertwined with many of the debates and questions that have been baffling African and development studies for decades.

Notes

1 Information according to email correspondence (July 2016) with an anonymous researcher following a fieldtrip to Angola.
2 Email exchange, anonymous researcher, July 2016.
3 Afonso Dhlakama, the leader of Renamo, the main opposition party, has been reported as being a shareholder in Socadiv Holding Lda., a company specialising in the exploitation of timber and set up in 2007. See *Indian Ocean Newsletter*, 2007.

References

Alden, Chris and Daniel Large. 2011. "China's Exceptionalism and the Challenges of Delivering Difference in Africa". *Journal of Contemporary China* 20(68): 21–38.

Alden, Chris, Sergio Chichava and Paula C. Roque. 2014. "China in Mozambique: Caution, Compromise and Collaboration". In *China and Mozambique: From Comrades to Capitalists*, edited by Chris Alden and Sérgio Chichava, 1–23. Johannesburg: Jacana.

Alden, Chris, Sérgio Chichava and Ana Christina Alves. eds. 2017. *Mozambique and Brazil: Forging New Partnerships or Developing Dependency?* Johannesburg: Jacana Media.

Alden, Chris and Ana Cristina Alves. 2008. "History and Identity in the Construction of China's Africa Policy". *Review of African Political Economy* 35(115): 43–58.

Alves, Ana Cristina. 2009. "Angola's Resources: From Conflict to Development". In *Africa and Energy Security: Global Issues, Local Responses*, edited by Ruchita Beri and Uttam K. Sinha, 155–175. New Delhi: Academic Foundation & IDSA.

Alves, Ana Cristina. 2013a. "Chinese Economic Statecraft: A Comparative Study of China's Oil-backed Loans in Angola and Brazil". *Journal of Current Chinese Affairs* 42(1): 99–130.

Alves, Ana Cristina. 2013b. "China's "Win-Win" Cooperation: Unpacking the Impact of Infrastructure-for-Resources Deals in Africa". *South African Journal of International Affairs* 20 (2): 207–226.

Bach, Daniel C. 2011. "Patrimonialism and Neopatrimonialism: Comparative Trajectories and Readings". *Commonwealth & Comparative Politics* 49(3): 275–294.

Bayart, Jean-Francois. 1985. *L'Etat au Cameroun*. Paris: Presses de la Fondation Nationale de Sciences Politiques.

Bayart, Jean-Francois. 2000. "Africa in the World: A History of Extraversion". *African Affairs* 99(395): 217–267.

Billon, Philippe le. 2001. "Angola's Political Economy of War: The Role of Oil and Diamonds, 1975–2000". *African Affairs* 100(398): 55–80.

Birmingham, David. 2015. *A Short History of Modern Angola*. New York: Oxford University Press.

Bratton, Michael and Nicholas van de Walle. 1994. "Neo-patrimonial Regimes and Political Transitions in Africa". *World Politics* 46: 453–489.

Brito, Luís. 1988. "Une relecture nécessaire: la genèse du parti-État FRELIMO". *Politique Africaine*, 29: 15–27.

Burgis, Tom. 2015. "Queensway Group Probed Over Use of 'Secrecy Jurisdictions'". *Financial Times*, 5 May. Available online: www.ft.com/content/a95e8252-f015-11e4-a b73-00144feab7de (retrieved 23 August 2017).

Callaghy, Thomas. 1984. *The State-Society Struggle: Zaire in Comparative Perspective*. New York: Columbia University Press.

Canby, Kerstin et al. 2008. *Forest Products Trade between China and Africa. An Analysis of Imports and Exports*. www.forest-trends.org/wp-content/uploads/imported/ChinaAfrica Trade.pdf (accessed 18 August 2017).

CARI Database: Chinese Loans to Africa (version May 2017), available online: www.sais-ca ri.org/data-chinese-loans-and-aid-to-africa (accessed 3 September 2017).

Chabal, Patrick and Nuno Vidal. eds. 2007. *Angola: The Weight of History*. London: Hurst.

Chabal, Patrick and Jean-Pascal Daloz. 1999. *Africa Works: Disorder as Political Instrument*, Oxford: James Currey.

Clapham, Christopher. 1996. *Africa and the International System: The Politics of State Survival*. Cambridge: Cambridge University Press.

Clapham, Christopher. 2008. "Fitting China in". In *China Returns to Africa: A Rising Power and a Continent Embrace*, edited by Chris Alden, Daniel Large and Ricardo Soares de Oliveira, 361–370. London: Hurst.

Clapham, Christopher. 1985. *Third World Politics: An Introduction*. Madison, WI: University of Wisconsin Press.

Cooper, Frederick. 2002. *Africa since 1940: The Past of the Present*. Cambridge: Cambridge University Press.

Corkin, Lucy. 2013. *Uncovering African Agency: Angola's Management of China's Credit Lines*. London: Ashgate.

Croese, Sylvia. 2012. "One Million Houses? Chinese Engagement in Angola's National Reconstruction". In *China and Angola: A Marriage of Convenience?* Edited by Ana Alves and Marcus Power, 124–144. Cape Town: Pambazuka Press.

Crook, Richard C. 1989. "Patrimonialism, Administrative Effectiveness and Economic Development in Cote D' Ivoire". *African Affairs* 88(351): 205–228.

EIA. 2013. "First Class Connections: Log Smuggling, Illegal Logging and Corruption in Mozambique." Available at: https://eia-international.org/wp-content/uploads/EIA-First-Class-Connections.pdf (accessed 6 April 2018).

Eisenstadt, S. N. 1973. *Traditional Patrimonialism and Modern Neo-patrimonialism*. London: Sage.

Evans, Peter. 1989. "Predatory, Developmental and Other Apparatuses: A Comparative Analysis of the Third World State". *Sociological Forum* 4(4): 561–582.

Forquilha, Salvador 2009. "Reformas de descentralização e redução de pobreza num con-texto de Estado-neo-patrimonial. Um olhar a partir dos conselhos locais e OIIL em Moçambique". In *Pobreza, desigualdade e vulnerabilidade em Moçambique*, edited by L. de Brito *et al.*, 19–48. Maputo: IESE.

Forquilha, Salvador and Aslak Orre. 2011. "Transformações sem mudanças? Os conselhos locais e o desafio da institucionalização democrática em Moçambique". In *Desafios Para Moçambique*, edited by L. de Brito *et al.*, 35–53. Maputo: IESE.

Hanlon, Joseph. 2004. "Do Donors Promote Corruption?: The Case of Mozambique". *Third World Quarterly* 25(4): 747–767.

Harrison, Graham. 1999. "Corruption as "Boundary Politics": The State, Democratisation, and Mozambique's Unstable Liberalisation". *Third World Quarterly* 20(3): 537–550.

Hodges, Tony. 2004. *Angola: Anatomy of an Oil State*, Oxford: James Currey.

Hodges, Tony. 2008. "The Economic Foundations of the Patrimonial State". In *Angola: The Weight of History*, edited by Patrick Chabal and Nuno Vidal, 175–199. New York: Columbia University Press.

Henriksen, Thomas. 1978. "Marxism and Mozambique." *African Affairs*, 77(309): 441–462.

Hwang, Jyhong, Deborah Brautigam, Janet Eom. 2016. *How Chinese Money is Transforming Africa: Is Not What You Think*, CARI, Policy Brief 11, available online: https://static1.squarespace.com/static/5652847de4b033f56d2bdc29/t/5768ae3b6a4963a2b8cac955/1466478245951/CARI_PolicyBrief_11_2016.pdf (retrieved 3 September 2017).

The Indian Ocean Newsletter. 2007. "Afonso Dhlakama goes into business in Mozambique" (1 December).

Jackson, Robert. 1990. *Quasi States: Sovereignty, International Relations and the Third World*, Cambridge: Cambridge University Press.

Jackson, Steven F. 1995. "China's Third World Foreign Policy: The Case of Angola and Mozambique, 1961–1993". *The China Quarterly* 142: 388–422.

Kohli, Atul. 2004. *State Directed Development: Political Power and Industrialisation in the Global Periphery*. Cambridge: Cambridge University Press.

Lindsay, Lisa A. 2014. "Extraversion, Creolization, and dependency in the Atlantic Slave Trade". *Journal of African History* 55: 135–145.

Mabunda, Lázaro 2014. "Partido FRELIMO financia-se com dinheiro de contrabando de madeira na Zambézia", CIP Newsletter, 11/2014. www.cip.org.mz/cipdoc%5C329_CIP-a_ transparencia_11.pdf (retrieved 2015).

MacauHub. 2017. "Angola Gives Tax Breaks to Real Estate Projects of Chinese Group CIF", 13 June 2017, available online: https://macauhub.com.mo/2017/06/13/pt-angola-concede-beneficios-fiscais-a-projectos-imobiliarios-do-grupo-chines-cif/ (accessed 3 September 2017).

Machel, Milton 2011. "Um negócio da China para a família Guebuza", *CIP Newsletter*, 11, CIP, Maputo (April).

Mackenzie, Catherine. 2006. "Forest Governance in Zambezia, Mozambique: Chinese Takeway!" *FONGZA*.

Mailey, J. R. 2015. *The Anatomy of the Resource Curse: Predatory Investment in Africa's Extractive Industries*. Washington, DC: Africa Center for Strategic Studies.

Médard, Jean-François. 1979. "L'État sous-developpé au Cameroun". In CEAN, *Anneé africaine 1977*, Paris: Pedone, 35–84.

Médard, Jean-François. 1991. "L'État néo-patrimonial en Afrique noire", In *États d'Afrique noire : Formation, mécanisme et crise*, edited by J-F Médard, 323–353. Paris: Karthala.

Médard, Jean-François. 1990. "l'État neopatrimonialisé". *Politique Africaine* 39: 25–36.

Médard, Jean-François. 1992. "Le "Big Man" en Afrique: Esquisse d'Analyse du politicien Entrepreneur". *L'Année Sociologique* (1940/1948-) 42: 167–192.

Miramar. 2015. "Dhlakama proíbe corte de madeira", 7 July. Available at www.miramar.co.mz/Noticias/Dhlakama-proibe-corte-de-madeira (accessed 20 August 2017).

Mosse, Marcelo 2007. "Carta Aberta a Hu Jintao. Canal de Opinião". *Canal de Moçambique*, Maputo (8 February).

Newitt, Malyn. 1994. *A History of Mozambique*. London: C Hurst & Co Publishers Ltd.

Nhamirre, Borges 2014. "A consumação do alerta do CIP. Migração Digital entregue à Empresa da Família Presidencial", *CIP Newsletter*, 2, CIP, Maputo (April).

Notícias. 2017. "Em apoio ao XI Congresso: Empresa chinesa oferece motorizadas à FRELIMO". Available at: www.jornalnoticias.co.mz/index.php/politica/71513-em-apoio-ao-xi-congresso-empresa-chinesa-oferece-500-motorizadas-a-frelimo.html (accessed 2 October 2017).

O País, 2009. "Empresa chinesa apoia campanha eleitoral da Frelimo" (4 August), Available at: http://opais.sapo.mz/index.php/politica/63-politica/2270-empresa-chinesa-apoia-campanha-eleitoral-da-frelimo-.html (accessed 2 October 2017).

Peiffer, Caryn and Pierre Englebert. 2012. "Extraversion, Vulnerability to Donors, and Political Liberalization in Africa". *African Affairs* 111(444): 355–378.

Pereira, Micael. 2017. "Angola's Dos Santos: One Last Gift from Father to Daughter", *News 24*, 3 September, available online: http://www.news24.com/Africa/News/angola-one-last-gift-from-father-to-daughter-20170903-2 (accessed 6 September 2017).

Pitcher, M. Anne. 1996. "Recreating Colonialism or Reconstructing the State? Privatisation and Politics in Mozambique". *Journal of Southern Africa Studies* 22(1): 49–74.

Pitcher, Anne, Mary H. Moran and Michael Johnston. 2009. "Rethinking Patrimonialism and Neopatrimonialism in Africa". *African Studies Review* 52(1): 125–156.

Rede Angola. 2016. "China recusa novo emprestimo a Sonangol", 16 October, available online: www.redeangola.info/china-recusa-novo-emprestimo-sonangol/ (accessed 3 September 2017).

Roth, Guenther. 1968. "Personal Rulership, Patrimonialism and Empire-Building in the New States". *World Politics* 20(2): 194–206.

Serra, Carlos 2007. "Carta para o Senhor Presidente da República, Armando Emílio Guebuza", 30 Janvier. http://oficinadesociologia.blogspot.com/2007/01/carta-para-o-senhor-presidente-da.html (accessed 1 October de 2017).

Soares de Oliveira, Ricardo. 2007. "Business Success, Angola-style: Postcolonial Politics and the Rise and Rise of Sonangol". *The Journal of Modern African Studies* 45(4): 595–619.

Soares de Oliveira, Ricardo. 2015. *Magnificent and Beggar Land: Angola Since the Civil War*. London: Hurst.

Vail, Leroy and Landeg White. 1981. *Capitalism and Colonialism in Mozambique: A Study of Quelimane District*. London: Heinemann.

van de Walle, Nicolas. 2003. "Presidentialism and Clientelism in Africa's Emerging Party Systems". *Journal of Modern African Studies* 41(2): 297–321.

Weimer, Marcus and Alex Vines. 2012. "China's Angolan Oil Deals 2003–2011". In *China and Angoa: A Marriage of Convenience?* edited by Marcus Power and Ana Cristina Alves, 85–104. Cape-Town: Pambazuka Press.

Zheng, Yang. 2016. "Witness to History." *ChinAfrica, 8.* www.bjreview.com/Nation/201607/t20160701_800061150.html (accessed 6 April 2018).

Zolberg, Aristide R. 1966. *Creating Political Order: The Party-States of West Africa*, Chicago, IL: Randy McNally & Company.

16

BETWEEN RESOURCE EXTRACTION AND INDUSTRIALIZING AFRICA

Mzukisi Qobo and Garth le Pere

An enduring paradox in Africa's post-independence development has been its inability to use its manifold resource endowments as an impetus for generating sustained industrialisation. In part this deficiency is rooted in the continent's colonial and post-colonial history: in the international division of labour, Africa has in the main been a supplier of raw materials, and an importer of finished products (see Rodney 1972). This heavy dependence on commodities accentuates Africa's vulnerabilities. That is why Africa's latent industrial potential has often been viewed as a prerequisite for guiding countries out of the twin challenges of structural poverty and underdevelopment. There has even been an element of utopianism attached to this project since the value-addition from industrialisation is considered salutary not only for broadening but also embedding the teleological goals of freedom and democracy, which are both constitutive of, and instrumental to, development (see Sen 1999).

Here industrial development should be conceived broadly. It is not only about using mineral resources as the sole basis for industrial development, but also entails value-addition in agriculture, developing production value chains on a horizontal basis, and creating the requisite ensemble of skills for enabling industrial growth. This process can also be conceived in Schumpetarian terms as "creative destruction" (Schumpeter 1994 [1942]), where entrepreneurial innovation rests on the introduction of new products and the continual improvement of existing ones. In short, it should be about freeing the productive capacities of society and enabling its progress.

The very idea of self-determination, which was the cornerstone of anti-colonial struggle, can be extended to the sphere of production as implying the capacity of a society's mastery over its economic evolution and the purpose for which this is shaped, similar to what Putnam has called bonding and bridging social capital (Putnam 2000). At the heart of this *problematique* is how the end of colonial rule in

Africa hardly generated substantive freedoms, expressed in the development of individual capabilities and nurturing broad-based industrial development. Seeking the political kingdom first, as Kwame Nkrumah exhorted, became an end in itself.

The entry of China onto Africa's development landscape and its consequent discourse thus presents an opportunity to re-examine the contours of the paradox referred to above as it relates to the tense dialectic between resource extraction and industrialisation. The shift from antiquated forms of Maoist egalitarianism to the "reform and opening up" period under Deng Xiaoping from 1978 set the stage for the conceptual and normative ordering of a new approach to economic development in China. This approach would engender shedding Mao-era anti-imperialist shibboleths by unleashing the social and economic forces behind China's drive to modernisation, thus enabling it to become a major geo-political player in the global circuits of trade, production, finance, and investment.

In the process of its own dramatic and unprecedented growth and development over the last five decades, China has always sought to burnish its developing country credentials on the basis of "actively promoting a preferred idea of what China is and what it stands for in international relations. This entails of form of Occidentalism where images and understandings of the 'West' are constructed against which non-western cultures, societies and states identify themselves" (Breslin 2011, 1339). Consider, for example, China's impressive performance: on the back of high growth rates, China managed to lift more than 700 million people out of poverty between 1978 and 2012, reducing extreme poverty rates from 84 percent to 10 percent during this period (Wu 2016). Its growth "miracle" has made it attractive to other developing countries, not least in Africa, as a possible model (see Alden, in this volume).

Concerns about the "China threat," mainly emanating from Western corridors of power and their academic acolytes, have been met with a vocabulary underscoring the language of change, such as China's "peaceful development" or "peaceful rise," which were further elaborated as China's quest for a "harmonious world" (Wang 2003). As a hermeneutic exercise, this language was also meant to set China apart from the Western idiom in the developing world, driven mainly by political expediency, an exploitative gestalt, and lack of ethnical restraint. As has been asserted "[i]n contrast to European imperialism, there is no overriding 'civilizing mission' driving China's approach to Africa ... and indeed, the Chinese government has demonstrated that it can quite happily work with any political and social environment it finds itself in" (Alden 2008, 355–356).

In the case of Africa, this language has framed the ideational parameters of economic cooperation centered on the principles of peace and friendship; non-interference; equality and mutual benefit; mutual support and sincerity; and the achievement of common progress. These principles, in turn, have informed a strategic logic toward Africa and the rest of the developing world that heralds a move toward "state-led pragmatic nationalism" by China. According to Zhao, this pragmatic calculus is "ideologically agnostic, having nothing, or very little to do with either communist ideology or liberal ideals. It is firmly goal-fulfilling and national-interest strategic behaviour, conditioned substantially by China's historical experiences and geo-strategic interests" (Zhao 2007, 39). At the core of such behaviour is tactical flexibility, strategic subtlety, and avoiding any appearance of being confrontational.

This chapter examines the role of China in Africa as a development partner through the prism of its involvement in Africa's resource-industrialisation complex and considers how meaningful this involvement has been from an African perspective in influencing the knowledge-generating landscape of Africa's growth and development.

Some discursive considerations

There has been no shortage of development initiatives in, and for, the African continent. Since the late 1970s, Africa has been a laboratory for development initiatives and plans. (Prior to this, in the 1960s, many African countries were grappling with the challenges and vicissitudes of newly won independence while others were still locked in anti-colonial and liberation struggles.) Notwithstanding this dense "ecosystem" of initiatives and plans, the few African countries that have managed to achieve high levels of growth and registered some gains in poverty reduction since the mid-1990s have mainly done so on the back of macro-economic reforms, fiscal prudence, and improving their governance systems. Building inclusive political and economic institutions, investing in infrastructure, developing human capital, and undertaking micro-economic reforms aimed at diversifying the economic structure are critical catalysts whose absence in the majority of countries continue to constrain Africa's development prospects.

China's emergence as a major actor in Africa has certainly been of major significance in debates about the trajectory of the continent's development. In Gramscian terms, China represents a counter-hegemonic movement against the dominance of essentially Western-defined models of development (Tu and Mo 2015). This has a lot to do with the mutually reinforcing interactions of knowledge and power. As Michel Foucault reminds us, all knowledge flows through circuits of power, and the contest between the two is fought through a discursive battle in terms of versions and logics of truth (Foucault 2002).

The historical dimension

Africa's stunted development has to be located in the epistemological pathologies of different paradigms of development thinking that have been tried but have failed Africa over the last five decades. Much development orthodoxy on Africa has emerged in the context of the Anglo-Saxon experience but has proven bankrupt when tested in the crucible of generating economic growth and promoting human welfare. When development issues became a subject of debate in the late 1950s, as many African countries gained independence, an increasing number were influenced by a heterodox mix of approaches, such as the mechanical application of

Keynesian demand management with reference to the post-war reconstruction of Europe, the efficacy of the Marshall Plan, and Soviet style *dirigisme* and central planning (see Lal 1985).

The Arthur W. Lewis model of growth (Lewis 1955) brought these together in a theoretical and policy synthesis for structural change based on modernisation and industrialisation, which found wide appeal in Africa, starting with the post-1960s period of independence. This was based on the assumption that the experience of the West could be replicated by moving away from subsistence-based agriculture to a modern industrial economy. As a result, agriculture as the mainstay of economic life in Africa declined in importance, while preference was given to capital formation for industrialisation, urbanization, and technological transformation. The result was rising food imports and a failure of export earnings to grow fast enough to generate the required industrial inputs. This was exacerbated by the poor absorptive capacity of the formal sector to cater for growing numbers of rural migrants and a burgeoning urban working class.

The next phase, in the 1970s, saw import substitution industrialisation (ISI) acquire increasing paradigmatic influence, inspired by the Argentine economist Raul Prebisch and his German collaborator, Hans Singer, under the auspices of the UN Economic Commission for Latin America (see Bloch and Sapsford 2000). Steeped in economic structuralism, their basic postulate was that terms of trade deteriorate because of a secular decline in the value of agricultural and commodity exports compared to manufactured exports. The logic of ISI was to use high tariffs and quotas on imports to stimulate domestic production, since this would encourage economies of scale and lower labour costs. However, reality proved to be different: protective tariff walls made products more uncompetitive and costly; government-subsidised imports of capital goods resulted in heavy debt burdens and serious balance of payments deficits; and commodity exports suffered because exchange rates were artificially over-valued so as to raise prices of exports and lower those on imports.

As African economies became even more fragile under the burden of balance of payment crises, and declining commodity demand due to the recession in developed countries in 1980/81, African countries embarked on a re-appraisal of their development strategies. The high water mark was the inward-looking Lagos Plan of Action (LPA) adopted by the Organisation of African Unity in 1980, followed by the Abuja Treaty in 1991 that envisioned the creation of an African Economic Community by 2028. Not only did the LPA encourage greater self-reliance and economic integration based on a Pan-African ethos and the promotion of African unity, but there was a greater focus on people-centered approaches to development such as addressing poverty, unemployment, and welfare challenges.

The LPA, however, was quickly eclipsed by the 1981 World Bank Report *Accelerated Development in Sub-Sahara Africa: An Agenda for Action*, which was a direct riposte to the LPA. This infamous "Berg Report," named after its author and American economist, Elliot Berg, attributed Africa's development failures to pricing policies, over-valued exchange rates, and excessive state intervention in the economy. The Report inaugurated the beginning of outward-looking structural adjustment policies in Africa with a focus on economic liberalization in the form of market-oriented reforms, export promotion, monetary and fiscal discipline, and greater economic laissez faire. This represented the apotheosis of the "Washington Consensus" dogma (van Waeyenberge 2006).

Alternative development theories flow from the dependency critique of modernisation, and include various strands and approaches that emphasise agency and a people-centered approach to development. Pieterse (2010) traces the first strands of this approach to the Dag Hammarskjold Foundation's 1975 report titled "What Now?", which articulated dissatisfaction with mainstream development thinking and practice. The central thesis of alternative approaches is that "development" is often an exogenous process for developing countries since its origins are in a Western epistemological frame, with no recognition of local conditions and context. Alternative development perspectives further problematise the notion of poverty itself. As Eichengreen (2007, 66) points out: "Countries accepting American aid had to sign bilateral pacts with the United States agreeing to de-control prices, stabilise exchange rates, and balance budgets. In effect, they had to commit to putting in place the prerequisites for a functioning market economy."

Alternative development also provides a critique of technology and science, emphasizing its non-neutrality and how it is steeped in uneven power-knowledge relations between the North and South, which perpetually leave the latter "lagging behind" (Rist 1990; Escobar 1995). Although emerging powers such as China and India have been climbing the technological curve faster, poorer regions of the world, Africa in particular, are still far from catching up, let alone owning the proprietary knowledge embedded in technology.

The solution, according to alternative development theory, is to radically change the worldview from which development is perceived. The current development paradigm is rejected because it is a westernization project. Developing countries must thus develop endogenous programs to deal with their problems, which reflect the will of their societies. This is what Rodrik (2007) refers to as "one economics and many recipes," suggesting that developmental approaches should be cognisant of diverse institutional qualities that exist in different countries, and that it might not be possible to universalise solutions that are appropriate in one context. The influence of alternative development has been evident in the way that international financial institutions and many current development programs have emphasised community involvement and participatory approaches.

An example of such an alternate approach took place in 1999, when African critics responded with their own systematic critique of structural adjustment with a study, *Our Continent, Our Future: African Perspectives on Structural Adjustment* (Mkandawire and Soludo 1999). The LPA's call for self-reliance was complemented by the imperative for African countries to compete in the global economy on the basis of comparative advantage. The study extolled the virtues of ISI as a means of promoting high-value added and labour-intensive manufactured exports and was a direct rebuttal of the neo-liberalism of the Washington Consensus. The study put a premium on bringing the state back in by according it a strategic developmental role, informed by the Asian experience, as the engine of sustained high growth rates and structural change in the system of production.

A major impulse that drove the next paradigmatic change in the 1990s came from "sustainable development," which drew its inspiration from the 1987 Bruntland Report, *Our Common Future*. The broad contours included not only economic growth but improvements in social, cultural, and political life anchored in environmental security. In this scheme, development had to confront the vagaries of drought, desertification, soil erosion, disease, and the extent to which human agency contributed to these (UNECA 2001).

A strong normative orientation defined the policy matrices: growth imperatives had to be balanced by equity principles with an emphasis on redistribution, meeting basic needs, integrating the formal sector, and empowering local communities. The concept of "sustainable development," however, was appropriated if not adulterated by the World Bank and the IMF, who reduced its ethos to mechanically programmed poverty reduction strategies grounded in functioning institutions, good governance, responsible leadership, fiscal discipline, better debt management, job creation and so on (Briseid et al. 2008).

Other complementary approaches evolved in the form of the heavily indebted poor countries (HIPC) initiative, to help African countries to overcome indebtedness and thereby ensure that they are on track toward growth and development. Such initiatives were narrowly focused on macro-economic and structural constraints that hindered growth in the African continent. These were supposed to lead to higher levels of growth and ensure poverty reduction. However, this post-Washington Consensus hardly ameliorated conditions as human development indicators declined, export revenues decreased, and shortages of development finance and slower growth caused many African countries to contract by more than 10 percent between 1990 and 2005.

The contemporary dimension

The current juncture has been shaped by another home-grown response in the form of the New Partnership for Africa's Development (NEPAD), which sought to address the negative impact of globalisation on Africa's political economy. NEPAD represented a normative call-to-arms for African leaders to address governance, peace, and security; to concentrate on infrastructure and human capital backlogs; to adhere to greater fiscal prudence; to mobilise resources for development; and to enhance conditions for more aid flows and market access to developed countries (Rukato 2010). However, the promise of NEPAD has quickly dissipated in the throes of bureaucratic paralysis in the African Union as the steering mechanism, compounded by programming incoherence and lethargic political leadership.

The global financial crisis that followed 2008 affected the African continent adversely. Even though the crisis had its origins in advanced industrial countries, the effects on the continent were particularly intense, given Africa's institutional vulnerabilities. As underlined by the 2010 Economic Commission Report on Africa, these effects included the slowing down of trade, increases in food and fuel prices (which had declined since late 2006), weakening demand for exports in goods and services, decreases in remittances, and reduced private capital inflows (UNECA 2010, 1). In turn, employment conditions became dire as trade finance dried up and production was scaled back. Indicative of hard times in the continent, the GDP growth rate in 2009 averaged 1.6 percent, a far cry from a year earlier where the continent grew by 4.9 percent (UNECA 2010, 2).

There has been a shift toward statism and industrial development amongst some of the key African organizations. In its report on "Governing development in Africa," UNECA (2011) sets out three critical challenges where state intervention would make policy sense in managing a country's competitive and comparative assets in any industrialisation agenda: diversification and structural transformation; an increasing role of the state in structural transformation; and the construction of a developmental state to enhance economic transformation. The kind of state intervention approach promoted by the UNECA emphasises the need to upscale industrialisation and look at geographic markets beyond traditional partners such as the EU and the US (UNECA 2016). The major argument here is that the state is best placed to drive economic development in Africa, given the pervasive failure of markets.

While in practice African countries have gained a better footing in their economic management through stabilizing their macro-economic environments and promoting market-based economic reforms, there has been a growing disappointment with the Washington Consensus models to development that emphasise deregulation, market-based economic management, and liberal trade and investment policies. The global financial crisis deepened the emerging global consensus that this model is no longer applicable. Nevertheless, it is not clear what innovative body of ideas would emerge to replace the Washington Consensus; moreover, there is also no particular consensus on the primacy of the state in economic management.

The main weakness with the UNECA's view is its failure to recognise and account for institutional weaknesses of the state. Consider its definition of such a state as "one that has the political will and the necessary capacity to articulate and implement policies to expand human capabilities, enhance equity, and promote economic and social transformation" (UNECA 2011, 118). This consideration remains absent in many African countries, revealing the chasm that exists between the ideal and reality. In their critique of institutional weaknesses in the African continent, Mazrui and Wiafe-Amoako (2016, 10) point out that: "African elites have constantly ignored creating sustainable institutions, not because those institutions are difficult to establish, but because of how those institutions fit into their [elite] personal values and those of the society in general."

Against this broad conceptual and normative alchemy, the strategic entry of China onto the African landscape has provided some developmental space for struggling economies, propelled by China's high demand for commodities and natural resources. However, herein lies an acute dilemma that requires qualification: on the one hand, this developmental space derives from China's norm of non-interference and its *de facto* unconditional terms of economic, investment, and commercial interaction. On the other hand, a reality-check is required about the extent to which this developmental space has delivered broad-based welfare enhancing gains, since by value two-thirds of Africa's exports to China in 2014 were crude oil and base metals that only originate in a certain cluster of countries (Romei 2015). That said, it must be recognised that the global financial meltdown and the recalibration of the Chinese economy as a consequence has resulted in a loss of export earnings due to the income elasticities associated with commodities. This has made the imperative for building industrial capacity all the more urgent (see UNECA 2016).

China's entry: the nature of debates

As early as 1967 the Ghanaian scholar Emmanuel Hevi wrote that "few subjects are as complicated as China's Africa policy and the motives behind it" (Hevi 1967, 2). This observation still has profound relevance since debates persist about whether China's motives are entirely altruistic as a trusted interlocutor in Africa's development after more than five decades of desultory progress, or whether these have a darker side as a resource predator driven by national interest. This dichotomous tension underpinning the dilemma referred to above has shaped much of the popular and academic literature on the subject (see Large 2008), and is well captured by the following sentiments:

> Africa sells raw materials to China and China sells manufactured products to Africa. This is a dangerous equation that reproduces Africa's old relationship with colonial powers. The equation is not sustainable for a number of reasons. First, Africa needs to preserve its natural resources to use in the future for its own industrialisation. Secondly, China's export strategy is contributing to the de-industrialisation of some middle countries ... . It is in the interests of Africa and China to find solutions to these strategies.
>
> *(Mbeki 2006, 7)*

These sentiments in many ways help to define the terms of an often tendentious debate and have also proved its heuristic value in helping to delimit and define the nature of empirical research, analytical frameworks, and conceptual refinements.

A more nuanced reading was provided by Alden in terms of three perspectival prisms. The first views China as a development partner committed to a win-win formula of mutual gains through trade, investment, and development assistance, which has injected a new-found dynamism into Africa's growth prospects and geo-strategic relevance. In the second formulation, China is an economic competitor whose national interests are concentrated on the extraction of Africa's resource abundance as a means of underwriting China's own modernisation and growth agenda. Here scant attention is paid to typical Western normative concerns relating to matters such as good governance, human rights, environmental protection, and labour standards. In the third, China is the embodiment of the new scramble for Africa and behaves no differently from other neo-colonial powers but whose ambition is to displace traditional Western spheres of influence under the remit of South–South cooperation. This style of "authoritarian capitalism" provides China with the long-term leverage and geo-strategic gravitas to re-shape the political economy of Africa (Alden 2007).

Another compelling framework has been provided by Kaplinsky (2007) who suggests that Africa and other developing countries will invariably be caught in the vortex of China's economic rise and that this will pose a "disruptive challenge" for the countries and regions concerned. His concern is to examine the impact of five motor forces responsible for China's spectacular growth. First, there is the sheer size and high trade intensity of the Chinese economy together with its comparative advantages to crowd out African and developing countries from moving up regional, let alone global, value chains. Second, China's massive foreign reserves provide it with a muscular resource base not only to dispense development assistance and investment but to gobble up bond markets and equity stakes in developing countries in a manner that undermines local private initiative and entrepreneurship. Third, China's state-owned enterprises have a competitive edge of easy access to capital and low-cost labour. They typically tend to ignore pressures for prudent corporate governance and social responsibility. Fourth, China's access to low-cost labour and increasing domination of regional and global value chains presents particular challenges to low- and high-tech manufacturers globally, most crucially since China is the second largest investor in research and development in the world after the US. And finally, China's appetite for raw materials reinforces African countries' dependence on commodity exports which fuels the "Dutch Disease" effect. This concentration tends to undermine other sectors of the economy by artificially inflating the value of a country's currency, thereby making other exports less competitive. Moreover, capital resources are drained from manufacturing and agriculture and diverted into services, transportation, construction and other non-tradeable sectors of the economy.

With variation on these themes, Mills and White offer several considerations suggesting that the calculus of decision for African countries will be determined by their collective response to whether China is a "threat or a boon"? (Mills and White 2006; see also Habib 2008). Their argument rests on the thesis that commodity booms are unsustainable over the long term and invariably result in declining terms of trade for African countries that rely on the extractive sector. China's industrial and trading capacity will, therefore, overwhelm African markets with cheap manufactured products and negate any attempts at nascent industrialisation, exacerbated by the breadth and depth of China's reach into third party markets that will inhibit Africa's ability to develop its own export alternatives. They repeat the concern that the thrust of China's strategic interests in Africa is to meet the demands of its own modernisation and relations with African countries will thus be hammered out on the anvil of cold, self-serving realpolitik rather than ethical restraint. Finally, they point out that the expansionary drive of Chinese companies "going out" into Africa could capture big slices of the natural resources sectors and undermine opportunities for economic diversification and structural transformation.

These perspectives cast an important spotlight on how the knowledge base of China's role in Africa has been shaped, albeit hardly from an ideologically agnostic point of view. The thematic menu has certainly grown as more empirical studies have sought to impose greater conceptual order and analytical rigor on the mounting ambiguities of China's African interface (see Broadman 2007; Alden et al., 2008; Bräutigam 1998; Bräutigam 2009). A major problem with this type of discourse is that it has often privileged Western exceptionalism and Eurocentrism in ways not dissimilar from the study of International Relations. This is a way of thinking that enables a modal moral "logic" since it "allows for the continued imagination and invention of Europe's intellectual and political superiority, treating the West as a perennial source of political and religious tolerance" (Kayaoglu 2010, 195–196).

One is thus left to ponder the contested epistemological claims that have crept into how and who shapes the research agenda in China–Africa studies and ultimately for what purpose? What is incontrovertible, however, is that:

> Given some of the more inflated claims about the impact of China in Africa, often contained within arguments about a "new scramble" or "new imperialism," there is a marked gap between the perceptions and exaggerated projections of an inexorable Chinese rise in Africa *and* knowledge of how this is actually playing out.
>
> *(Large 2008, 57; emphasis added)*

Modes of Chinese engagement

If we take the 1955 Bandung Conference as something of a point of departure in China's support for African countries' anti-colonial and counter-hegemonic struggles against Western domination, it established a moral basis for China to become "warm partisans of the cause of African freedom" (Snow 1988, 76). Bandung is often invoked as providing the legitimizing vector for China's expanding role in Africa. Such a role, it must be recalled, has shifted and changed with different rhythms with every passing decade or two as China's role has become more complex and heterodox. China supported Africa's proletarian revolutions in the 1960s; its counter-hegemonic leanings against the US and the Soviet Union in the 1970s; its quest for self-reliance in the 1980s (otherwise known as the "decade of neglect"); and then its more robust economic, diplomatic, and political relations ever since (see Eisenmann et al. 2007).

While often subject to polemical and rhetorical overkill, the "Beijing Consensus" is often contrasted with its antinomy, the "Washington Consensus" as a counterweight to Western conditional approaches to economic relations with African countries. For Ramo, the Beijing Consensus represents "the new physics of power and development ... and is as much about social change as economic change" (Ramo 2004, 4). However, some of the essential tenets of the Beijing Consensus as construed by Ramo have been challenged for peddling in "myths." Kennedy provides some examples: "... technological innovation has not been the centerpiece of China's growth ...; the evidence that China is pursuing sustainable and equitable development is highly limited; ... [and]. China's economic development strategy is [not] unique" (Kennedy 2010, 469–470).

Be that as it may and for what it is worth, the Beijing Consensus can be summarised in terms of five constitutive axioms: an emphasis on incremental reform; innovation and experimentation in promoting productivity and competitiveness; export-led growth and high global demand for China's manufactured goods; state-driven regulation of market forces and a leading role for state-owned enterprises; and governance through highly centralised and concentrated power structures (Williamson 2012, 6–7).

While perhaps a model of social and economic change (see Naughton 2010), the Beijing Consensus has not been an unmitigated blessing and has come with heavy costs for China's domestic environment. For example, public welfare provision and social safety nets have been sacrificed on the altar of economic rationality and efficiency; the aggressive push toward industrialisation and labour- and capital-intensive manufacturing have depressed wages and resulted in high levels of environmental pollution; and a highly competitive and profit-driven domestic market has encouraged corrupt business practices. And in the face of growing centrifugal political tendencies and increasing social stratification, the Communist Party of China (CPC) has been strengthening its authoritarian grip on the Chinese state and society.

In this context, and in terms of Zhao's "pragmatic state-led nationalism," commercial viability and not ideological issues propel China's interests in Africa. China has a "Going Global Strategy" in terms of which Chinese enterprises are expected to establish a multinational presence through the provision of soft loans and investment projects in areas of strategic significance for the Chinese economy. In Africa, the Forum on China-Africa Cooperation (FOCAC) was set up in 2000 in order to manage China's cooperation framework across a range of technical, economic, and political themes. The raft of commitments made at FOCAC have expanded exponentially and it has become a catalyst for shifting China's focus from narrow trade and investment relations to embody a wider range of development concerns. In addition, a new China–Africa Industrial Capacity Cooperation Fund was put in place as China's response to spurring industrialisation since many countries continue to feel the impact of the downturn in commodity demand. Moreover, the Fund is intended to be a partial corrective to the unbalanced trade and investment relationship and is an attempt to assist African countries' economic diversification and integration into regional and global value chains. This is because African countries continue to face the spectre of de-industrialisation that began in the 1980s; in 2014 value added manufacturing declined to less than 11 percent as a contribution to GDP, and Africa provides only 4 percent of global value added in manufacturing (UNECA 2015).

It has been estimated that as China rebalances it economy and moves to more capital-intensive processes, that it will shed more than 85 million jobs in manufacturing (Davies 2015). The challenge is whether African countries will be able to capitalise on this opportunity since even capturing a proportion of this number would have positive spinoffs for employment. Chinese investment has increased in the service and manufacturing sectors as well as in infrastructure, all of which could develop the potential for promoting growth and industrial transformation which has been described as "slow and volatile" (UNECA 2015, 81). African leaders have taken a strong interest in how they can replicate China's successful experience in using Special Economic Zones (SEZs) as platforms for industrial development, especially for attracting foreign investment and enhancing the competitiveness of the manufacturing sector. The establishment of SEZs also complements China's "going out" strategy as they improve domestic and regional markets for Chinese products and services besides leveraging trade and preferential market access arrangements. They have also assisted industrial restructuring in China by relocating well-established and labour-intensive enterprises to other countries. Moreover, they create economies of scale for investments that support clusters of Chinese companies, create supply chains, and spread technological innovation.

Ethiopia is an example of an offshore market where China has aggressively outsourced its rising labour costs: average pay in Henan rose 103 percent in five years from 2009; 80 percent in Chingqing and 82.5 percent in Guangdong for the same period. Ethiopia has thus become an attractive offshore destination because of its cheap labour and electricity as well as government efforts to attract foreign direct investment. China's Huajian Shoe's factory of 3,500 workers outside Addis Ababa produces two million pairs of shoes and the goal is to increase the workforce to 50,000 in eight years (Durden 2014). Moreover, Ethiopia follows a similar tightly controlled statist model as China and there are strong party-to-party links between the CPC and the Ethiopian Peoples' Revolutionary Democratic front. These political ties help to consolidate Ethiopia's cachet with the Chinese, especially in heavy public investment and the use of credit from Chinese state banks.

However, despite this great promise, Chinese-managed and operated SEZs in Africa (such as Ethiopia, Egypt, Mauritius, Nigeria, and Zambia) remain underdeveloped with low levels of investment and exports while their impact on job creation and integration with local economies remains limited (see Kim 2013).

The terrain of China–Africa relations has been strategically shaped and politically consolidated at the bi-lateral level. This has provided the chemistry for crafting modes of confidence-building with African leadership that has, for example, greatly enhanced the legitimacy and operational efficacy of the FOCAC process. If the bilateral formed the first generation of relations, then the challenges represented by the regional and continental matrices will be much greater since they will test China's bona fides as a partner in fostering integration, particularly in assisting

Africa's trans-boundary industrial development and market expansion (Schiere and Rugamba 2011, 91–101).

In celebration of the AU's 50th anniversary, in 2013 African heads of state and government adopted *Agenda 2063* as a transformative vision for "an integrated, prosperous, and peaceful Africa" which is meant to move the continent inexorably in the direction of enhanced growth and development over the next five decades (AUC 2014). The vision is based on a programmatic agenda of five 10-year plans and, while similar to NEPAD in some respects, it considerably raises the level of ambition as a collective charter for the continent's future growth and development. China has already pledged its support for *Agenda 2063*, which will unfold in the context of FOCAC as the primary vehicle of cooperation. To this end, at the 2015 FOCAC VI summit in South Africa, President Xi Jinping announced a $60 billion package in new financing for 10 major initiatives over the next three years. This included $10 billion for building industrial capacity in manufacturing, hi-tech, agriculture, energy, and infrastructure; $5 billion for aid and interest free loans; $35 billion for preferential loans and export credits; and expanding the China–Africa Development Fund to $10 billion.

There are some critical challenges, however, arising out of the strategic dialogue between China and Africa in this regard. First, there is improving access to the Chinese market, especially addressing non-tariff barriers. Chinese rules of origin requirements, for example, are much stricter than those of the EU and the US; and African countries must still compete with much more competitive Asian imports which enjoy easy preferential access to the Chinese market. Second, there is China's ability to plan and mount regional infrastructure projects that can better link many of Africa's small and undeveloped domestic markets and so assist in improving intra-regional trade. Various trade initiatives in Eastern and Southern Africa as well as the North–South corridor, can be examined in order to prepare a pipeline of investable projects. Third, Africa's integration could be greatly assisted with funding that could be unleashed from better coordinated debt relief. This theme forms part of the FOCAC process. Better China–Africa coordination with multilateral institutions and international lenders could result in improved debt sustainability. Finally, there is the matter of untying development assistance, which is not easily addressed since the Chinese approach is not only to assist the recipient country but also to support Chinese companies. However, if assistance was untied it could give African companies an opportunity to participate in tender projects across borders resulting from *Agenda 2063* projects, improve their technology and management skills, and enhance a sense of national ownership (see Pigato and Tang 2015).

The discursive contours of China–Africa relations are still being shaped and defined by China, albeit in the face of an Africa that is changing and becoming more assertive, with an emphasis on reciprocal partnerships and win-win formulae. In terms of agenda-setting, President Xi articulated "five major pillars" at the FOCAC VI summit that broadly find echoes in the historical lexicon: political mutual trust; solidarity and cooperation in international affairs; economic cooperation; mutually enriching cultural ties; and mutual security cooperation. These pillars rest on a new foundational positioning that brings the two sides together in a "comprehensive strategic partnership," fuelled by the potential functional efficacy of the 10 cooperation plans. However, this language and its underlying normative ethos masks an instrumentalism that harks back to Zhao's "state led pragmatic nationalism" in Africa. As with the history of China's interactions with Africa, the leitmotif that runs through its behaviour in Africa is based on tactical flexibility, strategic subtlety, and a non-confrontational appearance.

Ex hypothesi as it were, it would seem that a hybridization of development policy in China is emerging. This has features of both the Beijing and Washington Consensus in the sense that while much of China's relations with Africa will depend on state-led and state-to-state dynamics, there is an increasing reliance on market forces, private sector engagement, and privately directed capital to drive the next phase of engagement. This takes into account the shifts in China's domestic political economy and African countries' industrialisation imperatives away from resource reliance. The crowding in of market forces together with private investment will portend a diminishing role of Chinese state-owned enterprises in Africa whose returns to scale are increasingly being questioned. According to Davies, three considerations underpin this thesis: in China's industrial manufacturing sector, the cost of production now exceeds the gains in productivity; the rebalancing of the Chinese economy signals a move away from high rates of fixed asset investment toward a less resource-reliant growth model; and SOEs are no longer as dominant (Davies 2015, 4). This reorientation thus provides a fertile opportunity to rethink how the confluence of social, economic, and institutional forces as well as the political responses to these, which have so divided development thinking about Africa, can be incorporated into fresh and refreshing appraisals. This would entail bringing together different matrices of development theory, including Keynesian, Schumpetarian, and communitarian approaches, in an endeavour to reimagine what growth and development mean in terms of the collective China–Africa experiential models and what these imply for their opportunity sets and preferences (see Sen 1999).

Conclusion

What runs through this chapter is the extent to which Africa has been a reactive subject, rather than an objective agent, in setting the terms and conditions about the discourse and debates about its development. This goes to heart of both the epistemological and ontological deficits that still afflict China–Africa studies and, in particular, its path to development through industrialisation and more judicious usage of its resource abundance. That Africa still finds itself stuck on the cusp of the promise of an industrial revolution attests to the poverty of development thinking.

Here again, we encounter the problematic dialectic between knowledge and power. In terms of the logic of information, politics draws its inspiration from knowledge as a guide to action. In a critical vein, the domination of the China–Africa discourse and narrative development by Western scholars and policy-makers has privileged certain power relationships, which have contaminated the production of research and knowledge in violation of the Enlightenment dictate of reason (Foucault 2002). Such power shapes knowledge production and determines what the parameters are of legitimate knowledge, much to the detriment of establishing systems of truth. The politics of knowledge production thus trumps and impedes the construction of learning and the logic of information in order to advance partisan interests.

It is only with the entry of China that the hegemonic thrust that has accompanied much of the discourse and debates about Africa's growth and development started to shift, if only because the grand logic of China's intentions was being scrutinised with greater intensity compared to Africa's traditional interlocutors from the West. Much of this has to do with China's own modernisation and its strategic calculus in Africa which, by and large, has not followed the idioms of neo-classical development orthodoxy. Moreover, China's principle of non-interference and respect for state sovereignty draws its predicates from a world view that "… allows for plurality, built on China's historical cultural predilection for harmony, virtue and society and for solving problems peacefully" (Breslin 2011, 1339).

The hybrid frontier that mixes state-led initiatives with the dynamics of the market and private sector could herald a new beginning of thinking creatively about Africa's resource–industrialisation nexus and how the proverbial sow's ear could be turned into a silk purse. The landscape of over-determined state cooperation that has characterised most of the FOCAC process could yield to making more space for the emergence of new and different matrices of public–private partnerships that bring in African firms and enterprises, who might be better able and placed to define and take advantage of the terms and conditions of hybridization. This could take several forms: building the African private sector's confidence in economic processes and allocative mechanisms that come with incentives and coordination from the FOCAC process; African and Chinese companies identifying effective and acceptable distributional payoffs in any industrialisation endeavour so that gains and losses are shared equally; and encouraging mutual learning, problem solving, and compromises in dealing with the historical and atavistic obstacles to growth and development such as poverty, unemployment, and inequality. Surely, this would engender a stronger sense of African ownership and participation?

Importantly, China has aligned forces of the state, market, and bureaucracy to generate a model of authoritarian state-led development as the antithesis of "Anglo-Saxon" capitalism. Thus Breslin boldly asserts that "China provides an important example of an alternative to the neo-liberal project that had come to dominate developmental discourses in the first part of the millennium" (Breslin 2011, 1324). As such, China serves as a potent metaphor of "difference." This can be transposed to the need for Africa to chart a different development path, especially against the backdrop of global economic changes. There is no question about the need for the continent to break out of commodity dependence and undertake long-term structural change that could be expressed in the form of inclusive development and a diversified economic profile.

A development paradigm that could best drive this is clearly not that which derives from the now-discredited and essentially market-driven Washington Consensus. Yet, a blind statist approach could lead to destruction of value, if this is not underpinned by credible attempts to restructure the state–society nexus in ways that empowers citizens, develops human capital, builds resilient institutions, and nurtures strategic collaboration between government and the private sector. Hence, measured statism is intrinsic to the basic promise of hybridization and its experimental ethos. Indeed, Ha-Joon Chang has shown that the US and other developed countries prospered on the back of policies that invited appropriate levels of government intervention and the creation of robust public institutions (Chang 2005). For our purposes, hybridization entails managing the tensions between the institutional pre-requisites for development and the policy prescriptions that flow from neo-classical orthodoxy. This requires an openness to experimentation in all realms of development policy and practice in order to encourage Schumpetarian-style innovation and growth.

There is much to learn from China's experience, and there is certainly some value that African countries can derive from their relationship with China, but this on its own is not sufficient to drive broad-based industrial development. Just like the centuries old relationship with Western metropoles, the China–Africa relationship may not deliver development gains on a sustainable basis despite the inherent promise and potential. The major lesson to learn from China is the importance of building institutions to underpin development; experiment with heterodox economic policy approaches that are a function of an overarching and politically driven development strategy; and deliberately work toward industrial diversification.

Finally, the era where the West has the answers and African countries can only ask how high they should jump has reached an end point. China is not the savior either, since, as Breslin argues, "… it appears doubtful whether other developing states have the same conditions, factor endowments and social and historical backgrounds to be able to emulate what China has done" (Breslin 2011, 1324). Therefore, African countries should be the drivers of their own development programs and policies and the role of other external actors should be supportive and complementary. Where African leaders in both the public and private sectors feel a sense of developmental ownership, the burden of responsibility becomes inordinately high.

References

Alden, Chris. 2007. *China in Africa*. London: Zed Books.

Alden, Chris. 2008. "Africa without Europeans." In *China Returns to Africa: A Rising Power and a Continent Embrace*, edited by Chris Alden, Daniel Large and Ricardo Soares de Oliveira, 349–360. London: Hurst and Company.

AUC. 2014. *Agenda 2063: The Africa We Want*. Addis Ababa: AUC.

Bloch, Harry and David Sapsford. 2000. "Whither Terms of Trade? An Elaboration of the Prebisch-Singer Hypothesis." *Cambridge Journal of Economics* 24(4): 461–481.

Bräutigam, Deborah. 1998. *Chinese Aid and African Development: Exporting Green Revolution*. London: Macmillan Press.

Bräutigam, Deborah. 2009. *The Dragon's Gift: The Real Story of China in Africa*. Oxford: Oxford University Press.

Breslin, Shaun. 2011. "The China Model and the Global Crisis: From Friedrich List to a Chinese Mode of Governance?" *International Affairs* 87(6): 1323–1343.

Briseid, Marte, Laura Collinson *et al.* 2008. "Domestic Ownership or Foreign Control? A Content Analysis of Poverty Reduction Strategy Papers from Eight Countries." LSE: Crisis States Occasional Paper 5, March.

Broadman, Harry G. 2007. *Africa's Silk Road: China and India's New Economic Frontier*. Washington, DC: The World Bank.

Chang, Ha-Joon, 2005. *Kicking Away the Ladder: Development Strategy in Historical Perspective*. London: Anthem Books.

Davies, Martyn. 2015. "What China's Economic Shift Means for Africa." *World Economic Forum*, 11 March.

Durden, Tyler. 2014. "Where China Goes to Outsource its Own Soaring Labour Costs." *ZeroHedge*, 24 July.

Eichengreen, Barry. 2007. *The European Economy Since 1945*. Princeton, NJ: Princeton University Press.

Eisenmann, Joshua, Eric Heginbotham and Derek Mitchell. eds. 2007. *China and the Developing World: Beijing's Strategy for the Twenty-First Century*. Armonk: ME Sharpe.

Escobar, Arturo. 1995. *Encountering Development: The Making and Unmaking of the Third World*. Princeton, NJ: Princeton University Press.

Faucheux, Sylvie, David Pearce and John Proops. eds. 1996. *Models of Sustainable Development*. Cheltenham: Edward Elgar.

Foucault, Michel. 2002. *Archaeology of Knowledge*. London and New York: AM Sheridan Smith.

Frieden, Jeffry. 2006. "Will Global Capitalism Fall Again?" Bruegel Essay and Lecture Series.

Greig, Alastair, David Hulme and Mark Turner. 2007. *Challenging Global Inequality: Development Theory and Practice in the 21st Century*. New York: Palgrave MacMillan.

Habib, Adam. 2008. "Western Hegemony, Asian Ascendancy and the New Scramble for Africa." In *Crouching Tiger, Hidden Dragon: Africa and China*, edited by Kweku Ampiah and Sanusha Naidu, 259–277. Scottsville, SA: University of Kwa Zulu-Natal Press.

Hevi, Emmanuel. 1967. *The Dragon's Embrace: The Chinese Communists and Africa*. London: Pall Mall Press.

Kaplinsky, Raphael. 2007. *Asian Drivers and Sub-Sahara Africa: The Challenge to Development*. New York: Rockefeller Foundation.

Kayaoglu, Turan. 2010. "Westphalian Eurocentrism in International Relations Theory." *International Studies Review* 12(2): 193–217.

Kennedy, Scott. 2010. "The Myth of the Beijing Consensus." *Journal of Contemporary China* 19(65): 461–477.

Kim, Yejoo. 2013. *Chinese-led SEZs in Africa: Are they a Driving Force in China's Soft Power?* Stellenbosch: Centre for Chinese Studies Discussion Paper.

Lal, Deepak. 1985. *The Poverty of Development Economics*. Cambridge, MA: Harvard University Press.

Large, Daniel. 2008. "Beyond 'Dragon in the Bush': The Study of China-Africa Relations." *African Affairs* 107(426): 45–61.

Lewis, Arthur W. 1955. *The Theory of Economic Growth*. London: Allen and Unwin.

Mazrui, Ali A. and Francis Wiafe-Amoako. 2016. *African Institutions: Challenges to Political, Social, and Economic Foundations of Africa's Development*. London: Roman and Little.

Mbeki, Moeletsi. 2006. Editorial, *South African Journal of International Affairs*, 13(1): 7–9.

Mills, Gregg and Lyal White. 2006. "Africa Can Decide Whether China is a Threat or a Boon." *Business Day*, 18 January.

Mkandawire, Thandika and Charles C. Soludo. 1999. *Our Continent, Our Future: African Perspectives on Structural Adjustment*. CODESRIA: Africa World Press.

Naughton, Barry. 2010. "China's Distinctive System: Can it Be a Model for Others?" *Journal of Contemporary China* 19(65): 437–460.

Pieterse, Jan Nederveen. 2010. *Development Theory*. 2nd edition. London: Sage Publications Ltd.

Pigato, Miria and Wenxia Tang. 2015. "China and Africa: Expanding Economic Ties in an Evolving Global Context." *World Bank: Investing in Africa Forum*, March.

Putnam, Robert. 2000. *Bowling Alone*. New York: Simon and Schuster.

Ramo, Joshua Cooper. 2004. *The Beijing Consensus: Notes on the New Physics of Chinese Power*. London: The Foreign Policy Institute.

Rist, Gilbert. 1990. "Development as the New Religion of the West." *Quid Pro Quo* 1(2): 5–8.

Rodney, Walter. 1972. *How Europe Underdeveloped Africa*. Washington, DC: Howard University Press.

Rodrik, Dani. 2007. *One Economics, Many Recipes*. Princeton, NJ: Princeton University Press.

Romei, Valentina. 2015. "China and Africa: Trade Relationship Evolves." *Financial Times (London)*, 3 December.

Rukato, Hesphina. 2010. *Future Africa: Prospects for Democracy and Development under NEPAD*. Trenton, NJ: Africa World Press.

Schiere, Richard and A Rugamba. 2011. "Chinese Infrastructure and African Integration." In *China and Africa: An Emerging Partnership for Development*, edited by Richard Schiere, Leonce Ndikumana and Peter Waknehorst, 91–101. Tunis: The African Development Bank.

Schumpeter, Joseph A. 1994 [1942]. *Capitalism, Socialism and Democracy*. London: Roultedge.

Sen, Amartya. 1999. *Development as Freedom*. Oxford: Oxford University Press.

Snow, Philip. 1988. *The Star Raft: China's Encounter with Africa*. London: Weidenfeld and Nicolson.

Tian, Xuejun. 2016. "A New Era for China-Africa Relations." *The Thinker* 68: 22–26.

Tu, Xinquan and Huiping Mo. 2015. "China's Developing Country Identity – Challenges and Future Prospects." In *Handbook on China and Developing Countries*, edited by Carla P. Freeman, 89–108. Cheltenham: Edward Elgar Publishing.

UNECA. 2001. *State of the Environment in Africa*. Addis Ababa: UNECA.

UNECA. 2010. Economic Commission on Africa: Annual report. Internet: www.uneca.org /cfm/2010/documents/English/AnnualReport_2010.pdf#original.

UNECA, 2011. *Economic Report on Africa 2011: Governing Development in Africa – The Role of the State in Economic Transformation*. Addis Ababa: UNECA.

UNECA. 2015. *Industrialising Through Trade*. Addis Ababa: UNECA.

UNECA. 2016. *Transformative Industrial Policy for Africa*. Addis Ababa: UNECA.

Van Waeyenberge, Elisa. 2006. "From Washington to Post-Washington Consensus: Illusions of Development." In *The New Development Economics: After the Washington Consensus*, edited by K. S. Jomo and Ben Fine, 21–43. New York: Zed Books.

Wang, Hui. 2003. *China's New Order: Society, Politics, and Economy in Transition*. Cambridge: Cambridge University Press.

Williamson, John. 2012. "Is the 'Beijing Consensus' Now Dominant?" *Asia Policy* 13: 1–16.

Wu, Guobao. 2016. "Four factors that have driven poverty reduction in China." *World Economic Forum*, 21 October.

Zhao, Suisheng. 2007. "China's Geo-strategic Thrust: Patterns of Engagement." In *China in Africa: Mercantilist Predator or Partner in Development*, edited by Garth le Pere, 33–53. Johannesburg and Midrand: The South African Institute of International Affairs and the Institute for Global Dialogue.

17

A CHINESE MODEL FOR AFRICA

Problem-solving, learning and limits

Chris Alden

Conventional views hold that the notion of a "Chinese model for Africa" is either something entirely new and sought out by Africans or that, in the dissembling comments of Chinese officials, they are not promoting a model of development at all. Deng Xiaoping's admonishment to Ghana's leader, Jerry Rawlings, during an official visit to Beijing in September 1985 is often cited as evidence of the lack of ambition on the Chinese part to provide lessons to Africa. Deng pointedly said "Please don't copy our model. If there is any experience on our part, it is to formulate policies in light of one's own national conditions."[1] Contrary to these declared perspectives, however, the application of Chinese practices to the African environment has been an integral feature of virtually every phase of Chinese engagement with the continent. This can be seen as far back as the Maoist era, when ideas of adopting the Chinese approach to collectivizing agriculture took hold in some countries. It can also be seen in the present day, where the focus on replicating the industrialisation strategy introduced under Deng Xiaoping is gathering momentum. Indeed, promoting the packaged export of the Chinese experience is a necessary dimension of China's foreign policy, every bit as much aimed at affirming its identity as a global power as contemporary pre-occupations with the politics of expanding Beijing's access to international financial institutions or power projection in the South China Sea.

Moreover, influential African governments and policy makers have responded to China in terms suggesting that at least some of them see it as more than another foreign source of practical lessons and technical assistance. Indeed, Xi Jinping's affirmation in October 2017 that the Chinese experience offers "a new option" for developing countries cements the foreign policy posture of a more promotional kind of development model. In fact, there is an undercurrent in African interpretations of Chinese experiences, actively promoted by Beijing (if more muted in expression during Deng's era), which contributes to expectations that emulating

China will not only have a transformative impact on Africa but will further a larger emancipatory agenda for the continent as well (Cooper 2016, 12).

Understanding the significance of promoting Chinese models as proven instruments for resolving African problems underscores its centrality to Beijing's Africa strategy. Furthermore, it provides insights into the structure of China's relationship with Africa. At the same time, it underscores the difficulties in achieving these aims through the simplistic application of problem-solving approaches to complex development issues. This chapter, therefore, examines the historical place that Chinese models have in its foreign policy toward Africa, briefly summarises the implementation of the industrialisation and agricultural development models in Africa and, lastly, provides a critical assessment of that process.

Models, learning and meaning

The idea of models permeates the literature and experience of development as conceptualised and practiced in Africa (as elsewhere). Models, blueprints and other devices are seen to be particularly useful pedagogical tools distilling the key features of a given development experience into a set of policy prescriptions suited to being applied in the pursuit of achieving the same outcome by the recipient. Behind this notion of the salience of models to the process is the belief in a "logic of appropriateness," which finds in a conceptual rendering of policy and its actual implementation a source of relevance for countries confronting complex policy dilemmas with costly choices. Models, in short, based as they are on proven experiences, can help policy makers in similar circumstances conceptualise the problem they face, break down the steps necessary to address the issue(s) and offer a course of policy action aimed at overcoming these problems.

Transferring that lesson is a crucial aspect of the transformative process, and necessarily focuses on policy makers and implementing agents embedded within state institutions. Targeting the right individuals and departments, coupled to developing appropriate methodologies of policy transfer, is crucial to developing the conditions for learning to transpire. Commentators have pointed out that learning within organizations takes place at various levels, especially when authority is distributed across an organization, but broadly speaking follows a hierarchical logic of top-down or bottom-up (Majone and Wildavsky 1978; Matland 1995). The actual transfer of experience, sometimes referred to as "policy diffusion," is understood as organizational learning. Scholars like Argyris provide a picture of how institutions learn as expressed through changing routines (Argyris 1999). Becoming a "learning organization," that is, one which integrates learning into its very routines and practices, is often held up as the essence of institutional success (see Senge 2006). Government institutions as different as the military and aid agencies have embraced this approach by systematically applying it to their policy cycles through internal monitoring and evaluation of programs.

Reflecting on the nature of learning itself as it relates to the policy process, May divides it into four categories: instrumental, social policy, political and "mimicking"

(May 1992, 336–337). Deriving "lessons" from an analysis of past policy implementation forms a distinctive part of learning in the policy process. In this respect, the singularity of "failure" as a source of profound learning by organizations is notable and contrasted with the weaker impacts of positive lessons (Majone and Wildavsky 1978; May 2015).[2] For instance, as Meseguer highlights, the wave of embracing neo-liberalism by developing countries starting in earnest in the 1990s needs to be situated within the historical context of the recognition of previous policy failures of interventionism.

> These failures, coupled with the successes of other countries, seen and portrayed as champions of liberal economic reform, persuaded politicians of the virtues of liberal market policies.
>
> *(Meseguer 2005, 56)*

Recognizing the possibility of failure, and that such failures have shaped the very experiences which themselves became distilled as lessons, is imperative to reproducing a dynamic form of the lesson with genuine applicability to the target.[3] Reducing or removing the element of risk, the learning equivalent of eliminating the "moral hazard" in finance, weakens the saliency of the knowledge transfer process. Or, as Jack Ma, CEO of tech giant Ali Baba puts it: "You have to get used to failure. If you can't, then how can you win?"[4]

If models are best understood as vehicles for policy learning, it should nonetheless be recognised that they serve purposes beyond the content of policy. The narrative of transformation in China, often of an emancipatory character, plays a significant part in the cooperation strategies employed by China toward Africa.[5] In this respect, the Chinese offering of models for development – and here I am using the broadest sense of development as both a guide for the economy but also in political systemic forms as well – is founded on this discourse of similitude and possibility. This latter point is reinforced through the pairing of China as "the world's largest developing country" with Africa as "the region with the largest number of developing countries," a conventional rhetorical depiction of the close proximity between them coupled to the common assertion that they have a shared position as victims of colonialism. African governments and policy makers regularly recite these platitudes, explicitly recognizing that in China's own experiences there resides a source of policies that can readily be transferred across regions, cultures and time, to generate equivalent outcomes on African soil. This co-constitution of a shared identity is the binding substance that lays the foundations for consideration by Africans of the lessons to be learned from China's experience. For China, it provides the certitude that its status, both as a member of the developing world and a transcendent leader in that sphere, is recognised.

China–Africa and the role of models

The most ardent application of the Chinese model to Africa during the Maoist phase was found in Tanzania. As president of Tanzania, Julius Nyerere undertook

13 official state visits to China, and spoke admiringly of their collectivization of agriculture and its development (what today are widely seen to be its perceived rather than substantive) successes (cited in Hyden 1980, 100). Reflecting upon his own country's conditions on the eve of the launching of the *ujamaa* collective agriculture program inspired by China's experience, Nyerere declared:

> We are using hoes. If two million farmers in Tanzania could jump from hoes to the oxen plough it would be a revolution. It would double our living standard, triple our product. This is the kind of thing China is doing.
>
> *(cited in Brautigam 2009, 29)*

This inspirational discourse was, however, not consigned to the time of revolutionary fervor characteristic of the post-independence period in Africa. A generation later, the South African Chair of the African Union Commission, former Foreign Minister Nkosazana Dlamini-Zuma, spoke at a conference held in May 2015, which asked the question "Can Africa Emulate China's Development?" She answered unequivocally:

> We think it is possible to effect change in a generation [in Africa]. China has done it. If we want to do it, we can do it.[6]

This assertion reflected recognition at the extraordinary and rapid change in China's economic circumstances after 1980, inspiring a renewed belief amongst leaders that Africa could learn from its experiences and apply these lessons toward resolving the key development dilemmas of the continent. But what is especially revealing is that from 2004 onwards, the restored faith coincided with the promulgation by Beijing of a set of policies and institutions explicitly designed to transfer lessons derived from the Chinese experience of development to Africa. While not formally declared to be "models" at the time, they nevertheless are programmatic approaches rooted in the seminal reforms that kick-started the Chinese development process from late 1978 onward, and represent a distillation of these experiences into an African context. They focus on introducing innovative measures in two key sectors: agricultural reform, and industrial policy through the special economic zones (detailed in the next section).

The saliency of the Chinese experience for African development was affirmed by African governments, policy makers and practitioners.[7] The promotion of "Look East" policies across many African states accelerated after the global financial crisis of 2008. Seeking, in the first instance, to capitalise on China's huge financial resources and its disregard for regime type, such efforts came to include a conscious effort to imitate key aspects of the Chinese experience.[8] These focused on the aforementioned industrialisation strategy and agricultural development but extended further to include the management of state-owned enterprises, media control and enhancing the functioning of political party structures. Good governance and the role of institutions began to feature with regularity in the Chinese discourses toward African development, painting a picture of an alternative model of modernisation without necessarily subscribing to liberal democracy. This was encapsulated in the phrase the "Beijing Consensus" coined by foreign observers (Zhao 2010, 422).

China's influence was not, of course, ever relegated to the development sphere alone in Africa. During their struggles for independence, African liberation movements especially looked to the Chinese experience of revolutionary warfare as both an inspiration and a source of strategy and tactics for gaining power (see Chau 2014). Amilcar Cabral, leader of the *Partido Africano da Independencia de Guine e Cabo Verde,* relied on Chinese guerrilla warfare and its prevailing theory of revolution to structure his own campaign against the Portuguese colonialists; FRELIMO's guerrilla campaign drew partially from Chinese revolutionary doctrine after 1966 in colonial Mozambique, as did the approach adopted by Zanu-PF in the later stages of the fight in Rhodesia. Though eventually painted as a counter-revolutionary, Jonas Savimbi's UNITA movement owed much of its highly successful guerrilla campaign to the training that he and UNITA cadres received in the mid-1960s at the Nanjing military academy. Sometimes, borrowing from the model of revolutionary warfare was more explicit, such as the rather shameless reproduction of Mao's "Theory of Revolutionary Warfare" by Ghana's Kwame Nkrumah in his self-styled *Handbook of Revolutionary War,* which was written in exile two years after his ouster in a military coup in 1966. The longer-term impact of this revolutionary legacy, in spite of the ideological shifts that accompanied the "reform and opening" post-1978, is that it contributes to a sense of China's enduring association with an emancipatory agenda, even when reformulated into a developmentalist configuration, that can meaningfully be applied to African problems.

The Chinese development model as catalyst for a rising Africa

The focus in the contemporary period, underscored by the formal launching of these initiatives at the Forum on China-Africa Cooperation (FOCAC) Summit held in Beijing in November 2006, was the creation of "three to five" Economic Cooperation and Trade Zones (ECTZs) and "ten" (later extended to 20) Agricultural Technical Demonstration Centres.[9] The first was derived from the successful experience of the Chinese Special Economic Zones, which were launched in late 1978 in coastal southeast China, while the latter were the product of the cumulative experiences of agricultural reform started in the early 1980s.

With *industrial policy/special economic zones,* China's salutary experience of developing an export-oriented manufacturing sector built on FDI and technology transfer from the industrialised countries is replicated through its ECTZ initiative, which promises to combine Chinese state funding alongside public–private investment to "hot-house" a site for job creation, skills transfer and possible out-sourcing of increasingly costly Chinese industry to nurture a nascent African manufacturing sector. According to Brautigam and Tang, the "principles of profitability reign" at all levels of Chinese engagement in this process, from the conceptualization by Chinese provincial authorities, enterprises and developers, their financing by Chinese banks and the decisions to invest in local markets (Brautigam and Tang 2011, 49–51).

In the *agricultural sector*, the Chinese government established over 20 Agricultural Technical Demonstration Centres, including provisions for financing and technical expertise, whose primary purpose is aimed at raising agricultural productivity for local markets and, with that, improvements in rural incomes, bolstered by a range of technical cooperation programs in agriculture. A phased "public–private partnership" approach is used, commencing with Chinese designated provincial authorities partnering with local host government to set up the infrastructure of the center in the first year, provisions for training and experimental farms in the second year, and the handing over of the local government to manage in the third year. According to Jiang, one of the longer-term purposes is to create a platform for Chinese agricultural enterprises to obtain exposure to the local market in that African country, gain position and experience in globalizing their production (Jiang 2015, 16–17). Notably, despite their inclusion in the model, "public–private partnerships" were not a feature of the original Chinese experience in agricultural reform.

Collectively, it is clear that these Chinese initiatives being promoted in Africa are drawn in the main from the transformative policy approaches and implementation strategies that were behind the rapid development of the modern Chinese economy over the last four decades. They are grounded in the interest-based form of cooperation that has prevailed in China–Africa relations manifested in the solid commercial component devised for support and involvement of Chinese firms and their African counter-parts. While perspectives differ as to the role of the state and the private sector as catalysts in this process, they reflect an emerging consensus within the development community as to the importance of linkages between growth and poverty reduction in the case of China (see Ravallion 2009; China DAC Study Group 2009, 1–8). As such, these initiatives are central to the effort to bring a distinctive Chinese experience of development to the task of catalyzing African development.

However, while these Chinese-led initiatives in Africa are still very much in the process of being rolled out, there are some troubling indicators that they are not fully meeting expectations as catalysts for development. For instance, despite the publicity associated with the launch, a full decade later a number of ECTZs remain under-developed sites (Mauritius, Lekki) with scant evidence of Chinese investment or spill-overs to local economies while others are little more than the "rebranding" of existing Chinese investments as an ECTZ (Zambia) (Alves 2011; Alves 2012).[10] By way of contrast, the Ethiopian ECTZ outside Addis Ababa, host to the famous Haijian shoe factory, while reported to be uneconomical in 2012 by its Chinese developer has subsequently experienced an investment boom.

Unlike the ECTZs, the take up of Agricultural Technical Demonstration Centres was much higher across the continent though by 2011 actual FDI was limited to only $400 million, representing just 12 percent of total agricultural investment (Jiang 2015, 7). Equally concerning, however, is evidence from at least two of the Agricultural Technical Demonstration Centres that suggests there is a mismatch between the training programs on offer and the needs of local end-users, resulting in limited impact and poor delivery (Chichava et al. 2014). The flagship Agricultural Technical Demonstration Centre in Mozambique, considered by Beijing to be the most successful of the demonstration centres, has had great difficulties even in the hands of private Chinese companies in finding a local market that would sustain production at a cost effective level. This situation is compounded by the extended and ongoing series of disputes with local employees and local communities over labour conditions and property rights.[11]

The limits of Chinese solutions to African problems

An understanding of the problems experienced in realizing the transformative aims and practical outcomes of the Chinese development models is to be found in shortcomings that are both conceptual in nature and particular in content. These issues are derived from faulty assumptions about the model held by actors involved, lack of appreciation of the institutions and practices needed to achieve policy transfer, and finally a failure to account for the distinctive features of China's experience.

The nature of the model. The conceptual sources of the Chinese model reflect as much upon the assumptions underlying Chinese production of knowledge as the African interpretations of it. Problem-solving approaches to tackling the challenges posed by development dominate China's practical stance toward addressing everything from raising living standards to managing health delivery. Distilling this eclectic experience into a model tacitly integrates assumptions about the actors involved – in this case, that they will behave in ways that are comparable to their Chinese counterparts – and the conditions they find themselves in. Equally it makes assumptions about the exogenous environment as being roughly equivalent to the favourable trade and investment conditions experienced by China in its first three decades of "reform and opening."

These problems are compounded by the framing of a Chinese model of industrialisation and agriculture in Africa that precludes exposure to risk and, with that, potential failure which was integral in shaping the experimental and incrementalist approach adopted by Chinese officials. The stops and starts, the collapse of particular initiatives and the array of technical and political conundrums that had to be overcome to produce the outcomes in industry and agriculture cannot be replicated in an uncritical adoption of a model.

This is reflected in the lack of patent understanding on the part of African governments involved, which focus on the development outcomes these models are to bring rather than on a careful analysis of the evolution of the historical process of development in China itself. Writing at the outset of the launch of the Chinese initiatives (ECTZ and Agricultural Technical Demonstration Centres) for Africa, Li Zhibao noted:

(T)he intense trend of African countries studying China in most cases stops short at the surface, seldom getting down to a substantial level. After all, the most basic reason for this is that the majority of African nations do not really understand China – do not understand its politics, economy, history, culture or road to development. They only know a lot of China's success stories and a string of successful data that has dazzled the world, but rarely know the situation of its underlying policies. Even less so do they know that China has also taken a tortuous route and has made mistakes.

(Li 2008, 57)

For Li Zhibao, the failure of African policy makers and academics to have more than a superficial understanding of China is compounded by their unwillingness to "pay attention to the problems in China's development" including the overextension of the Chinese economic model, the growing gap between rich and poor, and what he calls the "illness" of the demand for the foreign investment (Li 2008, 65–69). Ding Xueliang picks up this theme, characterizing the obsession with the Chinese model of export-led growth as a "chronic illness" that promotes stability at the cost of long-term economic and environmental well-being (Ding 2010).

Policy transfer and Africa's institutions. A second set of problems is found in the realm of policy transfer where the model is reliant upon African institutions – generally an array of ministries, local government and party officials, though sometimes inclusive of tertiary education institutions – charged with facilitating all aspects of Chinese industrial and agricultural initiatives. In China, policy reforms that catalyzed private enterprise were underpinned by important changes to local governance structures that gave them a clear mandate to promote development. According to Breslin, this resulted in a development approach characterised by "gradualism" (渐进性 *jianjinxing*), "autonomy" (自主性 *zizhuxing*) and strong government (政府 *qiangzhengfu*) (Breslin 2011, 1329). Local government in China was given a significant degree of autonomy in areas deemed crucial to development, and the authority to raise revenue linked to local economic performance, all of which fostered a virtuous cycle of localised policies promoting self-sustaining growth. African institutions at the local level are, however, notoriously lacking in adequate human resources, sufficient fiscal autonomy and a clear mandate to drive development-oriented decision making in their local context (see Alence 2004).[12] Moreover, the flexibility that autonomy gave to local officials in China, by many scholars' reading, maximises "tinkering" that allows for a myriad local conditions and problems to be addressed is notably lacking in Africa (Heilmann 2009, 450). Without these features institutionalised as practices, it is proving difficult for local institutions in Africa to respond appropriately to the incentives and policy reform initiatives that have been inherent to the Chinese "developmental state" experience. African local government is in these cases not in a position to serve as a "development enabling" institution, a vital ingredient to the Chinese experience.

Learning. Under these sub-optimal circumstances (with the notable exception of Ethiopia), the learning that takes place is closer to "mimicking," as May calls it, rather than a process with meaningful impact on targeted actors and their policies (May 1992, 336–337). Here there is an adoption of policies not objectively related to the nature of the problem, or cognizant of the prevailing political environment, but effectively parachuted into an environment with limited to little cognizance of actors and conditions in place.[13] This experience of development and its human costs expended over the last generation, captured in the Chinese phrase "eating bitterness," is not reflected in the technical language of the model as presented, nor are the possibilities of failure alluded to or incorporated in the process. Knowledge production is, under these circumstances, designed less for learning and more as an emancipatory expression of solidarity that holds ideational meaning for the two sides. The gap between public declarations of the relevance of China's development model for Africa and private experiences on the ground widens, breeding cynicism amongst participants.

Conclusion

China's model for African development, founded on its own experience, offers a deceptively clear route to winning development gains. It is one that is built upon assumptions, conditions and contradictions that make its viability as an instrument of transformation limited without more active and innovative engagement by African hosts. Beyond the conceptual assumptions bedevilling this problem-solving approach, and the African institutional deficiencies crucial to the transfer of lessons and integrating these into sustained policy practices, a more intangible element is missing from the story of the Chinese model. Proponents of these Chinese models in Africa do not dwell on this aspect and the accompanying hardship, social upheaval and even environmental destruction, that is, the centrality of a process of experimentation and failure. This failure to do so detracts from the viability of any development model as a coherent and transferable set of policy prescriptions, causing it rather to assume the status of a utopian project.

Its contemporary entanglement with China's foreign policy aspirations for global leadership further threatens to confound its practical purposes with the imperatives of Beijing's wider ambitions. The essence of China's experience and its potential to contribution to African development is in danger of becoming just another blueprint for development rather than its catalyst.

Notes

1 Wei Wei Zhang, "The Allure of the Chinese Model," *International Herald Tribune*, November 1, 2006.
2 For a critique of lessons and policy transfer see Martin Lodge and Oliver James, "Limits of Policy Transfer and Lessons Drawing for Public Policy Research," *Political Studies Review* 1:2, 2003.
3 Toft and Reynolds (2005) characterise this as "isomorphic learning".
4 Jack Ma, cited in "South Africa: the Ma, the merrier", *The Africa Report*, 99: April 2018, p. 29.

5 Unlike the conventions of aid and technical assistance as promoted by Western sources at least since the 1970s, which emphasise more modest, technically bounded and measurable aims of improving outputs in target sectors.
6 Presentation by Chair of AU Commission, Nkosana Dlamini-Zuma as cited in "Africa Can Emulate China's Development," May 25, 2015, University of Cape Town. www. uct.ac.za/dailynews/?id=9160.
7 The general significance of a model with regional characteristics can be found as far back as the World Bank's publication *The East Asian Model* (Washington DC: IBRD, 1994), which lay the basis for its "export" to Africa. On the China-specific discourses, see Tian (2005); and Zhao, 2010.
8 For a description of the impact of the 2008 financial crisis on propelling the "China Model of Development" into prominence, see Joshua Kurlantzick, "China's Model of Development and the 'Beijing Consensus,'" April 29, 2013, www.cfr.org/china/china s-model-development-beijing-consensus/p30595.
9 FOCAC, "Forum on China-Africa Cooperation Beijing Action Plan (2007–2009)," 2006/11/16, www.focac.org/eng/zyzl/hywj/t280369.htm, accessed October 2, 2016.
10 Author's research visit to the ECTZ in Mauritius September 2013 confirmed no actual buildings had been constructed beyond a Chinese gate and locally funded Mongolian "hot pot" restaurant.
11 Interviews with Lu Jiang and Sergio Chichava.
12 For a specific comparative study on weak institutions in the natural resource sectors, see Oxfam America, *The Weak Link: The Role of Local Institutions in Accountable Natural Resource Management in Peru, Senegal, Ghana and Tanzania*, Oxfam Research Report, April 2016.
13 In a refreshingly candid outburst, Cui Jianjun, head of the semi-official Chinese Network for International Exchanges at the World Social Forum in Nairobi in 2007 declaimed angrily that if African wanted to develop they would have to make hard decisions on foreign investment or they would "lose, lose, lose" and remain "poor, poor, poor." Walden Bello, "China Provokes Debate in Africa," *Yes Magazine*, April 5, 2007, www.yesmagazine.org/issues/columns/china-provokes-debate-in-africa/.

References

Alence, Rod. 2004. "Political Institutions and Developmental Governance in Sub-Saharan Africa." *Journal of Modern African Studies* 42(2): 163–187.
Alves, Ana Cristina. 2011. "The Zambia-China Cooperation Zone at Crossroads: What Now?" SAIIA Policy Briefing 41. Braamfontein: South African Institute of International Affairs.
Alves, Ana Cristina. 2012. "Chinese Economic and Trade Cooperation Zones in Africa: the case of Mauritius." *SAIIA Policy Briefing 51*. Braamfontein: South African Institute of International Affairs.
Argyris, Chris. 1999. *On Organizational Learning*. 2nd edition. Malden, MA: Blackwell Publishers Inc.
Brautigam, Deborah. 2009. *The Dragon's Gift: The Real Story of China in Africa*. Oxford: Oxford University.
Brautigam, Deborah and Tang Xiaoyang. 2011. "African Shenzen: China's Special Economic Zones in Africa." *African Affairs* 49(1): 27–54.
Breslin, Shaun. 2011. "The 'China Model' and the Global Crisis: From Friedrich List to a Chinese Mode of Governance." *International Affairs* 87(6): 1323–1343.
Chau, Donovan. 2014. *The Influence of Maoist China in Algeria, Ghana and Tanzania*. Annapolis, MD: Naval Institute Press.
Chichava, Sergio, Jimena Duran and Lu Jiang. 2014. "The Chinese Agricultural Technology Demonstration Centre in Mozambique: A Story of a Gift." In *China and Mozambique:*

From Comrades to Capitalists, edited by Chris Alden and Sergio Chichava, 107–119. Auckland Park: Jacana.

China DAC Study Group. 2009. "Sharing Experiences and Promoting Learning about Growth and Poverty Reduction in China and African Countries." Paris: OECD.

Cooper, John F. 2016. *China's Foreign Aid and Investment Diplomacy, Volume III: Strategy Beyond Asia and Challenges to the United States and the International Order.* Basingstoke: Palgrave.

Covadonga, Meseguer. 2005. "Policy Learning, Policy Diffusion, and the Making of a New Order." *The Annals of the American Academy of Political and Social Science* 598: 67–82.

Ding, Xueliang. 2010. "Jingti zhongguo moshi de 'manxingbing'". *Nanfang Zhoumou* December 9, 2010.

Heilmann, Sebastian. 2009. "Maximum Tinkering under Uncertainty: Unorthodox Lessons from China." *Modern China* 35(4): 450–462.

Hyden, Goran. 1980. *Beyond Ujamaa in Tanzania: Underdevelopment and an Uncaptured Peasantry.* Berkeley: University of California.

Jiang, Lu. 2015. "Chinese Agricultural Investment in Africa: Motives, Actors and Modalities." Occasional Paper 223. South African Institute of International Affairs.

Li, Zhibiao. 2008. "How Should African Nations Draw Lessons from China's Development Experience?" *Contemporary Chinese Thought* 40(1): 49–55 [translation from Li Zhibiao. 2007. "Feizhou guojia ruhe jiejian Zhongguo de fazan jingyan." *Xiya Feizhou* (West Asia and Africa) 4: 49–55].

Majone, Giandomenico and Aaron Wildavsky. 1978. "Implementation as Evolution." In *Policy Studies Review Annual*, edited by in Howard E. Freeman, 437–462. Beverly Hills, CA: Sage.

Matland, Robert. 1995. "Synthesizing the Implementation Literature: The Ambiguity-Conflict Model of Policy Implementation." *Journal of Public Administration and Research Theory* 5: 144–174.

May, Peter. 1992. "Policy Learning and Failure." *Journal of Public Policy* 12(4): 331–354.

May, Peter. 2015. "Implementation Failures Revisited: Policy Regime Perspective." *Public Policy and Administration* 30(3–4): 277–299.

Nkrumah, Kwame. 1968. *Handbook of Revolutionary Warfare.* London: Panaf.

Ravallion, Martin. 2009. "The Developing World's Bulging (but Vulnerable) 'Middle Class'". *Policy Research Working Paper 4815.* Washington, DC: World Bank Group.

Senge, Peter. 2006. *The Fifth Discipline: The Art and Practice of the Learning Organization.* 2nd edition. London: Century.

Tian, Yu Cao. 2005. *The Chinese Model of Modern Development.* London: Routledge.

Toft, Brian and Simon Reynolds. 2005. *Learning from Disasters.* 3rd Edition. Basingstoke: Palgrave.

Zhao, Suisheng. 2010. "The China Model: Can it Replace the Western Model of Modernization?" *Journal of Contemporary China* 19(65): 419–436.

18

NEW STRUCTURAL ECONOMICS

A first attempt at theoretical reflections on China–Africa engagement and its limitations

Tang Xiaoyang

During the past decade, a substantial amount of research has examined various aspects of China–Africa relations, spanning geopolitical analyses, investigations of economic sectors, country-based studies, assessments of aid and loans to individual factories, overviews of peacekeeping and corporate social responsibility efforts, and case studies of individual migrants' experiences. A number of in-depth, fact-based studies have mapped an increasingly clear picture of the comprehensive relationship that has developed between China and Africa over the course of this century. At the same time, the Chinese government and Chinese enterprises have also made their activities in Africa more transparent to the international community. Consequently, researchers and the public have a much better understanding of the scale and areas of Chinese engagements in Africa today. In light of these new facts, and based upon the foundational research conducted over the past decade, a number of scholars have begun exploring theoretical models with which to explain the China–Africa relationship.

In particular, Chinese researchers have made various efforts to analyze this new South–South partnership from their own perspectives. The primary framework of interpretation so far has been borrowed from Marxist materialism, in line with the Chinese government's main ideology. Liu Hongwu and Luo Jianbo argue that China–African collaboration is an extended application of "Chinese Knowledge" of development. They put emphasis on defending the right of development of China and Africa, criticizing the West for using "double standards" (*shuangchong biaozhun*) and "hegemony of discourse" (*huayu baquan*) to judge other countries' cooperation (Liu and Luo 2011, 2). Zhang Zhongxiang's monograph on FOCAC investigated the Chinese government's strategy and cooperation patterns with Africa. Reviewing the changing trajectory of China's cooperation with Africa, he draws a "China model" of South–South cooperation, which bears the characteristics of developmentalism (Zhang 2012).

Since economic collaboration makes up the lion share of the China–African partnership, several Chinese scholars develop theoretical reflections, especially concerning the drivers and impacts of bilateral economic exchange. Li Xiaoyun and his colleagues attempt to construct China and Africa's rural poverty reduction models by comparison and experience sharing (Li 2010). Elsewhere, I analyzed the development of Chinese economic engagements in Africa through a global value chain angle (Tang 2014). Such efforts have not yet formulated a systematic theoretical framework. However, Justin Yifu Lin, former World Bank senior vice president and chief economist, has provided one of the first systematic theoretical analyses of China–Africa relations by proposing the use of New Structural Economics to explain China's unique and effective engagement in Africa.

Justin Yifu Lin's book *Going Beyond Aid: Development Cooperation for Structural Transformation*, coauthored with Wang Yan, a former World Bank senior economist, is based on many years of practical experience and theoretical research in the field of international development (Lin and Yan, 2016). The book takes development theory as a starting point for its inquiry. From the perspective of New Structural Economics, the authors analyze the past two centuries of successes and failures in the development of countries across the world. They suggest that every developing country must promote structural transformation in industry according to the principle of comparative advantage. They examine the East Asian "flying geese pattern" to show how, according to the comparative advantages they held at various times, East Asian countries proceeded step-by-step to complete the process of structural transformation. A country first starts with producing and exporting labour-intensive products such as textile and footwear, then begin to produce higher value-added products, like steel parts, and finally move to become exporters of sophisticated capital-intensive industrial products such as electronics and automobiles (Akamatsu 1962; Kojima 2000). The level of industrialisation and modernisation achieved in quick succession by East Asian countries like Japan, South Korea, and the People's Republic of China over the past several decades indicates the success of this path.

On this basis, the authors contend that instituting appropriate industrial policies to promote sectors in which a country has latent comparative advantages can accelerate a country's structural transformation. The authors argue that information scarcity, infrastructure backwardness and unwillingness of enterprises to invest in developing countries make it difficult for countries to rely on market mechanisms to allocate resources efficiently to make use of their comparative advantages. Therefore, governments ought to step in to facilitate structural transformation for the sake of countrywide development. Such governments should identify the sectors in which long-term comparative advantages are to be gained and direct resources toward these sectors in order to realise their comparative advantages by means of providing industrial information, coordinating infrastructure construction, offering incentives, attracting FDI and so on (Lin and Yan 2016, 28–29).

Reviewing traditional aid practices guided by the "Washington Consensus," they criticise a single-minded emphasis on free markets. Absent from this approach are investments in infrastructure or other sectors necessary for structural transformation and the stimulation of developing countries' potential. Western donors are said to focus on what developing countries do not have, such as good governance, institutional capital, favourable investment climate, viable infrastructure, diversified industries, human capital and implementation capacity. However, the weak record of structural adjustment practices has proved that relying on the free market may lead to idealistic and dogmatic policies toward developing countries.

Over the course of its own development, by contrast, China did not follow instructions from Western donors. It managed to take control of the policymaking process while receiving supplementary aid and loans from outside sources. In this manner, it has effectively combined foreign aid with homegrown policies geared toward domestic industry in order to achieve desirable results. This pragmatic spirit in turn characterises Chinese aid to other developing countries, in which China makes use of its current comparative advantages, such as in infrastructure construction and light industry. Through investment and/or commercial loans, China helps developing countries, both in Africa and elsewhere, climb the ladder of structural transformation and together form a new pattern of flying geese. The authors argue that initiatives like resource-backed infrastructure financing and the establishment of special economic zones and industrial parks in developing countries allow China and its partner countries to each benefit from their comparative advantages. Moreover, they provide an example of innovative development cooperation in line with New Structural Economic theory. The experience of successful structural transformation in China informs the direction and key sectors in which African countries should invest. Lastly, the authors ask international aid organizations to fully consider how similar commercial models have successfully promoted development. They call for an expanded definition of development financing to guide further global investment in international development.

Justin Yifu Lin's theory on international development hinges on a key principle of New Structural Economics, which states that structural economic transformation can be propelled through measures like industrial policies and special economic zones, with many examples from China's domestic reform and international practices cited as evidence. Among theoretical works that address international development from a Chinese perspective, this is the most comprehensive yet; it is sure to be extremely influential. Justin Yifu Lin and Wang Yan are not only familiar with the history of Chinese reform and opening, but they have also proven themselves to be highly capable of synthesizing a vast amount of primary source data to support their position. Moreover, their experience in high-level positions at the World Bank has given them a broad global vision and sound skills in economic theory. The result of their research is an original and helpful theoretical framework for Chinese scholars of international development, as well as a reference guide for policy and practice in government and business. Their call for transcending the existing international development aid system also reflects the rise of South–South cooperation and the critical reflections on the Washington Consensus being voiced by many countries today. Rich in theoretical insights, their book summarises and refines lessons learned from China's many years of flexible, pragmatic development aid practice. It provides an important reflection for China and other countries, as they continue to improve upon development aid models and achieve the UN Sustainable Development Goals. Overall, it has remarkable significance for the innovative reform of global development cooperation that is currently underway.

It is not surprising to see that economists have provided the first theoretical reflections on the China–Africa relationship, given that trade, loans and other commercial activities have been the most obvious driver of collaboration between China and Africa during the last decade. China has been Africa's largest bilateral trading partner since 2009. In 2014, Sino-African trade reached over $210 billion, which was more than Africa's combined trade volume with the United States, Japan, France, and the United Kingdom in the same year.[1] Chinese contractors have long dominated the construction market in Africa thanks to both reasonable pricing and abundant financing by Chinese banks. Chinese firms took a whopping 49.4 percent share of Africa's entire construction market in 2014.[2] The phenomenal growth of China–Africa economic ties looks even more remarkable given that Africa's development prospects were gloomy not so long ago. Despite policy reforms based on structural adjustment programs, sub-Saharan Africa did not realise many successful economic growth achievements in the 1980s and 1990s. As Rodrik (2006, 973) writes, "proponents and critics alike agree that the policies spawned by the Washington Consensus have not produced the desired results." China's economic engagement with Africa not only increased the market share of Chinese firms in Africa, but also significantly boosted Africa's economic performance while improving the continent's infrastructure. There is a clear trend in data and research indicating that the impacts of Chinese engagements in Africa are not limited to extractive sectors, but span from trade and construction to agriculture and manufacturing.

As Justin Yifu Lin writes, his theory has a main goal of serving as an alternative to the policy prescriptions of the Washington Consensus and mainstream Western economics theory (Lin and Yan 2016, 57–80). Both the Chinese model of economic development, and the Chinese approach to Africa, have gained increasing popularity among African countries, which desire to replicate China's success story and benefit from economic ties with the Asian giant. Theoretical research on the Chinese development model and its implication for other developing countries is thus in high demand. Justin Yifu Lin is an authority on China's economic development, and his work on international development bears clear marks of Chinese perspective. As the book points out, Chinese tend to view their development model as a dynamic process, not a static framework. While Western donors are said to focus on what developing countries "do not have," like governance and infrastructure, Lin and Wang call for more attention to what developing countries actually do have, including "the same 'dream' or aspiration like China had," and "the willingness to work hard and jumpstart growth" in addition to their existing natural endowments and comparative advantages.[3] It is interesting to see that both economists are no longer confined to objective analysis of economics, but call for attention to the aspiration and will of Africans. This special perspective is rooted in the authors' observation of China's development experience.

Even when Dani Rodrik, a professor at Harvard University's John F. Kennedy School of Government, agreed with Justin Yifu Lin's idea of using industrial policy to facilitate structural transformation in Africa, he did not agree to distinguish between static and dynamic comparative advantage (Rodrik 2006). In his eyes, the difference between static and dynamic comparative advantage is simply whether the advantage is calculated according to today's prices or intertemporal relative prices. For him, dynamic comparative advantage is nothing more than the overall advantage across an extended period. Whether industrial policy defies static advantage or dynamic advantage is just a secondary question. He considers industrial policy as anyway contradictory to market economy and criticised New Structural Economics for its inconsistency in this regard. As Lin favours both comparative advantages and industrial policy, Lin appears to follow market rules and defy market rules at the same time.

Yet the dynamic process that Lin and Wang talk about is not about predictable changes of comparative advantages. Since the overall challenges in developing countries are so high and the rewards for individual firms are so small, even over an extended period, no private investors are willing to initiate such ventures without the assistance of the government. Thus public agencies do not defy the market when they facilitate structural transformation, but instead bring the country from a situation in which market does not work to a stage in which market rules can function and enterprises can make use of comparative advantages. Because the market does not work in the initial stage, there is no market rule to be defied by the enabling government policy. Such a transition is not explicitly outlined by Lin and Wang but is precisely the path experienced by China in its transition from a command economy and agrarian society to an industrialised market economy.

It is within this context that the limitations of New Structural Economics can be seen as well. The reason why the authors may leave readers with the impression of defying market rules with policy tools can be traced back to the framework of this theory. First of all, their analysis of comparative advantage and structural transformation is based on economic theories that presuppose the existence of functioning market mechanisms. But because functioning markets are not a given in many developing countries in Asia and Africa, such a theoretical starting point immediately becomes questionable. Whether a country is transitioning from a socialist planned economy to a market economy, or from a traditional agricultural society to an industrial society, its development is not simply an economic question of transition from primary industry to secondary and tertiary industries. An agricultural society lacks all the necessary elements of a market economy, such as a free labour force, a clear system of property rights, a legal system to protect the operations of the market, contractual relations between community and individual, and an entrepreneurial spirit of pursuing capitalist profit. As discovered by Max Weber and Karl Polanyi in their investigations of industrialisation in European countries, deficiencies in societal structural or cultural elements can prevent the normal development of a market economy (Polanyi 1944, 41–42; Weber 1992, 21–25).

In many smallholder farming-oriented developing countries today, much of the institutional, social, and cultural reality does not conform to the requirements of a market economy. As was the case at the beginning of Chinese reforms over 30 years ago, transformation in these developing countries is not merely a matter of restructuring industry, to be understood only in economic terms. Rather, deeper and broader changes are necessary in political, social, and cultural structures, such that a market economy can begin to operate (Tang 2014, 17–24). If these challenges are not addressed, individual enterprises cannot expect reasonable profits from their operations, and market rules cannot be expected to work properly. Western donors perceived that developing countries indeed lack the political and social systems necessary for a market economy, but offered an incorrect solution to this problem. They sought to directly transplant to developing countries a ready-made socio-political system constructed according to the economists' perception of market economy. Unfortunately, this repeatedly proved to be incompatible with preexisting local political and social structures. In order to better analyze the development of these countries, New Structural Economics must also consider how to construct market economy systems and how to effect transformation toward a market economy. To be sure, Lin and Wang mention the importance of information, laws, market institutions and other "soft infrastructures" in structural transformation. Yet they viewed these factors merely as components affecting market prices and costs and did not see paradigm shift from non-market society to market economy due to the constraints preset by economic studies.

Second, in New Structural Economics much emphasis is placed on government policies, but there is insufficient discussion on the specific and diverse development mechanisms of society and markets. State industrial policies, financial support, and infrastructure projects alone are not enough to promote a country's sustainable development. Ideally, government policies must be coordinated between such areas as the supervision of government officials and strategies for enterprise management, or the education of society and technology transfer. In the development cooperation efforts it has launched in over 40 African countries, China has employed similar modes of financing and construction, but different countries obtain very different results. For example, in Mali, the Sukala sugar complex was transformed from an agricultural aid project to a successful investment (Lin and Yan 2016, 117–118). However, in many African countries, attempts to change what were originally industrial and agricultural aid projects into investment enterprises often meet resistance or failure. For the majority of Chinese-invested industrial zones in Africa, including the Yuemei Industrial Park mentioned in Lin and Wang's book, progress has been slow and difficult, departing markedly from original policy programs. The book mentions many examples of success in Chinese development aid, but it does not carefully analyze the many more cases of failure. Tianli Zone in Mauritius (see Cowaloosur and Taylor in this volume), Mulungushi and Urafiki Textile Mills in Zambia and Tanzania are some of the large projects in recent years that have suffered huge losses. A lot of other projects in agriculture, manufacturing and infrastructure construction are either slow, or encounter challenges in various phases. Experiences of failure and success are closely related to each country's specific practices; this proves that development is the result of an interaction between policy-making and the complexities of practice. Restricting theoretical analysis to the level of policy, and relying on a select number of successful cases, may mean neglecting questions of compatibility and adaptability to different environments, which would lead to biased conclusions and ineffective measures.

Third, New Structural Economics suggests that countries should focus on developing export industries according to their comparative advantages. Although exports are crucial for a country's integration into global production chains and for achieving high-quality development, industrialisation in underdeveloped countries should not be confined to export-oriented industries. On the one hand, many underdeveloped countries have a very weak industrial base; the products they manufacture cannot possibly meet the requirements for quality stability or the supply schedules of the international market. Under these circumstances, enterprises geared toward local market production can play a key role in industrialisation and modernisation. Some developing countries, of course, are quite small, in which case the local market would have to integrate with regional markets in neighbouring countries. On the other hand, even in countries like Ethiopia that have the conditions necessary for promoting export industry, a vast number of non-export-oriented industries still hold a major share in the national economy. Accelerating the development of non-export-oriented enterprises would benefit a country's overall growth and promote transformation of the entire socio-economic structure toward industrialisation. At the same time, it would reduce excessive dependence on the international economy and the risks posed by the fluctuations of global markets. China's own development process has been rooted in the simultaneous advancement of enterprises oriented toward the domestic market alongside those producing manufactured goods for export, thereby improving its chances for sustainability. If new theories of development aid also addressed non-export-oriented industries, they would be more applicable to developing countries. For example, foreign investments in manufacturing sectors in Nigeria and Ghana are mainly driven by domestic and regional market demands.

In sum, New Structural Economics not only offers a perspective from China, but also constructs an innovative theory that addresses new approaches and initiatives currently under experimentation in developing countries. Africa is a particular focus area of this theory, as China–Africa economic cooperation has witnessed strong growth for over a decade. The theory has pioneering significance in its field, and its mode of inquiry can serve as a guiding inspiration for future researchers. It deserves careful study and discussion by scholars and students conducting research on South–South cooperation, as well as those researching international development more broadly. I believe that as we continue to explore and add to the theory and practice of international development, the theories of New Structural Economics will be gradually perfected and deepened. More data collection and detailed case studies will help economists correct and improve the theory to make it more precise to explain the reality of development. Meanwhile, the theory, as an experimental effort, cannot cover all phenomena in China–Africa relations. But in preparing a foundation of theoretical reflection on this subject, it will undoubtedly serve to stimulate a more comprehensive theoretic analysis on socio-economic transformation related to China–Africa engagement. I myself am particularly eager to see researchers from fields beyond economics (for instance, Sociology or Anthropology) begin to offer their own theoretical analyses on China–Africa relations, because the transformation processes toward modernity of both partners through their interaction are too comprehensive to be grasped by one discipline. Multidisciplinary research can reveal the interactions between Chinese and Africans in more depth. It will also enrich our understanding of the interrelated impacts of various aspects in this important relationship between developing countries.

Notes

1 United Nations Comtrade Database, https://comtrade.un.org/, accessed July 10, 2016.
2 Peter Reina and Gary J. Tulacz. "The Top 250 International Contractors." *Engineering News Record,* 2015, August 24/31: 40.
3 Wang Yan, presentation in People's University, Beijing, December 16, 2016.

References

Akamatsu, Kaname. 1962. "A Historical Pattern of Economic Growth in Developing Countries." *The Developing Economies* 1(s1): 3–25.
Kojima, Kiyoshi. 2000. "The 'Flying Geese' Model of Asian Economic Development: Origin, Theoretical Extensions, and Regional Policy Implications." *Journal of Asian Economics* 11(4): 375–401.
Li, Xiaoyun. ed. 2010. *Zhongguo he Feizhou Fazhan Pinkun yu Jianpin* [Development, Poverty and Poverty Alleviation in China and Africa]. Beijing: Zhongguo Caijing Chubanshe [China Financial and Economic Publishing House].
Lin, Justin Yifu and Wang Yan. 2016. *Going Beyond Aid: Development Cooperation for Structural Transformation*. Cambridge: Cambridge University Press.
Liu, Hongwu and Luo Jianbo, 2011. *Zhong-Fei Fazhan Hezuo: Lilun Zhanlue yu Zhengce Yanjiu* [Sino-African Development Cooperation: Studies on Theories, Strategies and Policies], Beijing: Zhongguo Shehui Kexue Chubanshe [China's Social Sciences Press].
Polanyi, Karl. 1944. *The Great Transformation*. Boston, MA: Beacon Press.
Rodrik, Dani. 2006. "Goodbye Washington Consensus, Hello Washington Confusion? A Review of the World Bank's Economic Growth in the 1990s: Learning from a Decade of Reform." *Journal of Economic Literature* 44(4): 973–987.
Tang, Xiaoyang. 2014. *Zhongfei Jingji Waijiao Jiqi Dui Quanqiu Chanyelian De Qishi* [China-Africa Economic Diplomacy and Its Implication to Global Value Chain]. Beijing: Shijie Zhishi Chubanshe [World Knowledge Publishers].
Weber, Max. 1992. *The Protestant Ethic and the Spirit of Capitalism*. Translated by Talcott Parsons. London: Routledge.
Zhang, Zhongxiang. 2012. *Zhongfei Hezuo Luntan Yanjiu* [Study of Forum of China-Africa Cooperation]. Beijing: Shijie Zhishi Chubanshe [World Knowledge Publishers].

19

CHINA, AFRICA, AND GLOBAL ECONOMIC TRANSFORMATION

Alvin Camba and Ho-Fung Hung

This chapter explores China's strengthening links with Africa within the political economy of global transformations. As a result of complex shifts in manufacturing in the world economy from the West to the East, however, we now find China's emergent role as an exporter of capital, some of which has succeeded in promoting natural resource extraction and industrial manufacturing in Africa and other developing regions. We might be moving into a new global political economy in which Chinese foreign direct investment (FDI), as an example of South–South economic integration, can serve as a new engine of development in Africa.

Most research on Chinese engagements in Africa has focused on the dynamics of investment at the national level between China and individual African states, while recent work on sectoral variation analyzes the dynamics of various kinds of Chinese and African actors in subnational contexts. The former examines the types of loans, terms of exchange, and impact on national development, whereas the latter looks at sectoral linkages, resource flows, and commodity chains in different states across the African continent. Both research trajectories, however, tend to treat China–Africa relations in isolation, ignoring and obfuscating the dynamics of China's domestic political economy, as well as China's role in other developing regions and in the global political economy at large.

This chapter argues that we need to conspicuously trace the dynamics of Chinese investments at global, national and sub-national levels by emphasizing the context of China's developmental trajectory, geopolitics, and the different responses of developing states. Rather than obscuring China's multiple and simultaneous engagements in various regions by a singularly narrow African focus, the chapter unpacks China's role and actions in the continent in connection with the rise and fall of the China boom, together with a comparison of how different regions in the world respond to the active overtures of Chinese state and private actors. We first outline Chinese investments in different sectors and countries in Africa, and then connect this outward investment to China's domestic political economy. We then discuss the geopolitical implications of Chinese investment, and compare China's presence in Africa with that of other developing regions.

Contours of Chinese outward investment in Africa

China's practice of extending developmental assistance to African countries in exchange for their political allegiance dates back to at least the 1960s. At that time, China was active in supporting revolutionary movements and governments in developing regions, Africa in particular. After economic reform in the 1980s, China's attention toward Africa abated. It was renewed with greater vigor in the 2000s, when rapid economic growth in China urged Beijing to get "back to Africa" as a strategy to secure the supply of oil and other raw materials and export surplus capital in China.

In this vein, China's general approach to African natural resource exports is to befriend whoever is in power with loans, aid, and infrastructure investment projects. China has not been discriminatory in dealing with any type of regime, courting both democratic and authoritarian governments. In comparison with US investments, Chinese investments in the continent are more evenly spread across different countries and carry more generous terms from the perspectives of the host country recipients (see Bräutigam 2011). Though the amount of Chinese economic assistance has trailed those offered by traditional Western powers, most of all the US (Hung 2015: Table 5.8), China's assistance generally brings new and positive gains to the continent, creating competitive pressures for other developing and developed countries to offer better terms in dealing with African nations.

Chinese investments in Africa can be summed up in terms of natural resource extraction, infrastructural investments and manufacturing. First, China's national oil companies (NOC) have taken the lead to invest in the oil-rich states of Africa and even dominate some selected countries. The China National Petroleum Corporation, China Petroleum and Chemical Corporation and the China National Offshore Oil Corporation invested in ten overseas acquisitions globally worth $18.2 billion in 2009 (Jiang and Sinton 2011). Numerous authors have noted the tendency of NOCs to pay overinflated oil prices to compensate for the initial market lead of Western energy companies (Frynas and Paolo 2007; Lee and Shalmon 2008; Zhang 2007). While NOCs competed in oil, China Nonferrous Metal Mining Group (CNMG) became the largest overseas non-ferrous Chinese mining company (Haglund 2008). CNMG, a Chinese SOE, operates in Zambia through NFC Africa Mining plc (NFC-A), which acquired the Chambishi mines in 1997 (Kragelund 2009). As Zambia's Consolidated Copper Mines were privatised in 1997, Chinese mining companies capitalised by becoming major shareholders of several strategic mines.

Second, in infrastructural development, which has often included transportation and energy (see Figure 19.1), Nissanke and Söderberg (2011) note that China has on average focused more on infrastructure than OECD Development Assistance Committee (DAC) countries. China has been able to acquire most of the civil works contracts to build infrastructure in Ethiopia, Tanzania, Democratic Republic of Congo, Zambia, Mozambique, yet the size of infrastructure has been widely debated. While most projects were worth less than $50 million, there have been several mega projects worth over US$1 billion (Renard 2011). Through a new methodology that analyzes press reports and the China Export–Import Bank database, some have estimated that Chinese infrastructural investment inflows have been worth US$500 million annually since 2000 (Foster 2009).

Finally, Chinese interest in agriculture has not led to substantial financial investments (Bräutigam and Zhang 2013). Though there have been agricultural acquisitions in a range of countries, including Ethiopia, Ghana, Mali, Madagascar, Mozambique, and Tanzania (Cotula et al. 2011), most of these assets were built in the 1960s by Mao's government. While the Chinese government does have programs to encourage agricultural investment overseas (Buckley 2013), poor infrastructure, bureaucratic patronage, and multiple, overlapping land ownership claims have generally discouraged huge financial commitments in the continent (Bräutigam and Zhang 2013). China's meager agricultural acquisitions have been used to produce agricultural goods for local African markets or raise biofuel for exports.

Though not as significant as resource extraction sector or infrastructure investment, China's investment in manufacturing in Africa has also started. Indeed, manufacturing FDI in the region occupies only 4.33 percent of the overall amount.

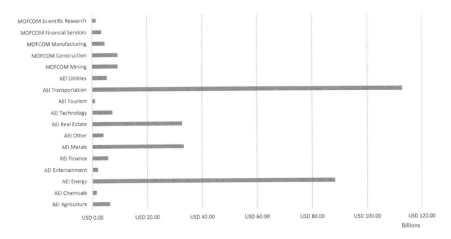

FIGURE 19.1 Sectoral Distribution of Chinese Overseas Direct Investment in Africa, Second Quarter of 2016

Source: Authors' Modification based on American Enterprise Institute's Chinese Global Investment Tracker (CGIT) and the Ministry of Commerce's 2015 Statistical Bulletin of China's Outward Investment Database

Zambia created a Multifacility Economic Zone that allowed smaller and privately owned producers to relocate in the manufacturing sector. The Chinese Ministry of Finance and Commerce also established a special fund to encourage the offshore relocation of textile companies in low cost countries. As Eliassen (2012) points out, the textile and clothing industry has the advantage of the low startup cost, labour-intensive requirements for employment, knowledge and technology transfer, trade-intensive links, and multiple linkages for different import and exporting sectors. However, Elliassen and others argue that China's desire to bypass WTO textile and clothing quotas motivated China to invest in Zambia's textile and clothing industry (Bräutigam and Tang 2011; Eliassen 2012). After China's WTO accession, there was little reason to follow up on the commitment.

Domestic dynamics of China's capital export

There is often a strain of methodological nationalism and unreflective empiricism in literature on China–Africa economic relations, and research on Chinese engage-ments in Africa is often framed in isolation from transformation within China. We propose three solutions in this context. First, China's economy is not just the lar-gest in the world, but also occupies a central position in the global economy as an exporter of goods and capital. Indeed, after China's currency devaluation and generous policy preferences to foreign manufacturers, particularly after China's accession to WTO in 2001, East Asia manufacturers shifted their labour-intensive manufacturing from the US and Europe to China. Manufacturing sectors, origin-ally located in the major markets of the West, transformed China into the "work-shop of the world" by drawing on a huge reserve army of rural migrant labor. In addition, apart from exporting manufactured goods, fixed asset investment is the other dominant source of Chinese economic growth. Since most investment pro-jects by state-owned companies or local governments have been fueled by cheap credits from state banks, the expansion of the state-owned investment sector is grounded in China's export growth. The growing liquidity fostered by China's cen-tral banks created the conditions for an explosion of loans, which in turn relies on the rapidly expanding foreign exchange reserve originating from the export sector to back up its lavish money supply (Hung 2015).

What is the connection between the different projects in Africa and China's recent readjustment from fixed-asset investment to consumer-driven growth? China's rapid economic growth over the last two decades has been driven mostly by export-oriented manufacturing and fixed-asset investments but the export engine has weakened in the past few years. The reckless investment floodgate during the rebound of 2009–2010 has created a gigantic debt bubble no longer matched by a commensurate expansion of foreign exchange reserves. We continue to gather different kinds of empirical data on China's natural resource projects in Africa, but know very little about the connection of those projects to the changes in the Chinese economy and the broader global economy. For example, we know much about the different types of loans from China – grant aid from the Ministry of Commerce, Export-Import Bank of China's loans, and commercial loans and joint ventures – that have flourished in Africa, but we know very little about the connection of those loans to the viability and recent troubles of the PRC's fixed-asset investments.

Furthermore, cheap manufactured goods that maintained debt-fueled consumption in the Western world have been largely dependent on natural resource inputs from Africa and the rest of the developing world. We have accumulated knowledge on the activities of Chinese resource extraction in Africa and the rest of the world, but know very little about the dependence of these development strategies and economic processes to debt-fuel consumption in the US and the EU. Put simply, the study of Chinese engagements in Africa could advance further by analyzing the indispensable connection among, and changing dynamics of, export manufacturing, financial repression, and natural resource extraction in Africa.

In addition, conditions in China and the rest of the world have already changed. Between 2008 and early 2015, outstanding debt in China skyrocketed from 148 percent of GDP to 282 percent, exceeding the level in the US and most other developing countries. China's foreign exchange reserve ended its long rise and started to shrink in 2014 (Hung 2015). As foreign exchange reserves continued to fall, the many redundant constructions and infrastructural investments resulting from the debt-fueled economic rebound appear to be unprofitable in the coming years. To make matters worse for global and Chinese capital, the escalation of peasant resistance and labour unrest in China since the 1990s forced the Communist party-state to make concessions by improving rural economic conditions (that finally curtailed flow of rural migrant labour to coastal export sectors) and labour conditions in the manufacturing sector. These concessions increase wage levels and put further pressure on the profitability of capital. Furthermore, manufacturing capacity and infrastructural expansion of apartments, coal mines, steel mills, and others that expanded rapidly during the boom time and the post-2008 rebound have reached excess capacity with falling profit rates.

How do we conceptualise Chinese economic engagements within the post-China boom context today? The literature has taken into account the variety of Chinese actors, and the multiple kinds of economic engagements, but has largely remained faithful to a homogenous and timeless characterization of the Chinese economy. In its most sophisticated version of Chinese engagements (Gallagher and Porzecanski 2010), scholars derive China's impact from the independent effects of Chinese manufacturing exports, commodity imports, and as a foreign investment competitor of Latin American and African states, whereby some conducted econometric analyses on the trade-offs of Chinese trade and investment relations with states in the global South. This version, however, remains solely applicable to Latin American states, and appears to ignore the link between China's export-manufacturing production, primary commodity imports, and fixed-asset investments in the Chinese economy.

If Chinese investment in Africa depends on, and remains tied to, China's economic model, what happens in the post-China boom world economy? If repayment and debt servicing is dubious, and a big debt time bomb has been formed, how are these finances linked to Chinese investments in the African context? If China's fixed asset investment as a development strategy has run out of room for growth, while the export sector continues to struggle, what happens to the loans, migrants, communities and workers in Africa? Research needs to empirically study the post-boom plans of Chinese companies, the lives of the migrant Chinese communities, and the response of host states to China's potential to scale back. Given that we are now in the waning years of the China boom, many have pointed out that conditions for sustained development will be difficult. Continuing the assessment of Chinese projects in Africa and other parts of the world, while linking the processes to the Chinese economic conditions, becomes more pressing than ever.

From geoeconomics to geopolitics

As Lenin diagnosed long ago in *Imperialism, the Highest Stage of Capitalism,* an overaccumulation crisis is as old as capitalism itself. An overaccumulation crisis within a national economy drives capitalists to export capital overseas in search of places with higher profit rates. It was how manufacturing capital from the core relocated to Asia and China after the 1970s in the first place. Now it is China's turn to become a victim of overaccumulation and feel the urge to export capital. Ever since the early 2000s, China's capital export soared. Stock of China's outward foreign direct investment jumped from USD$28 billion in 2000 to $298 billion in 2012, though it is still small in comparison with smaller advanced capitalist economies like Singapore (see Hung 2015: Table 5.4). The accumulation of foreign exchange reserves and trade surplus leads China into a typical overaccumulation crisis, epitomised by the many ghost towns and shut-down factories across the country. However, as Lenin also argued that exporting capital overseas necessitates the need to protect it, pushing the Chinese government to project its political and military power overseas, leading to numerous territorial disputes, and instigating inter-imperial rivalry with other capital-exporting states.

As a result of capital export, China's increasing presence as a new source of financial support and opportunities offers these African countries new autonomy to resist political demands from the US and other Western powers. At the same time, many African states are reciprocal in their relations with China, returning China's economic favours by supporting Beijing in such political issues as the status of Taiwan or Tibet and the Dalai Lama. Just like many of China's Southeast Asian neighbours, who feel insecure with their increasing dependence on the Chinese economy, some African leaders have voiced concerns about "Chinese colonialism," with opposition movements across the continent seeking to capitalise on the growing popular anti-China sentiments by attacking incumbent governments for becoming subordinate to Chinese interests.

During the rundown to the BRICS summit in Durban in March 2013, attended by leaders from Brazil, Russia, India, China, and South Africa, NGOs and activists in Africa organised a counter summit and employed the concept of "sub-imperialism" to refer to the domination of China and other BRICS countries in the continent. Some go as far as claiming that BRICS' enthusiasm in expanding their presence in Africa resembles the "scramble for Africa" among European imperial powers after the Berlin Conference of 1884–1885 (Bond 2013). This concern about China's growing influence in Africa has been so powerful and widespread that even current governments with close relations with China voice their anxiety about China's economic presence openly.

The limit of China's political influence in other countries, however, ultimately constrains the expansion of China's economic influence. The transformation of China's global economic role, the waning of the commodity boom, and the rebalancing from investment-fueled growth to consumer-driven growth all point to a change in capital export. While China has previously lacked the will and capability to project hard power, it may be changing very soon. China's National Defense White Paper in 2013 stated explicitly for the first time that protecting overseas economic interests is now one core goal of the People's Liberation Army.

In this vein, three interesting research trajectories emerge from geopolitics and the post-China boom context. First, while China will not be a new hegemonic or dominant power in the world anytime soon, its increasing presence across the developing world, most notably in Africa, is already changing the dynamics of global politics by empowering other developing countries to resist Western political influence. As many studies have pointed out (e.g. Kentor and Boswell 2003), the political subjugation of developing countries to Western countries was not caused by trade with, and investment from, developed countries per se, but has been a result of Western countries' monopoly role as sources of investment and major trading partners. As such, research connecting the implications of the post-China boom to the shifts in alliance formation would be most interesting. As China's desire to project political and military power in other regions of the world becomes more apparent, it will significantly affect Chinese investments and political relations in Africa. China might mobilise African states for specific political purposes, which might impact Africa's relations with the West. Put simply, the study of China–African relations cannot be disassociated from China's central and evolving position in geopolitics.

Second, what are the implications for state-led development in the developing world? With the rising prevalence of China as a major new trade partner and alternative source of investment, many developing countries can more readily reduce their one-sided reliance on the West for investment and markets. With a wide range of developing countries competing for investment from a limited number of developed countries, or exporting similar low-value-added products to a smaller number of developed world markets, developing countries lack bargaining power. This renders these countries less capable of resisting political demands from developed countries in a bilateral setting or in multilateral organizations like the WTO. Thus, China as an alternative improves their bargaining position in bilateral and multilateral negotiations. What remains to be explored is how much this new geopolitical dynamics of development is contributing to the recent economic renaissance of the African continents beyond the commodities boom (e.g. Ali and Dadush 2015) over the last decade or so. It is also important to see how the more recent economic slowdown of China is going to take a toll on Africa's development, which has become increasingly reliant on Chinese investment.

Finally, the implications on global economic and security governance cannot be disregarded. With the origins of the Asian Infrastructure Investment Bank (AIIB), numerous countries joined from all over the world despite Chinese economic troubles. In Southeast Asia, China's Regional Comprehensive Partnership with ASEAN has been active in forming networks and groups to counter the possible emergence of the Transpacific Partnership.

China's capital export in regional-comparative perspective

Besides Africa, China has been active in exporting capital to other developing regions as well. In 2011, China bypassed the US as Latin America's largest trading partner. Chinese investment in Latin America, worth US$121.85 billion, occurs between SOEs and focuses on natural resource extraction. In most segments of the world economy, SOEs have waned in influence, often due to government decisions to embrace privatization and market competition. But in Latin America, the role of state enterprises in the oil, gas and mining industries has remained stronger than ever. This is especially the case in Brazil and Chile, but also to a greater degree in the wider Latin American region. As Chinese investments advance in Latin America and elsewhere, Chinese companies have partnered up with the dominant state-owned enterprises. Chinese capital has focused on extractive industries, which export commodity minerals to the Chinese factories, and agro-businesses, which ship soybeans to Chinese markets. Chinese investment also went into services and other sectors, but the Latin American geography hinders further incorporation into the Sino-Centric global production network.

There are also the differentiated impacts of Chinese partnerships across Latin American states. States with strong SOEs, particularly in the extractive sector, benefited from China's resources demand (Gallagher 2016). As a result of extraordinary high commodity prices, the steady improvement in technology, innovation, and transfer of know-how in the mining sector, the commodity boom in Brazil and Chile, states with strong SOEs, translated into windfall profits during the early 2000s. However, most Latin American exports to China, even from Brazil and Chile, have been natural resources, which has recently slowed down due to China's economic adjustment. In addition, the manufacturing sector of the most competitive Latin American states – Brazil, Chile, Argentina, and Mexico – has lost to, and stagnated because of, Chinese manufacturing exports. Experts in the region argue that Mexico, which has traditionally exported to the US, has suffered the most because of Chinese manufacturing (see Gallagher 2016). In addition, other resource-dependent countries did not benefit as much. For example, Peru and Venezuela also implemented quite similar macro-economic reforms aimed at attracting Chinese investment, but the outcomes appear to be divergent from Brazil and Chile's experience. As a result of the weakness of their tax authorities, Peru and Venezuela lack the mechanisms for direct fiscal transfer to social programs from the SOEs. Conversely, Brazil and Chile have produced stable tax regimes and created "enabling" political environments, which allowed successive governments to implement a policy of continuity in order to retain Chinese investments and resource demands in their domestic markets.

In Southeast Asia, Mao's China exported communism by supporting revolutionary movements in ASEAN during the Cold War. In contrast, Deng Xiaoping marked a turn in China–ASEAN relations, tapping into the long and rich history of the two dating as far back as the pre-colonial period. With respect to China's overtures, diplomatic ties were normalised in the 1990s, starting in Thailand but followed by all ASEAN states. China's bilateral Overseas Development Aid (ODA) to ASEAN states also significantly increased. Trade volume between China and ASEAN trade reached US$20 billion in 1995, but eventually ballooned to $480 billion in 2014 (Siriphon 2015). While ASEAN's failure in the Asian Financial Crisis led to China's tremendous rise in the 1990s, ASEAN states eventually adjusted to the Sino-centric global production network to remain competitive in the global economy (Hung 2015). While huge state initiatives, such as the Belt and Road Initiative, remain uncertain in the bigger picture, China's "going out" strategy has been more relevant to Southeast Asian development. In the $1.5 trillion Chinese capital across the world, there is an estimated amount of $144 billion in Southeast Asia (see data in the American Enterprise Institute's (AEI) Chinese Global Investment Tracker (CGIT) database).

While research on Chinese investments in ASEAN is still at the early stages, there are several interesting patterns. The heterogeneity of state capacity and economic development in ASEAN has attracted different kinds of Chinese investments. High-income countries like Singapore attracted knowledge-based, service sector companies. In extractive industries, national SOEs, similar to the ones in Africa, invested heavily in Indonesia's large-scale mining, but only regional Chinese mining companies went to the Philippines. Thailand and Vietnam appeared to attract most Chinese manufacturing due to geography. However, the economies of Myanmar, Laos, and Cambodia appear to be largely dependent on Chinese investments across a whole variety of sectors. While Latin America has some degree of variation too, the variety of developmental stages and geographic proximity of Southeast Asian states made Chinese investments more varied. Furthermore, ASEAN also differs because economic transactions and joint ventures occur between Chinese investors and the Southeast Asian Chinese communities in region. Southeast Asia has a long history of Chinese migrants since the ninth century (Yeung 2006), amounting to an estimated population of 33 million ethnic translocal Chinese.

The pursuit – or retreat – of extractive Chinese investments in Southeast Asia was not only because of the external pressures of export manufacturing growth and fixed asset investments, but also partially a consequence of their geopolitical realities. Unlike other regions in the world, Chinese investors had to deal with states strongly aligned to the West. Hence, when territorial issues were used against China, the West quickly branded them as "exploiters" or "bullies," something that also happened in 2016 when The Hague tribunal supported the Philippines' South China Sea claim. In this context, the Philippines is perhaps illustrative of how territorial issues might have compelled Chinese investors to redirect their development strategies and concentrate their fiscal resources in elsewhere. Under Rodrigo Duterte's administration, the Chinese and Philippine governments seem to be rebuilding the friendly diplomatic relationship that the Gloria Macapagal–Arroyo government started in 2001. With impressive levels of human capital and continuing growth in the services industry, the potential opening of the Philippines and various Southeast Asian states to Chinese capital makes us ask what holds for Chinese investments in Africa in the coming decades. One might hypothesise that the changing geopolitical situation and the post-China boom context might change the direction and patterns of Chinese investments across the world.

China has invested the most amount of money in north and sub-Saharan Africa. As seen in Figure 19.2, Africa has $305 billion in the AEI's CGIT dataset. As discussed earlier, Chinese investments in the extractive sector, infrastructure, and manufacturing comprise the bulk of capital. The absence of strong domestic SOEs in the region, as well as the more homogenous kind of states in the terms of capacity, mean that China's impact has been more uniform.

FIGURE 19.2 Regional Distribution of Chinese Overseas Direct Investment in the World, Second Quarter of 2016

Source: Authors' Modification based on American Enterprise Institute's Chinese Global Investment Tracker (CGIT) and the Ministry of Commerce's 2015 Statistical Bulletin of China's Outward Investment Database

The regional brief comparison has two conceptual implications for the future of Chinese engagements in Africa. First, with regard to the study of Chinese investments at the end of the China boom, it appears that research needs to move beyond the specification of loans and project implementation. This is a technocratic exercise promoted by international financial institutions when discussing "capacity-building" in specific regions, exemplified by an obsession with assessing the impact of projects, working to improve the revenue capture abilities of tax agencies and streamlining of royalties and taxes. This type of research needs to find patterns and connections between Chinese investments across regions, unbounded by a priori area specifications. Chinese investors and their projects face a complex political calculus of economic constraints to contend with. Only through research efforts geared toward connecting multiple and similar processes at the global level can we become more adept at designing and conducting research projects that capture these complicated patterns. In short, a focus on Africa might miss inter-regional processes that trespass "a priori" area studies boundaries and affect long-term economic outcomes. Regional specialists need to build credible inter-regional research partners, which means not simply co-writing comparisons but also gaining legitimacy and academic support from various areas and inter-disciplinary social science traditions.

Second, with China's economic slowdown, the developmental spaces for poor, resource-rich African countries might be narrowing as the pull from China weakens. This is also true to different extents in other developing regions. Research on Chinese engagements in Southeast Asia is still at an early stage, while there is now growing consensus on the variation of impact in Latin America. At the same time, Chinese investment in different developing countries is becoming politicised. For example, some resource-rich African countries want more power to make political decisions about *when* and *at what pace* investments should take place in their countries. As the experience of the Philippines shows during the Aquino period (2009 to 2016), a pro-American administration with some degree of civil society support rejected China's overtures and instigated an international case in the South China Sea. In this context, competing foreign investors may move to take some of China's place. China's economic inroads into other developing countries have woken up other big emerging economies to pursue capital export. Continuing the assessment of Chinese projects vis-à-vis their competitors in Africa and across the world, while linking the stories to the Chinese economic conditions, has become more pressing than ever.

Conclusion

While previous works painstakingly analyzed China's engagements in Africa at the national and sectoral levels, we suggested, as the main argument of the chapter, an alternative research angle that carefully considers China and Africa within the political economy of global transformations. After synthesizing the literature on Chinese investments in Africa, we argued three points. First, we discussed the importance of treating China's capital export dynamically. The rise of Chinese FDI and ODA has depended on the availability of cheap credit, the export of cheap manufactures, and the demand from the credit-driven consumption in the West. As Pettis argues (2013), China's developmental model begun to run into problems since the 2009 financial crises. From growth rates of 10 to 12 percent in the early 2000s, China has slowly but surely slowed down since the end of the commodity boom in 2011. The buyers from the West, as consumers of last resort, started to experience class bifurcation, shrinking an already declining middle class. As the commodity boom ended and resource demand from Indonesia and Africa declined, we suggested that Chinese FDI projects in Africa, particularly the ones in extractive industries, should not only be treated within the backdrop of China's economic adjustment but also be analyzed within these global economic changes. In order to do so, research that pursues a within-case design analyzing projects before and after the China boom can illustrate these dynamics. In addition, comparatively studying different types of Chinese FDI projects can better situate these processes.

Second, in the realm of geopolitics, we asked about the intersection between China's geopolitical ambitions and economic goals. We discussed China's economic engagement in the developing world as a way to dissuade them from pursuing territorial claims or affirm China's political interest. Simultaneously, China is also empowering these states to resist Western influence. Our argument dovetails well with Swaine (2014), who posited that China under President Xi Jinping has moved increasingly toward rewarding allies and punishing enemies in territorial disputes. We argued that the study of China's engagements in Africa must not lose sight of Beijing's political goals. At the same time, it will be beneficial to be equally cognizant of Africa's bargaining position with the West regarding political and economic matters. These can be analyzed through the attempt of various states to pursue state-led development and their participation in global political governance, such as the AIIB, the UN, and many others.

Finally, we argued that Chinese FDI and ODA can be best studied in a comparative-regional perspective. Regional specialists in Latin America, Africa, and Southeast Asia have produced considerable amounts of work regarding China's involvement in their various regions. While more collaboration with China specialists will be needed, future research projects examining these Chinese FDI projects *in relation to one another* will help cast Chinese FDI from a global perspective. With the announcement and the continuation of China's Belt and Road Initiative, Chinese FDI seems to be expanding further in Central Asia and the Middle East. As Chinese SOEs move forward in different parts of the world, the study of the patterns and connection of Chinese FDI projects can better illustrate the logic and dynamics of Chinese capital. These research projects can be done through the study of comparable Chinese projects across different countries or the comparison of different types of Chinese capital within a certain geographic space. Furthermore, as the global economy transforms some countries to export more capital, Chinese FDI will be competing with other investors. Emerging markets need to export capital to safeguard their economies from global economic shocks (Prasad 2015). As China was once a challenger to the West, other emerging economies will be challenging China this time around. How Africa and the developing world respond to these capital exporters will shape the lives of millions.

References

Ali, Shimelse and Uri Dadush. 2015. "Is the African Renaissance for Real?" *Carnegie Endowment for International Peace.* http://carnegieendowment.org/2010/09/30/is-africa nrenaissance-for-real (accessed May 19, 2016).

American Enterprise Institute. 2016. Chinese Investment Tracker Database. www.aei.org/ch ina-global-investment-tracker/ (accessed December 2, 2016).

Bond, Patrick. 2013. "Sub-imperialism as Lubricant of Neoliberalism: South African 'Deputy Sheriff' Duty Within BRICS." *Third World Quarterly* 34(2): 251–270.

Bräutigam, Deborah. 2011. *The Dragon's Gift: The Real Story of China in Africa.* New York: Oxford University Press.

Bräutigam, Deborah and Tang Xiaoyang. 2011. "African Shenzhen: China's special economic zones in Africa." *The Journal of Modern African Studies* 49(1): 27–54.

Bräutigam, Deborah and Haisen Zhang. 2013. "Green Dreams: Myth and Reality in China's Agricultural Investment in Africa." *Third World Quarterly* 34(9): 1676–1696.

Buckley, Lila. 2013. "Chinese Agriculture Development Cooperation in Africa: Narratives and Politics." *IDS Bulletin* 44(4): 42–52.

Campell-Mohn, E., 2015. "China: The World's New Peacekeeper?" *Duke Political Review.* http://dukepoliticalreview.org/china-the-worlds-new-peacekeeper/.

Carmody, Pádraig and Ian Taylor. 2010. "Flexigemony and Force in China's Resource Diplomacy in Africa: Sudan and Zambia Compared." *Geopolitics* 15(3): 496–515.

Cotula, Lorenzo, Sonja Vermeulen, Paul Mathieu and Camilla Toulmin. 2011. "Agricultural Investment and International Land Deals: Evidence from a Multi-country Study in Africa." *Food Security* 3(1): S99–S113.

Di John, Jonathan. 2011. 'Is There Really a Resource Curse? A Critical Survey of Theory and Evidence." *Global Governance* 17: 167–184.

Eliassen, Ina Eirin. 2012. "Chinese Investors: Saving the Zambian Textile and Clothing Industry?" Stellenbosch: Stellenbosch University Centre for Chinese Studies.

Foster, Vivien, William Butterfield, Chuan Chen and Nataliya Pushak. 2009. *Building Bridges: China's Growing Role as Infrastructure Financier for Sub-Saharan Africa* (Trends and Policy Options Vol. 5). Washington, DC: World Bank Publications.

Frynas, Jedrzej Frynas and Manuel Paulo. 2007. "A New Scramble for African Oil? Historical, Political, and Business Perspectives." *African Affairs* 106(423): 229–251.

Gallagher, Kevin P. and Roberto Porzecanski. 2010. *The Dragon in the Room: China and the Future of Latin American Industrialization.* Stanford, CA: Stanford University Press.

Gallagher, Kevin P. 2016. *The China Triangle: Latin America's China Boom and the Fate of the Washington Consensus.* Oxford: Oxford University Press.

Haglund, Dan. 2008. "Regulating FDI in Weak African States: A Case Study of Chinese Coppermining in Zambia." *The Journal of Modern African Studies* 46(4): 547–575.

Hung, Ho-fung. 2015. *The China Boom: Why China Will Not Rule the World.* New York: Columbia University Press.

Information Office of the State Council, PRC. 2013. *The Diversified Employment of China's Armed Forces.* http://news.xinhuanet.com/english/china/2013-04/16/c_132312681.htm (accessed May 19, 2016).

Jiang, Julie and Jonathan Sinton. 2011. "Overseas Investments by Chinese National Oil Companies: Assessing the Drivers and Impacts." *International Energy Agency Information Paper*.

Jiang, Wenran. 2009. "Fuelling the Dragon: China's Rise and Its Energy and Resources Extraction in Africa." *The China Quarterly* 199: 585–609.

Kentor, Jeffrey and Boswell, Terry. 2003. "Foreign Capital Dependence and Development: A New Direction." *American Sociological Review* 68: 301–313.

Kragelund, Peter. 2009. "Part of the Disease Or Part of the Cure? Chinese Investments in the Zambian Mining and Construction Sectors." *European Journal of Development Research* 21(4): 644–661.

Lee, Henry and Dan Shalmon. 2008. "Searching for Oil: China's Oil Strategies in Africa." In *China into Africa: Trade, Aid, and Influence*, edited by Robert Rotberg, 109–136. Washington, DC: Brookings Institution Press, 2008.

Mohan, Giles and May Tan-Mullins. 2009. "Chinese Migrants in Africa as New Agents of Development? An Analytical Framework." *European Journal of Development Research* 21(4): 588–605.

Nissanke, Machiko and Marie Söderberg. 2011. "The Changing Landscape in Aid Relationships in Africa: Can China's engagement make a difference to African development?" UI Papers 2011/2. *Swedish Institute for International Affairs*.

Pettis, Michael. 2013. *The Great Rebalancing: Trade, Conflict, and the Perilous Road Ahead for the World Economy*. Princeton, NJ: Princeton University Press.

Prasad, Eswar S. 2015. *The Dollar Trap: How the US Dollar Tightened its Grip on Global Finance*. Princeton, NJ: Princeton University Press.

Renard, Mary-Françoise. 2011. "China's Trade and FDI in Africa". Working Paper 126. *African Development Bank Group*.

Siriphon, Aranya. 2015. "Xinyinmin, New Chinese Migrants, and Influence of the PRC and Taiwan on the Northern Thai Border." In *Impact of China's Rise in the Mekong Region*, edited by Yos Santasombat, 147–166. London: Palgrave.

Swaine, Michael D. 2014. "Chinese Views and Commentary on Periphery Diplomacy." *China Leadership Monitor* 44(1): 1–43.

Yeung, Henry Wai-chung. 2006. "Change and Continuity in Southeast Asian Ethnic Chinese Business." *Asia Pacific Journal of Management* 23(3): 229–254.

Zhang, Zhongxiang. 2007. "China's Hunt for Oil in Africa in Perspective." *Energy & Environment* 18(1): 87–92.

20

CHINA AND AFRICAN SECURITY

Lina Benabdallah and Daniel Large

For China, meaning a variety of Chinese state and non-state actors, Africa has become known as an arena of foreign policy experimentation in general (including economic and business ventures) and, more recently, in the realm of security. Africa stands out in China's global security foreign policy. Approached from the perspective of African security, however, China represents a comparatively new and ever more influential external power. At the same time, many of the issues the subject of security relations raises are historically familiar and represent variations in continuity.

This chapter aims to provide an overview of context, treatments so far and possible directions for future studies concerning security in China's relations with Africa. It first introduces background, including key drivers of China's security engagement with Africa, before surveying academic treatment to date. The relationship between economic development and security, a crosscutting theme that has attracted theoretical and applied interest, is then explored. It concludes by suggesting that while analysis of China's security engagement is necessary, this needs to be framed more in terms of African security within a global perspective on China's evolving security engagement.

Context

Histories of China's military ties with Africa remain present today, reformulated into discourses mobilised to enhance solidarity on the basis of selective use of African liberation history. These underline how different current relations are from the late- and post-colonial periods. Africa had appeared to be, and was represented by China as, a fertile theater to promote revolutionary Chinese foreign policy (see Van Ness 1971). More recent trends emphasise not the positive value and potential of strife but instead the spectre of insecurity, risk, conflict and political instability. While often regarded as threats to Chinese interests and longer-term prospects in the continent, these very challenges also represent opportunities for "learning by doing" in security policy, thus carrying potential benefits for China as well. Such historical and discursive changes across time, connected by the common theme of domestic Chinese economic and political change and the corresponding evolution of China's Africa relations, are themselves worthy of further exploration. A further refrain, variations on the China threat in Africa, has been periodically rejuvenated (e.g. Fergusson 1964; Mallaby 2005) and challenged. Discursive and representation issues in a social-media age play out as an important part of China's official security engagement in Africa, notably UN peacekeeping, a fairly clear case of Beijing's effort to mobilise positive soft power. At the same time, less official information emanating from the spread of Chinese migrants can be at odds with such official messaging. The upshot is different streams of engagement and types of knowledge, meaning that many of the sources used to inform understandings are bound up in or designed for other ends.

Security arrived as an official policy area in China–Africa relations only in 2012. Amidst turmoil in North Africa, and at the end of President Hu Jintao's period in office, FOCAC V formally declared that a China–Africa Cooperative Partnership for Peace and Security would be established. This was further elaborated in 2015 at FOCAC VI, and in China's second Africa policy.[1] Security was elevated to one of five foundational "pillars" upholding the China–Africa "comprehensive strategic partnership." Beyond FOCAC, and bilateral military relations between the Chinese and African governments, China's security engagement has been expanding. This involves a notable UN role. China's UN peacekeeping remains high profile in Africa, seen as a means to enable China to raise its international status, improve relations with host countries and Western governments, protect interests abroad and give the PLA the opportunity to gain operational experience.[2] Africa is thus one part of China's evolving global security engagement. Besides China's increasing arms sales to many African countries, and its role in fighting piracy off the Gulf of Aden, recently, China has been seeking to establish itself as a "maritime great power." The leasing of a naval logistics base in Djibouti, formally announced in December 2015, marks a significant step in the use of military power projection in foreign policy.[3] Given that much has been happening in a relative short period, what has been driving this mainstreaming of security in China's engagement?

Drivers of China's security engagement in Africa

Security may seem to be a comparatively recent, more manifest direction in official China–Africa relations but it has been integral to the development of relations over the past two and a half decades. First, China's current engagement with African security is proceeding in and partly shaped by a historically new phase in relations. Unlike past relations, and in some African countries more than others, the Chinese government is pressured by economic stakeholders as well as its nationals aboard to protect against violent conflicts. Given such pressures, a more politicised role of China is evolving: despite efforts to uphold non-interference, the practical reality of greater involvement in politics and exposure to conflict, insecurity, terrorism and risks of political instability has raised the importance of investment protection for state-owned as well as private corporations.[4]

The African continent may be prominent but is by no means unique in terms of Chinese foreign policy dealing with the protection of Chinese nationals and global interests. However, two somewhat paradoxical effects of this have emerged. On the one hand, the recent global commodity price downturn and reduced domestic growth have triggered calls for China to rethink its approach to investing in Africa. One the other, Chinese firms began engaging in hiring and training private security forces in African countries in order to navigate the boundary of non-interference and secure Chinese investments. Indeed, non-interference in internal affairs remains the dominant thinking and declaratory policy at the strategic level, but China's policy makers have been gradually adopting a more flexible view of the principle in tactical terms (see Large 2009). The former Director-General of the Ministry of Foreign Affairs' Department of African Affairs suggested in 2013 that China's inadequate involvement in African peace and security affairs was a weakness in China's Africa policy and urged greater engagement (Lu, 2013).

Second, a less tangible dimension as to why China is increasing its support for peace and security efforts in Africa concerns efforts to exercise, and gain credit for exercising, international responsibility by contributing to peace and security (Alden and Large, 2015). Such reputational drivers serve additional purposes, notably doubling as an opportunity to gain operational military experience for the PLA, meaning bilateral interests can be advanced under the banner of multilateralism (Gill and Reilly 2000; Taylor, 2014). A decade ago, China was criticised by some as a security free-rider whose economic role was not commensurate with security burden sharing. Now China is the largest provider of peace-keeping troops among permanent UN Security Council members. There is a realization that peacekeepers can represent China as a responsible international actor that does not only care about its own self-centered economic interests but also genuinely engages in peace promotion (Large, 2016). The future logic and consequences of a more involved military role are thus starting to take shape. Indeed, emergency rescue, disaster relief, counter-terrorism, stability maintenance, as well as international peacekeeping are presented and rationalised in China's Military Strategy (2015) in terms of China doing its "utmost to shoulder more international responsibilities and obligations, provide more public security goods, and contribute more to world peace and common development." In the context of "national rejuvenation" under President Xi, the PLA's evolving role in Africa is presented as China supporting existing international institutions by being involved in "more international responsibilities and obligations," providing "more public security goods," and contributing "more to world peace and common development." (ibid.).

Third, the shift from China to Chinese engagement in Africa, extending beyond state dominance through a diversification of actors, has introduced new dynamics of influence on the role of the Chinese state. Migration has added the imperative for Beijing to protect Chinese nationals abroad, a key driver of efforts to enhance force projection capability (Parello-Plesner and Duchatel, 2015). This ontological change has seen the previously dominant state-based logic shift to one augmented by individual, social and corporate aspects, and interactions between these. Finally, there has also been interest by African states and organizations led by the AU in enhancing security cooperation with China. Since the early 2000s the AU, which replaced the Organization of African Unity (OAU), has been prioritizing an African roadmap for Africa's Security Architecture (ASA). The ASA is grounded on the firm belief that Africa should take ownership of finding solutions to the continent's security problems. More importantly, the AU views ASA as an amalgamation of governance, security, and development. As a result, relaxing the boundaries between security and development, and viewing these as necessarily interdependent, is compatible with the Chinese state's views on peace and security (see Benabdallah 2017). For this reason, one notable driving factor is the role of African states, regional organizations and the AU in seeking to forge relations with China in the field of security. While security arrived late as an official policy area in China's relations with Africa, following on from the cultivation of political ties and promotion of economic interests, it now stands out as a dynamic area of striking importance.

China–Africa security relations in theoretical context

While security as a policy issue travelled from a relatively marginal to a more prominent position in a fairly short space of time, much coverage has emanated from media or policy quarters, including think tanks and NGOs (van der Putten 2015; Saferworld 2012). The relative paucity of scholarly analysis thus far in the face of the evident growing importance of the theme suggest that this is a new frontier in China–Africa and Africa–China Studies (Huang and Ismail 2014; Alden et al. 2017). The present juncture is thus notable for a contrast between the recently growing but still comparatively underdeveloped studies of China compared to other literature on African security, and the wide-ranging policy debates this has given rise to in China (e.g. see Williams 2016; Wang 2012). As well as notable for China's evolving African engagement, these are testament to the importance of security as a barometer for wider evolution of China's global relations.

There are three widely recognised theoretical approaches to International Security Studies: traditional, non-traditional, and human security paradigms. The discourse on African security has, for a long time, reflected the first two approaches, although Human Security has surfaced along with efforts to democratise the continent and incentivise states to promote human and civil rights. Likewise, in the context of China, traditional and non-traditional security have been dominant. The Human Security approach has limited relevance given its liberal peace and democratic governance premises, which contradict China's governance model. When put together, China–Africa security relations have been, so far, mostly approached from traditional and non-traditional security perspectives. This reflects the challenges of understanding security, arguably an inter-subjective concept that gives rise to multiple, sometimes contending approaches; how one perceives security depends very much on how one answers such questions as who merits security (the state, institutions, individuals?), or how security should be promoted and can be achieved (armament, alliances, economic growth, democratization, human rights?). This section explores these broad approaches and applies them to the context of China–Africa security relations.

Before doing so, it is imperative to reflect that these different paradigms of security studies have been designed primarily from the perspective of Western powers and global North states. IR in general, and security studies as a sub-field, have yet to take seriously theorizing security and other concepts from the perspectives of the global South and emerging powers (Ayoob 1995). By examining these three approaches, this section highlights the ways in which global South perspectives to security (both African and Chinese understandings of security) fit within the traditional, non-traditional, and Human Security discourses.

Traditional security studies are associated with the Cold War, having been shaped primarily by U.S.–Soviet relations. Interest in the Third World was largely based on whether this mattered to Washington–Moscow relations. Traditional security approaches were dominated by interest in state security. When exploring foreign policy making, these operate through the prism of national security and state interests, and tend to rely on a hard power definition of security capability. In the context of China–Africa relations, scholars have explored questions related to China's interests in a naval base in Djibouti (Shinn 2015), use of UN veto power on peacekeeping mission deployments to thwart Taiwanese influence (Duchatel et al. 2014), combat troops deployed in Mali and South Sudan UN peacekeeping China's interests in negotiating peace agreements in African conflicts, or arms sales (Shinn and Eisenman 2012). One challenge of this approach is the narrow means by which it conceptualises and examines security. As seen in China–Africa relations, and more broadly, there are pressing security issues, such as transnational criminal networks of crime or environmental threats that are not captured by traditional security approaches and that identify the limits of military/hard power strategies.

Non-traditional security (NTS) also privileges states but departs from traditional security approaches by expanding understandings of threat to include aspects such as terrorism, disease, environment, or drug trafficking (Buzan and Hansen 2009). It emerged and gained influence after the Cold War. Issues of infectious diseases or the environment had previously been located within Development Studies, only discussed in the security context if they had palpable security consequences on developed states. The elevation of non-traditional issues to security studies can be explained by the rise in threats to state borders by non-state actors and transnational risks.

NTS is a defining characteristic of China's policy commitment and discourse on security in Africa (see FOCAC 2015). Likewise, the AU and other regional organizations across Africa approach security from NTS lenses and make it a priority to address issues such as trans-border migration, transnational crime, drugs, etc. In turn, a growing body of scholarship on China–Africa has sought to approach security from this perspective (see, for example, Anthony et al. 2015). NTS is a more appropriate framework to study security in China–Africa relations than the traditional/realist paradigm (at least for the time being; yet, in the likelihood that China may get more assertive with naval power projection in Africa, the traditional security paradigm may resurface as a useful framework to understand China–Africa security relations). There are several reasons for this. First, the non-interference principle has hitherto been an influence upholding the primacy of development over military security in the foreign policy making of many global South states. Non-interference recognises and upholds states' sovereignty over domestic issues in their territories. While foreign policy making in China and African states is careful in formally articulating this principle, and ensuring mutual interest in development cooperation and not in military intervention, China's adherence to the principle has shifted toward more pragmatic engagement (Alden and Large 2015; Aidoo and Hess 2015). At the same time, there has been a shift by some African states and the AU toward more interventionary practices, despite the continuation of the sovereignty framework that non-interference is rooted in.

Second, from the perspective of discourses on African Security, it is well-documented that issues of international terrorism and external conflicts are valid security concerns for many African states, but for the most part it is issues such as HIV, malaria, poverty, and other non-traditional security concerns which are viewed as threats to regime stability and state security. The AU and its ASA framework recognise that civil conflicts in Africa are mostly instigated by lack of or limited access to food, water, and health. Additionally, climate change and its impact on fluctuating agricultural yields are further concerns to many African states. These reasons make NTS a highly relevant framework and indicate, in part, why the Human Security paradigm is missing from this scholarship.

Finally, Human Security (HS) had its origins in post–Cold War efforts, including by UN agencies like UNDP, to reinforce the role of development in security and to move beyond the state-centrism of traditional and non-traditional security paradigms. By seeking to deemphasise state borders and states as the referent object of security, the aim was instead to foreground "people." Indeed, governments supportive of the concept, such as Canada, Japan, or Norway, associated human security with the "progressive values of the 1990s: human rights, international humanitarian law, and socio-economic development based on equity" (Suhrke 1999, 266).

Distinguishing HS from NTS might seem straightforward. Whereas NTS takes a state-centric approach and views citizens' security to be a part of the bigger picture of state security, in HS discourse the state is obligated to ensure citizens' security. In this way, NTS is the reverse of HS since it gives state security primacy over that of individuals. Even more, in HS discourse, the state's security apparatuses can be viewed as a source of insecurity for individuals if civil rights and freedoms are constrained, threatened or violated by the state. In considering NTS and HS in the context of China–Africa relations, however, the differences between the two are less stark given the state centrality in the governance model of both China and most African countries. As many scholars assert, despite being distinct in the mainstream literature, NTS and HS are not necessarily in opposition in Chinese security discourses (Breslin 2015).

Applying HS to the study of authoritarian governments and developmental states presents a number of challenges. First, in HS discourse, states are responsible for providing the environment necessary for citizens to flourish and develop. By contrast, in developmental states (such as China), development is state-led and centrally planned. In other words, HS's emphasis on development and security for people before the state clashes with developmental states where development comes from the state. Second, the idea that HS revolves around achieving "freedom from fear and want" may not seem achievable in the context of authoritarian regimes found in many African states and China. The authoritarian state is viewed, in HS discourse, as a source of fear. Lastly, a challenge in both human and non-traditional security approaches is that it is difficult to draw a line between what counts as a security issue and what does not. Indeed, when issues as broad as human rights, food, environment, energy or health can all be interpreted as human security issues, it can be difficult to know what human security is not. Thus, where the traditional security paradigm may seem to be too constraining, the NTS and HS approaches may be too broad.

Yet, given that the bulk of the relationship between China and many African states is one where economic growth and development projects take priority over hard security and military considerations, a fruitful way of approaching China–Africa security relations is through the prism of the development–security nexus. This nexus is narrower than the overwhelmingly broad scope of NTS and HS, but much more encompassing than the traditional approach. The security–development nexus also allows for an interesting intersection of questions of theory and practice because it depends on states' views on security and the extent to which economic growth is considered as part of security. Indeed, the amalgamation of security and development in China–Africa relations has been present from the early stages of encounters making it all the more appropriate to further examine this nexus.

Development–Security

The notion of a development–security nexus points at the centrality of the interdependent connection between improving socio-economic living conditions and the prevention of conflict (Duffield 2010; Jensen 2010). This view of security rests on two principle assumptions: a negative understanding of security, taking the absence of conflict to mean security; and a view that conflict is rooted in economic underdevelopment (Reid-Henry 2011). This understanding takes improvement of living conditions via job creation, infrastructure development, and poverty reduction as the means to achieving lasting conflict resolutions and, thus, a state of realised security. In recent years, international conflict resolution and peacekeeping efforts have also reflected a shift away from traditional military and political missions to missions that enhance economic and social development. Post–Cold War peace-keeping missions are thus a good example of the development–security nexus. The Brahimi Report (2000) proposed that achieving sustainable peace required more than a ceasefire and peacebuilding, galvanizing the UN to restructure its peace-keeping operations to include economic and social development programs. This view of achieving sustainable peace through development is congruent with contemporary Chinese perspectives on peace and security.

While development–security was a Western coinage as such, the concept is not without equivalents in the Chinese context. China has had over thirty-five years of domestic experience with strengthening security and stability through focusing on economic development (Wang 2010, He Yin 2014). China's history produced a strong belief in the necessity of economic growth to maintain internal order. The CPC became a fervent adherent of its approach to development–security. Under Deng Xiaoping's leadership, it put a premium on enhancing development, reducing poverty, and providing jobs as measures to ensure relative stability. Under President Jiang Zemin, "development was regarded as the key to solving all problems in China" (Wang 2010, 5) and identified as the main objective of the CPC. Both the "sci-entific development" associated with President Hu Jintao, and the "China Dream" of Xi Jinping, identified economic development as the central task of the party.

The "new security" concept, initiated by Chinese officials after the Cold War, was based on a shift from understanding security as self-help to a more mutual/common security.[5] From the late 1990s, China embraced a progressive view of security that was not based on a Cold War zero-sum mentality but instead on common interest and mutual security based on dialogue and win/win development.[6] More recent versions of this concept hone in on NTS; a 2015 update of China's military strategy paper, for instance, reiterates Beijing's adherence to an approach to security in which mutual peaceful development, trust, and win-win relations are central.[7]

Emphasis on domestic economic growth as the basis of security for China, and CPC legitimacy and power, has been prominent in the domestic Chinese context, but it has also been externalised in China's Africa relations to form a prominent aspect of security relations founded in the centrality of economics. President Xi has been a strong exponent of economic development as the foundation of security and, indeed, development as the highest form of security, not merely within China but also in a global setting. The Chinese government's emphasis on economic development as the basis for security is thus by no means unique in its Africa relations. Development, as China's foreign minister stated, "holds the key to numerous problems in the world" (Wang 2013).[8] Such a view is compatible with that of many state leaders in global South countries.

China's support for security–development involves not merely material or eco-nomic engagement but also ideational dynamics and points of normative contention. China's approach has hitherto not been politically disinterested, as simplistic binary characterizations of a division of labour between Chinese hardware and Western software claim, but has consistently stressed the role of economic growth, and criticised democracy and human rights. Finally, however, for all the faith in the efficacy of economic growth in delivering lasting peace, where such intent has been applied to actual settings, such as Sudan and South Sudan, the results to date have not vindicated this faith. The ascribed and assumed efficacy of economic development in overcoming conflict should be problematised, backward in terms of a historicised analysis exploring earlier precedents, and in terms of Africa relations going forward.

Future directions

Some of the (un)intended consequences of rapid economic growth include potential controversies regarding environmental degradation, rapid depletion of natural and wildlife resources, as well as social justice concerns including issues of indigenous people's rights. Although the development–security nexus seems like an appropriate lens to evaluate African security issues, it should also be noted that given the relatively emergent status of security questions in China–Africa relations, it is still too early to have a full grasp on the consequences of China's development and security policies in Africa. As an emerging theme that has become more prominent in recent years, security and security-related themes appear set to feature more in various engagements with China's African relations. Five of these might be noted here.

First, the interconnections between domestic dynamics, and debates about security and human security within China will continue to shape and influence the various aspects of Chinese security engagement with Africa. From terrorism, social media, to private Chinese security companies and profit-driven engagement seeking new markets abroad, domestic dynamics exert formative influence.[9] In the context of China's "new normal" economics, and concerns about the impact on Africa of reduced economic engagement and waning of China's developmentalist foreign policy, a more salient set of issues arose from the links between economic interests and security.

Second, and in terms of the role of the Chinese state and military, the expanding role and reach of the PLA, and other state and non-state security actors, looks set to be a prominent research area. Following China's earlier, predominantly economic "go global" drive, the PLA's own equivalent and related process of going global is underway, albeit in uncertain ways that remain subject to political sensitivities. Further moves, from maritime security to land deployments, are likely. China's multilateral security role, including UN peacekeeping and how this overlaps with and feeds into China's bilateral engagement, is one aspect of this. As an extension of state foreign policy intended formally to demonstrate China's contribution to world peace and security, how China adds substance to exercising "responsibility" as a new type of great power looks set to remain an issue.

Third, widening the framework of analysis to consider China's role in African security in global security terms will enable China's engagement in relation to others, such as Japan and the US, to be better considered. This reinforces other calls to analyze African security in global perspective (Abrahamsen 2013). From China's perspective, is Africa different in terms of its security engagement? While prominent, Africa is one part of China's evolving global security engagement; it is important to approach Africa in this context and not in isolation from other foreign policy trends under President Xi. This involves reconsidering established ways of thinking about and understanding China's core interests in a global context. As Tang suggested, for Chinese foreign policy going forward, Africa, together with Latin America and the Middle East, may prove to be "a more challenging core foreign policy issue for China in the next decade" (Tang 2011, 187). The Belt and Road Initiative (BRI), depending on how it proceeds, is likely to hammer this home. The continent may not feature prominently in recent debates about China's "core interests" and how to secure these (Zeng et al. 2015). However, there are good reasons to argue that core interests need to be reinterpreted to better align with and reflect the globalisation of Chinese interests and the concomitant diversification and diffusion across space and political terrain of security-related challenges for the Chinese state. Approaching Africa in this global context, through transcaler analysis rooted in defined local settings, would help scholarship go beyond the dominance of methodological nationalism and inter-state analysis to encompass a range of non-national spaces of global security like cities, cyberspace, or the globe itself (Adamson 2016). In view of China's emerging security roles, Africa is prominent but one notable area in a wider evolving global Chinese role.

Fourth, there are a host of questions and areas pertaining to China from diverse African perspectives, regions and actors. These include the process of how China is fitted into the AU's ASA, a policy and political work in progress, in which not merely political and military cooperation is involved but also adjustments to the normative framework of African security. This includes the evolution of regimes of intervention in Africa away from non-interference and toward greater acceptance on behalf of key states in military intervention in neighbouring states. One notable theme, tested in action over Darfur, has been the question of China's adaptation and possible socialization into new norms, such as the AU Constitutive Charter's non-indifference, and efforts to define and mobilise Chinese alternatives to intervention, such as R2P (Ruan 2012).

Most important in political terms, however, are a range of Africa–China security dynamics involving not so much the exercise of Chines military power as part of a story of the global expansion of the PLA as much as the incorporation of China into regional and domestic African state agendas. In examining proactive and successful elite securitization strategies by Uganda, Rwanda and Ethiopia, for example, Fisher and Anderson only bring China in at the end of their analysis as a factor likely to grow in importance or that will need to be further taken into account (Fisher and Anderson 2015). From this perspective, China is a relative newcomer but is likely to conform to established patterns of extraversion (Clapham 2008).

There is much to support such an effort to emphasise the deeper, more complex African security context. Going forward, however, perhaps there is more to this than a simple apparent divide between studies privileging Chinese foreign policy or African politics. Rather than privileging China as "different", it can be inserted into established analytical frameworks seeking to understand relations between external actors and various kinds and levels of African political protagonists. This suggests that China, naturally in some places more than others and in mixed, uneven ways, has become or is becoming constitutive, raising the question of how local and external, non-state and state forces "interact to produce order and authority in various different kinds of social and political space" (Latham et al. 2001, 4).

Conclusion: toward Africa–China security studies

Further changes are underway that are indicative of the deepening complexity of relations. These involve an evolving shift from a state-based China engagement to one involving multiple Chinas, involving the need to move beyond a state-based ontology to better reflect the transition from state-dominated relations to more plural, dispersed interests and growth of Chinese social formations in parts of the continent. These also necessitate recalibrating the hitherto dominant spotlight on China's role in African security, toward one that allows a better grounding in African security contexts and changing the privileging of China's agency and foregrounding of geopolitical realism to enable greater attention to multi-level African engagement and local agency. In turn, this points to the need to position China as one aspect of broader issues concerning Africa in global security strategy studies. Other changes in the international sphere, moreover, may have (potentially mixed) effects on Africa–China security engagement. Declining economic growth in China, the expressed interest in cutting U.S. foreign aid to African states, and major initiatives such as the BRI should all be taken into account when analyzing China–Africa security relations. While the slowdown of China's economy might suggest a scale back from China's commitments to supporting African peace and security apparatuses, the isolationist foreign policy of the latest U.S. administration seems to favour a rapprochement between China and African counterparts. Similarly, with the establishment of BRI, China's security interests in the continent might seem to revert to the traditional military build-up of naval and replenishment bases, yet in the longer run BRI is formally also supposed to involve creating more economic integration and connectivity in terms of trade, energy, as well as intelligence sharing. In this vein future China–Africa security scholarship is better served by following developments in PLA reforms and exploring the impacts of these on China's foreign policy engagement in Africa.

China's engagement with Africa's security should not be approached in isolation from other regional engagements and Beijing's evolving global role. At the same time, the continent is a revealing theater showcasing foreign policy experimentation in the realm of security. For the AU and the continent in general, just how

China can be best fitted into the existing peace and security architecture, remains a work in progress.

Approached from the perspective of China's ruling party-state, Africa is a frontier of foreign policy experimentation and its security engagement is often presented as new, innovative, dynamic and unusually salient in China's global security engagement (e.g. Shen 2011; Shinn 2016). Approached in terms of the history and political economy of African security, however, China may enjoy a more prominent position by virtue of interest in its wider continental engagement, but many of the issues it raises are historically familiar and represent variations in continuity. This reflects both a basic divide in efforts to examine security dynamics, and an ongoing challenge in the study of security in the Chinese–African context, suggesting that a reorientation toward a better grounding in African security contexts is required in order to deepen and advance scholarship.

Notes

1 See Declaration of the Johannesburg Summit of the Forum on China-Africa Cooperation (2015), para. 25.4; The Forum on China-Africa Cooperation Johannesburg Action Plan (2016–2018), section 6. China's Second Africa Policy, December 4, 2015 at Section 6.
2 It has evolved from largely support roles into force protection; South Sudan saw the first deployment of Chinese combat troops under a UN Chapter VII mandate.
3 Although China has been participating in anti-piracy patrols in the Gulf of Aden since 2008, the establishment of the naval base is much more assertive.
4 In late November 2015, for example, an assault on the Radisson Blu hotel in Bamako, Mali, killed three senior Chinese employees of the China Railway Construction Corporation. Afterwards, China's Foreign Minister pledged to fight extremism and strengthen counter-terrorism cooperation with Africa.
5 "China's Position Paper on the New Security Concept" July 21, 2002; at www.fmprc. gov.cn/mfa_eng/wjb_663304/zzjg_663340/gjs_665170/gjzzyhy_665174/2612_665212/ 2614_665216/t15319.shtml, accessed June 21, 2016.
6 White Paper on China's National Defense, 1998, p. 6. http://csis.org/files/media/csis/p rograms/taiwan/timeline/sums/timeline_docs/CSI_19980727.pdf.
7 The State Council Information Office of the People's Republic of China, China's Military Strategy www.chinadaily.com.cn/china/2015-05/26/content_20820628.htm.
8 Wang Yi's remarks at the World Peace Forum 2013 http://calgary.china-consulate.org/ eng/lgxw/t1053966.htm.
9 The future impact of recent legislation further down the line, for example the December 2015 Anti-Terrorism Law, which permits PLA become involved in anti-terrorism operations overseas, remains to be seen.

References

Abrahamsen, Rita. 2013. "Introduction: Conflict and Security in Africa." In *Conflict and Security in Africa*, edited by Rita Abrahamsen, 1–12. Woodbridge and Rochester, NY: James Currey and Boydell & Brewer.

Adamson, Fiona B. 2016. "Spaces of Global Security: Beyond Methodological Nationalism." *Journal of Global Security Studies* 2016: 1–17.

Aidoo, Richard and Steve Hess. 2015. "Non-interference 2.0: China's Evolving Foreign Policy towards a Changing Africa." *Journal of Current Chinese Affairs* 44(1): 107–138.

Alden, Chris and Daniel Large. 2015. "On Becoming a Norms Maker: Chinese Foreign Policy, Norms Evolution and the Challenges of Security in Africa." *China Quarterly* 221: 123–142.

Alden, Chris, Laura Barber, Abiodun Alao and Zhang Chun. eds. 2017. *China and Africa: Building Peace and Security Cooperation on the Continent*. London: Palgrave Macmillan.

Anthony, Ross, Harrie Esterhuyse and Meryl Burgess. 2015. "Shifting Security Challenges in the China–Africa Relationship." *SAIA Policy Insight Paper 23*, September 2015.

Ayoob, Mohammed. 1995. *The Third World Security Predicament: State Making, Regional Conflict and the International System*. Boulder, CO: Lynne Rienner.

Benabdallah, Lina. 2017."Explaining Attractiveness: Knowledge Production and Power Projection in China's Policy for Africa." *Journal of International Relations and Development*. Online first. doi:10.1057/s41268-017-0109-x.

Breslin, Shaun. 2015. "Debating Human Security in China: Towards Discursive Power?" *Journal of Contemporary Asia* 45(2): 243–265.

Buzan, Barry and Lene Hansen. 2009. *The Evolution of International Security Studies*. Cambridge: Cambridge University Press.

China's Military Strategy. 2015. Beijing: State Council Information Office of the People's Republic of China.

Clapham, Christopher. 2008. "Fitting China in." In *China Returns to Africa: A Continent and a Rising Power Embrace*, edited by Chris Alden *et al.*, 361–369. London: Hurst.

Duchâtel, Mathieu, Oliver Bräuner and Zhou Hang. 2014. "Protecting China's Overseas Interests: The Slow Shift away from Non- interference." Stockholm: Stockholm International Peace Research Institute (SIPRI). Policy Paper 41, June.

Duffield, Mark. 2010. "The Liberal Way of Development and the Development–Security Impasse: Exploring the Global Life-Chance Divide." *Security Dialogue* 41(1): 53–76.

Fergusson, A. 1964. "The Chinese Threat to Africa." *Statistics*, December 11.

Fisher, Jonathan and David M. Anderson. 2015. "Authoritarianism and the Securitization of Development in Africa." *International Affairs* 91(1): 131–151.

FOCAC. 2015. Declaration of the Johannesburg Summit of the Forum on China-Africa Cooperation (2015), para. 25.4; The Forum on China-Africa Cooperation Johannesburg Action Plan (2016–2018), section 6. *China's Second Africa Policy*, December 4, 2015.

Gill, Bates and James Reilly. 2000. "Sovereignty, Intervention and Peacekeeping: The View from Beijing." *Survival* 42(3): 41–59.

He, Yin. 2014. "United Nations Peace building and Protection of Human Security." *Guoji Anquan Yanjiu* (Journal of International Security), 3, 75–91.

Huang, Chin-hao and Olawale Ismail. 2014. "China." In *Security Activities of External Actors in Africa*, edited by Olawale Ismail and Elisabeth Sköns, 15–37. Oxford: Oxford University Press.

Jensen, Steffen. 2010. "The Security and Development Nexus in Cape Town: War on Gangs, Counterinsurgency and Citizenship." *Security Dialogue* 41(1): 77–97.

Large, Daniel. 2009. "China's Sudan Engagement: Changing Northern and Southern Political Trajectories in Peace and War." *China Quarterly* 199(3): 610–626.

Large, Daniel. 2016. "China and South Sudan's Civil War, 2013–2015." *African Studies Quarterly* 16(3–4): 35–54.

Latham, Robert, Ronald Kassimir and Thomas M. Callaghy. 2001. "Introduction: Trans-boundary Formations, Intervention, Order, and Authority." In *Intervention and Transnationalism in Africa: Global-Local Networks of Power*, edited by Thomas Callaghy, Ronald Kassimir and Robert Latham, 1–20. Cambridge: Cambridge University Press.

Lu, Shaye. 2013. "Grabbing the Opportunities and Overcoming Difficulties: Promoting New Developments in China-Africa Relations" (zhuazhu jiyu yingnan ershang tuidong zhongfei guanxi xinfazhan), *International Studies*, 2.

Mallaby, Sebastian. 2005. "The Next Chinese Threat." *Washington Post*, August 8; Page A15.

Parello-Plesner, Jonas and Mathieu Duchâtel. 2015. *China's Strong Arm: Protecting Citizens and Assets Abroad*. London: IIAS.

Putten, Frans Paul van der. 2015. "China's Evolving Role in Peacekeeping and Africa Security: The Deployment of Chinese Troops for UN Force Protection in Mali." *Clingendael Report*, September 2015.

Reid-Henry, Simon. 2011. "Spaces of Security and Development: An Alternative Mapping of The Security-Development Nexus." *Security Dialogue* 42(1): 97–104.

Ruan, Zongze. 2012. "Fuzeren de baohu jianli geng anquan de shijie" ["Responsible Protection: Building a Safer World"], *International Studies* 34 (June).

Saferworld. 2012. *China and Conflict-affected States: Between Principle and Pragmatism*. London: Saferworld.

Shen, Dingli, "Nansudan shi zhongguo suzaoxing waijiao de shiyantian" ["South Sudan is A Testing Ground for China's Proactive (Shaping) Diplomacy"], *Oriental Morning Post*, July 12, 2011.

Shinn, David H. 2015. "China's Growing Security Relationship with Africa: For Whose Benefit?" *African East-Asian Affairs* 3(4): 124–143.

Shinn, David H. 2016. "Africa: China's Laboratory for Third World Security Cooperation." *China Brief* 16(11): 6 July.

Shinn, David H. and Joshua Eisenman. 2012. *China and Africa: A Century of Engagement*. Philadelphia: University of Pennsylvania Press.

Suhrke, Astri. 1999. "Human Security and the Interests of States." *Security Dialogue* 30(3): 265–276.

Tang, James T.H. 2011. "Chinese Foreign Policy Challenges: Periphery as Core." In *New Frontiers in China's Foreign Relations*, edited by Allen Carlson and Ren Xiao. Plymouth: Lexington Books.

Taylor, Ian. 2014. "China's Role in African Security." In *Routledge Handbook of African Security*, edited by James J. Hentz, 245–257. London: Routledge.

Van Ness, Peter. 1971. *Revolution and Chinese Foreign Policy: Peking's Support of Wars of National Liberation*. Berkeley: University of California Press.

Wang, Xuejun. 2010. Paper presented at the conference on "China, South Africa and Africa," *South African Institute of International Affairs*, Zhejiang Normal University, November.

Wang, Xuejun. 2012. "Zhongguo canyu feizhou heping yu anquan jianshe de huigu yu fansi " [Review and Reflections on China's Engagement in African Peace and Security], *International Studies*, 1: 72–91.

Wang, Yi. 2013. "Exploring the Path of Major-Country Diplomacy With Chinese Characteristics." June 27. Beijing: Tsinghua University.

Williams, Paul. 2016. *War and Conflict in Africa*. Cambridge: Polity (2nd edition).

Zeng, Jingzhang, Yuefan Xiao and Shaun Breslin. 2015. "Securing China's Core Interests: The State of the Debate in China." *International Affairs* 91(2): 245–266.

Conclusion

21

CONCLUSION

Chris Alden and Daniel Large

The collection of chapters that make up this book, while aspiring to represent the diversity of scholarship and disciplinary engagements found in the study of Africa and China, will naturally always fall short of that aim. Nevertheless, we believe that this assemblage sheds light not only on new directions in academic work on China and Africa but more broadly on some key themes, conditions and challenges facing the social sciences. Some of these themes and their meanings are developed below.

Western Topiary in the Garden of Solidarity: post-colonialism revisited

> I'll love you, dear, I'll love you
> Till China and Africa meet
> And the river jumps over the mountain
> And the salmon sing in the street.
> *(W.H. Auden 1937, in Auden 1978, 61)*

In what could be called an ode to surrealism, W.H. Auden sought to contrast the absurdity of finding China and Africa in proximity with other outlandish and decidedly unnatural happenings. In this, he is of course recasting the iconic phrase in one of the quintessential poems of empire, Rudyard Kipling's 'East is East and West is West, and never the twain shall meet'.[1] If Kipling's poem is framed by a stern admonishment of impending Judgement Day befitting the Victorian reign, Auden's is an evocation of the witticism typical of popular culture during late colonialism in the 1930s, of the Noel Coward's 'mad dogs and Englishmen' variety. Unlike Kipling, Auden's poem does not foresee an eventual meeting between these polar oppositions (and an implied equality in death if not life); rather, he uses

China–Africa as a device to describe a hurly burly world of Bretonian surrealism where time undoes every human promise of faithful adoration.

How far are contemporary Western portrayals of 'China–Africa' from that classic binary of self and other, captured in Edward Said's scholarship on 'Orientalism' (Said 1995)? Is the frisson that drives media imagery and academic research merely a replaying of this trope, with the added dimension of the 'peculiar' engagement between two Western-designated 'others'? For evidence of this assertion, one need look no further than the proliferation of media reports, photographic essays, films, book titles and covers that feature across the majority of the published and web-based outputs. Animal imagery, with the red dragon wrapping its tail around the African continent, typifies the resort to a twinning of exoticism and threat that abounds in giving visual meaning to the relationship in terms familiar to post-colonial scholarship. Far from never meeting, as Auden would have it, encounters between Africa and China are not only possible but are to be understood as evocations of fear, violence and loss that threaten to shake the foundations of African civilization. Never mind the fact that this portrayal of dragons is taken from a distinctly Western cosmology and denies the Sino cultural reading, which might otherwise cause one to see this embrace in more positive terms for Africans. Equally, the appearance of classic colonial tropes of Africans as innocents in an unspoiled haven of nature, on display in posters put up for the FOCAC III Summit in Beijing in 2006 but reappearing in marketing produced by Chinese companies aimed at Chinese tourists, underscores how easily this image has found its way into popular Chinese perspectives on Africa (Alden 2007). Surely the idea of the 'native as savage' does not lurk far behind, and we already see that representation flitting in and out of the Chinese commercial advertising and social media, as van Staden and Wu, and Huynh and Park highlight in their chapters.

As parts of this book have sought to demonstrate, post-colonial studies – with its focus on discourses, social hierarchies and enduring structures of power – is of great relevance to the emerging scholarship on China–Africa. Frantz Fanon's own work, which emphasised the dehumanizing impact of colonialism on colonised populations, the subjugation of identity to a belittling one imposed from outside and its psychological damage, is legend (Fanon 1967). Said's pointed accusation of cultural imperialism as the defining element of 'Orientalism' which perpetuates the subjugation of once-colonised societies through the more insidious production of knowledge is another (Said 1995). As chapters on the politics of knowledge and race suggest, the China–Africa relationship is necessarily framed in terms which reflect Western epistemologies, are recreated through its discourses and ultimately serve its interests.

At the same time, one cannot help but reflect upon how the diffusion of global power (of which more is said below) is being experienced today, if not eventually upending this particular post-colonial narrative, certainly overlaying it with a set of alternative narratives grounded in their own interpretive experience. Some scholars have tended to see the racist (or at least anti-Chinese) positions and imagery articulated in the African media and some academic work as ultimately being derived from the dominance of Western knowledge production prevalent in a variety of forms across the African continent (Sautman and Yan 2016). Certainly, as Sautman and Yan point out, there is strong evidence to suggest that racist discourses play an important part in shaping perceptions of Chinese conduct. At the same time, the idea implicit in such depictions of Africans presented in these critiques is one of passivity, devoid of ideas and agency, willingly adopting Western positions for lack any originality of their own. If African societies, as well as the governments which preside over them, did once 'borrow' the second-hand visual topiary of the West, they are in the midst of adopting, interrogating, shedding and remaking their understanding of 'China–Africa' today. The periodic surveys of African opinions of China as generally positive are testimonial to the fact that the deterministic expectations of Western influences by scholars routed through everything from academia to art are being regularly buffeted by African opinions and sensibilities (Afrobarometer 2016).

What this points to is another revelatory feature emerging in the wake of the study of the 'China–Africa' relationship. The simple binaries of 'orientalism' – settler and native, coloniser and colonised, etc. – are as saturated with their own assumptions regarding the characteristics of the oppressor societies as they are when exposing the brutalizing impact of the coercive relationship on the oppressed. In part, these assumptions are derived from the stultifying solidary rhetoric of the Cold War, a kind of 'red-washing' of all distinctive features of a relationship and context into a single 'self and other' emancipatory narrative to be subsumed in the larger task of political mobilization. While this solidarity discourse is ritualised in diplomacy and expanded to include notions of the cultural proximity of developing societies, there are other shared elements that might equally (or better) explain the evolving content of a complex relationship. In addition, surely some allowance must be made for contemporary experiences of engagement, its parochial character and messiness as sources of imagery of the 'other' which is far removed from the high table solidarity rhetoric invoked by political elites.

What is perhaps underlying all of this is a fear prevalent in diplomatic and scholarly circles that the African perspective on China and the Chinese perspective on Africa are not in fact embracing the given rhetoric of solidarity but trotting down an all too familiar path that mirrors traditional 'Africa–West' discourses. The chapters by Cowaloosur and Taylor, as well as Qoba and Le Pere, see much that resembles the asymmetries of Africa's experiences with the West. Are subjected societies stubbornly unable to shake off colonial caricatures of the exploitative outsider? Do they replicate, perhaps unwittingly, the prevailing patterns? Kwame Anthony Appiah cautions against the trend towards substituting one set of binaries, with its simplistic assumptions, for another in describing Africa (Appiah 1992). Can it be that Chinese and Africans have their own sources of experience, codified as history or derived from contemporary anecdote, which holds sufficient interpretive ballast to create a new narrative with the outside world? In short, is it truly possible to forge the 'new type of strategic China–Africa partnership' that Xi Jinping spoke of at FOCAC VI in 2015 in the face of what post-colonial scholars tell us is a structurally determined response by damaged colonial societies to the external world? The undercurrents of xenophobia sweeping anew across the globe, with their hardened political manifestations, suggest the magnitude of the challenge.

Lost voices found and agency discovered: a new archeology of knowledge

'Cherchez la science, jusqu' en China'
('Seek knowledge even as far as China')
(Prophet Sidna Mohammed cited in
Oualalou 2016, 6)

Fathallah Oualalou's publication *La Chine et nous – responder au second depassement* is but one of a number of recent books written that are not in English originating from the African continent (and beyond), that attempt to tell the story of China and Africa (Oualalou 2016). Indeed, the gap between the original Arabic and the French language is an expression of the distance that the quotation on China from the hadiths (sayings attributed to the Prophet) had to travel before finding its way into this particular English-language publication. It also reminds us how much of this relationship is examined through the narrow lens of Anglo-American scholarship, with only limited concessions to other voices, as Anthony explores in his chapter.

These concerns were echoed in a heated debate within the Chinese in Africa–Africans in China Research Network Google group, a loose affiliation of researchers sharing information and publications about China–Africa matters, as to the dominance of English as a discourse of exchange as opposed to French, which underscored one aspect of this problematic.[2] The regular output of Chinese scholarship has received little attention until recently in English language publications, as Li Anshan's chapter clearly demonstrates. Memoirs by Chinese diplomats who spent time on the continent almost constitute a minor sub-field of Chinese-language accounts of Africa.[3] African contributions are dominated by South Africans, employing an epistemology that is rooted in Western practices, and a perspective that can at times strike many other Africans as being at a remove from the rest of the continent's concerns (see, for example, Shelton et al. 2015). Lusophone Africa's most active researchers on this topic have published in Portuguese (French and English) on aspects of China–Africa, China–Angola and China–Mozambique relations, with the journal *Daxiyangguo* playing a prominent role in leading the way on this theme. African journalists from across the continent, epitomised by the work of Kenyan Bob Wekesa, are amongst the mostly widely published on China–Africa topics, many of them under the auspices of the Wits University's Africa–China Reporting Project. African language publications on this topic, about which little is known, are rarely if ever consulted.

Non-Western observers of the China–Africa phenomenon, most often writing in their home language, have also weighed in on China–Africa but their positions have not featured much in English-language publications. These include Japanese scholars, such as Katsumi Hirano, which Kweku Ampiah notes in this chapter (see, for example, Hirano 2007; and Marukawa 2007). Historical sources are barely consulted as yet, especially from materials outside of the West. Researchers in China are beginning to amend this through archival studies, as are their counterparts in African and Western universities. Luis Camoes' *Lusiads*, published in the sixteenth century, provides some tantalizing hints of a world 'discovered' that included Africans and Chinese, while the writings of European missionaries such as Francis Xavier, and the Dutch merchants in Batavia all provide potential insights. Investigations of the work of figures such as Ibn Battuta's journeys in the fourteenth century in Africa and the Middle East offer clues as to more primordial sources of diplomatic, commercial and social interactions between the two regions. In more contemporary terms, anti-apartheid activist and poet Denis Brutus' exile from South Africa took him on a three-month tour of China in 1973 that inspired his largely forgotten collection called 'China Poems' (which were translated into Chinese as well). Angolan writers like Jose Eduardo Agualusa and Mozambican poets like Mia Couto describe the Chinese presence in the shadow of colonialism, civil war and its aftermath (Agualusa 2007; Couto 2010).

Beyond the blank spot that is the linguistic divide, there is a further and more damning critique again emanating from post-colonialist literature. Gayatri Spivak tells us that the subaltern's voice is muted, and even lost altogether, in a process that privileges the language of power and concurrently reifies its hegemonic context (Spivak 1998). Subaltern voices as such exist in the shadows of the piazza of imperialism, found in sometimes faint echoes of historical archives, snatches of protest songs threaded through society or, more obviously, in the transactional life of the market. The paradoxical invisibility of the visible is the defining expression of power and reflects what Spivak labelled 'epistemic violence', the systematic erasure of other voices as part and parcel of the construction of the dominant narrative. The most obvious absence is that of women from the literature, though other identity-based groups, such as the LGBT community, have yet to find much of a hearing in Africa–China studies (though there will be those who argue that the incorporating of these particular voices into the research agenda is not an extension of a putative universal emancipatory agenda but rather yet another demonstration of the forcible imposition of the Western liberal canon on non-Western societies).[4] Graham Harrison goes so far as to warn that the 'force of liberal ideology produces a kind of epistemic authoritarianism, as Westerners impose liberal frames of analysis … on African realities' (Harrison 2002, 19).

Insofar as Western knowledge production in the realm of the social sciences is bound up with power, however, the dislodging of that power should be – and is – freeing up discursive space for some of these alternative voices to be heard. In this respect, the story of China and Africa/Africa and China is not being scripted exclusively by Western sources anymore but rather is being produced from a range of disparate sources. These voices are rendering the relationship in different ways and perspectives, which correspond with their experiences and are mediated

334 Chris Alden and Daniel Large

through the cultural lens. In some cases, as described in Sheridan's chapter on Tanzania, the community of rumour carries greater currency than any other sources of meaning. In this respect, an abiding fallacy adopted by scholars and policy makers alike is that sustained and deepening engagement necessarily produces more nuanced understanding between peoples, thus laying the foundation for better ties between countries and their citizens. Greater exposure may in fact reinforce noxious attitudes and images, playing on prevailing fears and feeding these through instruments like social media. The controversy over the decimation of rhinos in South Africa inspired the circulation of racist postings on Facebook by South Africans, despite the fact that the Chinese had been a part of South African society for over hundred years. An equally startling example is an Arab publishing house's particularly gruesome depiction of the relationship on the cover of its Arabic edition of *China in Africa*, the African zebra's throat slit with red blood pouring out, demonstrating that non-Western cultures can draw from their own disturbing stereotypes in representing ties.[5]

Wrapped up in these linguistic concerns is a larger issue of agency. African agency and its expressions has pre-occupied Western and African scholarship over the last decade, drawing inspiration in part from the interpretive historical scholarship of Jean-Francois Bayart and more recently realised through the work of William Brown and Sophie Harman (2013). It draws on the ideational terrain of Said's critique of the 'moral geography on international relations', emerging out of the productive systems of nineteenth-century colonialism and reproduced in the procedural rules and practical conduct of elites, states and international finance in subsequent decades which fundamentally reifies inequalities of power distribution and social exclusion in the contemporary international system. In so doing, it patently adopts a normative position on IR, which manifests variously as an emancipatory agenda (voicing the subaltern), an existentialist stance (seeking ontological security for states, societies and peoples of Africa and the South), and even exclusionary elite *revanchism* (preserving predatory regimes and practices in the South) in international affairs.

However, to what extent is the very idea of agency bound (and indeed even gagged, to complete the imagery of voices lost) to a rigid, deterministic structuralism that does not recognise or account for change and power transformation in the international system? Brown and Harman, as well as a growing body of African and Africanist scholars like Lucy Corkin, have sought to lift agency from these interpretive constraints and apply it to the conditions and experiences found in Africa–China relations. As presented in these works, incipient expressions of agency are found in the negotiation strategies of African elites as they bargain over loan packages, redefine the terms of agreements and assert in effect their parochial interests through state instruments in engaging with the Chinese (Corkin 2013). This book extends that narrow depiction of agency, in chapters by Alves and Chichava, Soulé-Kohndou and Procopio amongst others, to bring in a richer and more textured presentation of African agency, found in dimensions as diverse as bureaucratic obstruction and strategies of partial compliance adopted by local communities.

What this implies is that the structure of power, at least as experienced at the micro and mesa-levels in Africa, sees in China conditions that replicate power asymmetries, which if not echoing the discourses of protest and challenge utilised by Africans in describing engagement with the West in the colonial and post-colonial era, certainly chime with them. Is there a new critical vocabulary to express these circumstances, one which does not hide under the bedcovers of solidarity rhetoric but nevertheless does not feel compelled to revert to ahistorical application of the demonizing tropes of the other? In dismissing the determinism of scholars like Octave Mannoni – whose book, *Prospero and Caliban: The Psychology of Colonization*, offered no escape from the damage done by colonialism upon its subjects – Fanon argued that liberation from this mindset was not only possible but a vital component of de-colonization (Fanon 1967; Mannoni 1990). Put another way, are the discourses of solidarity and South–South cooperation generated by developing countries of sufficient depth and co-constituted meaning to peoples in Africa and China to account for difference and disputes?

Beyond and beneath Area Studies

Implicit in the interrogation of 'China–Africa/Africa–China' studies is, as noted above, the problematic of how to research this topic and, in particular, the controversial role Area Studies occupies within that process. African and Africanist scholars like Sabelo Ndlovu-Gatsheni and Siphamandla Zondi argue that decolonizing the university requires a wholesale shift away from the Eurocentric ontologies and epistemologies that define both scholarship in African universities and the study of Africa (see Ndlovu-Gatsheni and Zondi 2016; and, for earlier critiques, see Dunn and Shaw 2001; Cornelissen et al. 2012; Death 2013; Brown and Harman 2013). Echoing this sentiment, Rita Abrahamsen hones in on the problematic of IR and Area Studies, which is 'unable to capture the historical specificity of the postcolonial African state, to perceive of difference as anything but deviance from a norm, and therefore also unable to capture the continent's globality'. According to her, under these circumstances, Africa represents 'IR's permanent "other", serving to reproduce and confirm the superiority and hegemony of Western knowledge, epistemologies, and methodologies' (Abrahamsen 2017, 125).

For Zondi, the counter-approach moves away from the false God of Western universality, and involves embracing belief systems such as *ubuntu* and *sumak kwasay*, which are truly humanist in perspective and, as such, offer ontological and epistemological alternatives that can lay the basis for a new social science that is not only inclusive but no longer destructive of communities and the environment (Zondi 2016).[6] Abrahamsen's solution to these issues facing the Western-centric field of International Relations is to pursue a scholarly approach based on 'assemblage' as the foundation of its 'meta-narrative'.

> (T)he notion of an assemblage invariably draws attention to the multiplicity of actors, their various forms of power, and their struggles over influence.

Thinking politically with assemblages accordingly demands constant attention and vigilance towards how the political orders of contemporary Africa come into being, what forms of agency and power different actors, actants, norms, and values have, in order to ensure that scholarship is not simply serving the powerful but instead seeks to uncover new political possibilities.

(Abrahamsen 2017, 139)

If these debates are consuming scholars in African studies, many of the same concerns are being discussed by academics working in Asian studies and, more specifically, Chinese foreign policy studies. As in the case of Africanist critiques, the field of IR and its relationship to Area Studies comes under much withering scrutiny with the quintessential post-colonial critique, Edward Said's 'orientalism', serving as the defining cardinal point (see, for instance, Ling 2002; Chowdhry and Nair 2002; Bilgin 2008j). Amitav Acharya (2014, 649) goes further in highlighting the problematic linkage between Area Studies and IR theory:

The problem of ethnocentrism in IR theory will not disappear by using case studies from the non-Western world primarily to test theories generated in the West. Instead, it will merely reinforce the image of area studies as little more than a provider of raw data to Western theory.

This contribution builds on the critique levelled by scholars like Ching-Chang Chen, Pinar Bilgin, L.H.M. Ling, Geeta Chowdhry and others, all of whom have taken the study of Asia as a starting point (Chowdhry and Nair 2002; Ling 2002; Chen 2011; Bilgin and Ling 2017). Similar to the calls from Africanists, Acharya proposes the adoption of a 'Global IR' as an alternative to the Eurocentric depictions of international politics that is centred on pluralistic universalism, employs world history, subsumes IR theory and methods, integrates Area Studies and regionalism, avoids 'exceptionalism', and recognises multiple forms of agency beyond mere material power (Acharya 2014). Ling offers a re-founding of the cultural and accompanying normative basis of the discipline through compatibility of the archetypes found in depictions of the Westphalia system and the Chinese classic, *Romance of the Three Kingdoms* (Ling 2017). Nele Noesselt provides a useful review of the efforts to construct a Chinese theory of IR, which not unlike Africanist counterparts, aspires to an alternative approach that draws from the past and shapes this with contemporary ideas (Noesselt 2015). It is notable that some of these efforts, unlike the 'plurilateral world' and 'Global IR' movements described above, seem more closely fitted to the conventions of geo-politics.

Intertwined within both critiques of the origins and lingering purposes of Area Studies and IR as an instrument for Western dominance is a parallel exploration of Afro-Asian agency, coupled with efforts to develop an alternative approach to the Western social sciences. The contemporary work on African agency has been largely situated within a statist or regime context, though some scholarship does examine subaltern perspectives ranging from the peasantry to that of bureaucrats (for examples, see van Binsbergen and van Dijk 2004; Brown 2012; Brown and Harman 2013; Tieku 2013). Interestingly the most recent writings on agency feature Africa–China engagement as their subject, suggesting a shift in the meaning of power asymmetries in the relationship (Corkin 2013: Mohan and Lampert 2013; Gadzala 2015). By way of contrast, scholars working on Asian studies focus on a much more historical account of 'Eastern' agency, usually in the context of colonialism and the Cold War, and often as a necessary prerequisite for reclaiming contemporary meaning and standing (Hobson 2017).

Detecting and elucidating upon the presence of agency is only possible through detailed rendering of aspects of bilateral relations if not other kinds of relationships. This points to the critical scholarly advances garnered in Africa–China studies through the pursuit of empirical case studies. Starting with Taylor's work on China and states in Southern Africa in the 1990s, this was picked up in research produced on Sudan by Large and Patey producing a succession of in-depth studies of bilateral ties. Flowing from these were a range of commissioned bilateral studies produced by the South African Institute of International Affairs in Johannesburg, the Centre for Chinese Studies at Stellenbosch and African Economic Research Consortium based in Nairobi. Deborah Bräutigam's blog *China in Africa: The Real Story* – tended to see the relationship through a detailed sectoral lens, and exposed fallacies in the press and scholarship. These developments helped move scholarship away from the inevitable critiques of meta-statements about Africa and China ('Africa is not a country' and 'one country, many Chinas'), introducing context and nuance to blunt assertions that dominated much of the first generation research.

China as champion of globalisation: a world after its own image?

> The cheap prices of commodities are the heavy artillery with which it batters down all Chinese walls, with which it [*the Western capitalists, eds.*] forces the barbarians' intensely obstinate hatred of foreigners to capitulate. It compels all nations, on pain of extinction, to adopt the bourgeois mode of production; it compels them to introduce what it calls civilisation into their midst, i.e., to become bourgeois themselves. In one word, it creates a world after its own image.
>
> *Karl Marx and Friedrich Engels,* The Communist Manifesto *(1848)*

Writing against the backdrop of the upsurge of incipient rebellions across continental Europe, Marx and Engels attempted to mobilise the workers' outrage and budding nationalist-tinged sentiments and direct them towards the unleashing of class revolution that would topple the old order and usher in proletarian rule. On full display, of course, are the racial and cultural biases that featured in Marx's nineteenth-century scholarship but, if one can set that aside, the description put forward in the 'Communist Manifesto' is one of rampant globalisation and its impact on traditional societies undergoing forcible integration into the Western capitalist system.

While few contemporary scholars are writing fully in this polemical tradition when analyzing Africa and China, there are emerging strands in the literature of the left which is beginning to revisit these old chestnuts in light of the ascendancy of emerging powers and their position as beneficiaries of globalisation and architects of state-led capitalism. Achin Vanaik's work examines the role of emerging powers as represented by the BRICS countries and sees the key states (China, India and Russia) as adapting to an informal elite consensus with the US and the EU aimed at stabilizing the world for global capital (Vanaik 2014, 5; see also Vanaik 2013). He dismisses claims of China's ascendancy to global leadership (admittedly Vanaik is writing in 2014) in preference for a continuation of US hegemony through a more collective power-sharing via what he calls the 'quintet'. Patrick Bond and Alicia Garcia also criticise BRICS as less emancipatory in their agenda but closer to being 'sub-imperial' collaborationists with the prevailing capitalist order, offering policies that are designed primarily to further great power aspirations and promote national mercantilist agendas as evidenced in their conduct in Africa (Bond and Garcia 2015, 1–14). Howard French's argument in *China's Second Continent* adopts a different tack, building his case for China's hegemonic aspirations in Africa upon the steady flow of hundreds of thousands of Chinese migrants to Africa over the last decades and their impact on local economies and societies (French 2014).

What is certain, with Xi Jinping publically adopting the mantle of leadership of globalisation in early 2017 in the aftermath of Donald Trump's abandonment of this traditional US role, is that China has become the 'face of globalisation'. Heavily dependent on exports, Beijing's fear that Western protectionism will shut it out of key markets coupled to continuing weak domestic demand is driving China's expansion policy towards the financing of trade and investment on a global scale. The Asian Infrastructure Investment Bank, the BRICS New Bank and the widely touted Belt and Road Initiative are all part of China's effort to generate demand for its products and services abroad in what might be called a 'development as grand strategy'.

In this context, the most salient point for Africa is not so much Chinese companies' participation in resource extraction and infrastructure development across the continent, but rather China's unprecedented position as a significant holder of African sovereign debt and its stated interests in aligning itself with a 'rule-based system that does facilitate necessary restructurings and, in so doing, discourage over-lending and mispricing of risk'.[7] Whether it is Zambia's $642 million shortfall on a $3 billion Chinese loan payment, or Nigeria's agreement to a currency swap and floating of 'panda bonds' to raise cash for debt payments, Beijing's handling of the debt problem is reshaping the China–Africa relationship in profound and possibly enduring ways.[8] Such was the fear in Beijing of losing their money loaned to local corruption – and the continued inability of Zimbabwe to pay back its outstanding US$60 million debt to China – that in 2015 they insisted on the placement of Chinese officials directly into positions with the Zimbabwean government ministries and parastatal offices to provide oversight and engineer reforms to management procedures, stirring resentment within elements of ZANU-PF.[9] That these steps are reminiscent of classic IMF conditionalities is not lost on African governments and civil society.

For Africans, China's leadership of globalisation is an evident reality and increasingly influences their understanding of the relationship in all its facets (Alden 2007, 128–129). Debates amongst African elites, and in mainstream African media, periodically use terms like 'colonialism' as a proxy for all manner of growing Chinese ascendancy in different sectors of daily life.[10] The larger point is *not* that twenty-first-century China as global leader is the new architect of colonialism with Africa as a primary target, but rather, in an era where the locus of global power is shifting decidedly to China, that societies will be confronted by both the substantive and interpretative challenges of explaining Chinese economic and political dominance. Appropriating the historical memory of colonialism – Africa's own 'hundred years of humiliation', a parallel to the Chinese dictum – to make sense of this contemporary condition, if not inevitable, is certainly understandable. Suspicions voiced by some Africans as to Beijing's role in the military coup in Zimbabwe are lingering expressions of this colonialism trope.

Fanon's observation that 'to speak … a language means above all to assume a culture, to support the weight of civilization' is relevant in this context (Fanon 1967, 8). Perhaps the regular repetition of Chinese mantras like 'win-win' to describe the relationship by African officials and some academics heard these days, coupled to the expansion of Chinese language studies amongst Africans through Confucius Centers and other educational outlets, is an illustration of that truth on course to being realised. African language studies in Chinese universities are also on the increase, though as yet tellingly it has not produced an equivalent sounding of African terminology in Chinese official circles to describe ties between them.

Broader trends can be adduced as well, with academic scholarship beginning to display signs of a shift towards adopting Sino-centric concepts as interpretive tools. Yuen Foong Khong's application of imperial China's 'tributary system' to describe US hegemony in the twentieth century is an early sign that Western knowledge production, like Western power, may be gradually assimilating to the Chinese worldview (Khong 2013). Reimagining the present in terms that echo China's past is crucial to reconstructing new narratives that put knowledge production on a Sino-centric footing. In the end, insofar as knowledge production in the realm of the social sciences is bound up with power, the study of Africa–China is both a harbinger of a world undergoing profound transformation and has proven to be at the frontier of the challenges facing the social sciences in understanding that emerging world.

Concluding thoughts

The nature and scale of China's current ties with Africa constitute an unprecedented phase in the historical terms of these relations. The challenges in studying China–Africa, however, are by no means unique and exist in radically evolved and changed circumstances mediated by more extensive and consequential global connections. One of the directions for studies of China and Africa that this books points to is the growing interest in framing and approaching this in Africa–China terms (Bodomo 2012; Wekesa 2017). There are compelling reasons to support this. After all, as the late Calestous Juma noted:

> The story of Africa's relations with China is ultimately an African story. It is up to Africa to write the plot.
>
> *(Juma 2015, 118)*

However, the very pluralism evident in fluid and dynamic forms of research engagement should not be discounted. The eclecticism of work on China and Africa is one of its key strengths. Taken forward, it also suggests that beyond established disciplinary engagements, or work in specialist sub-fields, the very lack of an easy fit into a defined, established field of study demonstrates how pursuing questions concerning Africa and China has the potential to open up and contribute towards new ways to grapple with global changes. In this way, as George Yu suggests here, whether as an interdisciplinary field of studies, such as China studies and African studies, or as a sub-field in an existing disciplinary field, there is great promise in such scholarship which, given its complexity, requires 'multiple approaches *and* particularised attention'.

Notes

1 A view which, most reviewers will point out, Kipling himself uses the poem to challenge.
2 See Africa–China–Africa research Network Google Group correspondences, 8–9 December 2016, email Folashade Soule-Kohndou.
3 Thanks to Ambassador Shu Zhan for pointing this out.
4 Solange Chatelard's work on Chinese women working in business in Zambia is a notable exception.
5 As authors the world over discover again and again, the publishing rights once sold do not afford the author *any* say in the presentation of the author's work.
6 *Sumak kwasay* is a Quenchua phrase for living in harmony with the world, while *ubuntu* reflects an African ethic that one's humanity is affirmed from and is an expression of one's relationship with the community.
7 Domenico Lombardi, 'Sovereign Debt Restructuring: current challenges, future pathways', presentation at 2015 Money and Banking Conference, CIGI, Buenos Aires, 4–5 June 2015; also see Skylar Brooks, Domenico Lombardi and Ezra Suruma, 'African Perspectives on Sovereign Debt Restructuring', 19 August 2014, www.cigionline.org/p ublications/african-perspectives-sovereign-debt-restructuring.htm.
8 *Lusaka Times*, 'Zambia Will Struggle to Repay Debt that has Increased by 176% since 2011', 8 January 2016, www.lusakatimes.com/2016/01/08/zambia-will-struggle-to-repa y-debt-that-has-increased-by-176%-since-2011.htm; *This Day*, 'Nigeria Offered $6bn Chinese Loan, Agrees Currency Swap to Shore up Naira', 13 April 2016, www.thisda ylive.co/index/php/2016/04/13/nigeria-offered-6bn-chinese-loan-agrees-currency-swa p-to-shore-up-naira.htm
9 'China Puts Screws on Zim', *Mail and Guardian*, 23 January 2015, www.mg.co.za/a rticle/2015-01-23-china-puts-screws-on-zim.htm. The awarding of a Chinese version of the Nobel Peace Prize to Robert Mugabe in 2015 for his role in promoting world peace and its $65,000, perhaps an effort by Beijing to assuage him in the aftermath of the failure to provide expected loans, was even rejected by its recipient.

10 See, for example, Mpumelelo Mkhabela, 'ANC is Selling our Country to China', *Sowetan Live* 5 October 2015, www.sowetanlive.co.za/2015/10/05/anc-is-selling-our-country-to-china1.htm.

References

Abrahamsen, Rita. 2017. "Research Note: Africa and International Relations: Assembling Africa, Studying the World." *African Affairs* 116(462): 125–139.

Acharya, Amitav. 2014. "Global International Relations and Regional Worlds." *International Studies Quarterly* 58(4): 647–659.

Afrobarometer. 2016. http://afrobarometer.org/blogs/heres-what-africans-think-about-chin as-influence-their-countries (accessed July 19, 2017).

Agualusa, Jose Eduardo. 2007. *As Mulheres do Meu Pai*. Alfragide, Portugal: D Quixote.

Alden, Chris. 2007. *China in Africa*. London: Zed.

Appiah, Kwame Anthony. 1992. *In My Father's House: Africa in the Philosophy of Culture*. Oxford: Oxford University Press.

Auden, W. E. 1978. *Selected Poems*. London: Faber and Faber.

Bilgin, Pinar. 2008. "Thinking Past Western IR." *Third World Quarterly* 29(1): 5–23.

Bilgin, Pinar and L. H. M Ling. eds. 2017. *Asia in International Relations: Unlearning Imperial Power Relations*. London: Routledge.

van Binsbergen, Wim and Rijk van Dijk. eds. 2004. *Situating Globality: African Agency in the Appropriation of Global Culture*. Leiden, Boston: Brill.

Bodomo, Adams. 2012. *Africans in China: A Sociocultural Study and its Implications on Africa-China Relations*. New York: Cambria Press.

Bond, Patrick and Ana Garcia. eds. 2015. *BRICS: An Anti-Capitalist Critique*. London: Pluto Press.

Brown, William. 2012. "A Question of Agency: Africa in International Politics." *Third World Quarterly* 33(10): 1889–1908.

Brown, Willam and Sophie Harman. 2013. *African Agency in International Politics*. Abingdon: Routledge.

Brutus, Dennis. 1978. *Stubborn Hope: Selected Poems of South Africa and a Wider World, including China Poems*. Portsmouth, Hampshire: Heinemann.

Chatelard, Solange. 2016. 'Rien ne vaut son chez-soi': Pérégrinations d'une famille chinoise en Zambie. *Hommes & Migrations*. 71–75. 10.4000/hommesmigrations.3640.

Chen, Ching-Chang. 2011. "The Absence of Non-Western IR Theory in Asia Reconsidered." *International Relations of the Asia-Pacific* 11(1): 1–23.

Chowdhry, Geeta and Sheila Nair. 2002. *Power, Postcolonialism and International Relations: Reading Race, Gender and Class*. London: Routledge.

Corkin, Lucy. 2013. *Uncovering African Agency: Angola's Management of China's Credit Lines*. Farnham: Ashgate.

Cornelissen, Scarlett, Fantu Cheru and Timothy M. Shaw. eds. 2012. *Africa and International Relations in the 21st Century*. London: Palgrave.

Couto, Mia. 2010. "A China dentro de nos". *Pensageiro frequente*. Editorial Caminho.

Death, Carl. 2013. "Governmentality at the Limits of the International: African politics and Foucauldian theory." *Review of International Studies* 39(3): 763–787.

Dunn, Kevin C. and Timothy M. Shaw. eds. 2001. *Africa's Challenge to International Relations Theory*. Palgrave: London.

Fanon, Frantz. 1967 [1952]. *Black Skin, White Masks*. New York: Grove Press.

Fanon, Frantz. 1963. *The Wretched of the Earth*. New York: Grove Press.

French, Howard. 2014. *China's Second Continent*. New York: Alfred A. Knopf.

Gadzala, Alexandra ed. 2015. *Africa and China: How Africans and their Governments are Shaping Relations with China*. New York: Routledge.

Harrison, Graham. 2002. *Issues in the Contemporary Politics of Sub-Saharan Africa: the dynamics of struggle and resistance*. Basingstoke: Palgrave.

Hirano, Katsumi. 2007. Executive Summary. *Perspectives on Growing Africa: From Japan and China*, IDE-JETRO Seminar, 10 September.

Hobson, John M. 2017. "The Postcolonial Paradox of Eastern Agency." In *Asia in International Relations: Unlearning Imperial Power Relations*, edited by Pinar Bilgin and LHM Ling, 109–120. London: Routledge.

Jamba, Sousa. 1992. *Patriots*. London: Penguin Books.

Juma, Calestous. 2015. "Afro-Chinese Cooperation: The Evolution of Diplomatic Agency.", In *Africa and China: How Africans and Their Governments are Shaping Relations with China*, edited by Aleksandraa W. Gadzala. Lanham: Rowman and Littlefield.

Khong, Yuen Foong. 2013. "The American Tributary System." *Chinese Journal of International Politics* 6(1): 1–47.

Ling, L. H. M. 2002. *Postcolonial International Relations: Conquest and Desire between Asia and the West*. Basingstoke: Palgrave.

Ling, L. H. M. 2017. "Romancing Westphalia: Westphalian IR and Romance of the Three Kingdoms." In *Asia in International Relations: Unlearning Imperial Power Relations*, edited by Pinar Bilgin and L. H. M. Ling. 184–198. London: Routledge.

Mannoni, Octave. 1990. *Prospero and Caliban: The Psychology of Colonization*. Ann Arbor, NY: University of Michigan Press [2nd edn.].

Marukawa, Tomoo. 2007. "Afurika ni shinshutsu suru Chūgoku". [China's Advance into Africa] http://web.iss.u-tokyo.ac.jp/~marukawa.

Mohan, Giles and Ben Lampert. 2013. "Negotiating China: Reinserting African Agency into China–Africa Relations." *African Affairs* 112(446): 92–110.

Ndlovu-Gatsheni, Sabelo and Siphamandla Zondi. eds. 2016. *Decolonising the University, Knowledge Systems and Disciplines in Africa*. Durham, NC: Carolina Academic Press.

Noesselt, Nele. 2015. "Revisiting the Debate on Constructing a Theory of IR with Chinese Characteristics." *The China Quarterly* 222: 430–448.

Oualalou, Fathallah. 2016. *La Chine et nous – responder au second*. Casablanca: La Croisee des Chemins/OCP.

Said, Edward. 1995. *Orientalism*. Harmondsworth: Penguin.

Sautman, Barry and Hairong Yan. 2016. "The Discourse of Racialization of Labour and Chinese Enterprises." *Ethnic and Racial Studies* 39(12): 2149–2168.

Shelton, Garth, Yazini April and Li Anshan. eds. 2015. *FOCAC 2015: A New Beginning of China-Africa Relations*. Pretoria: Africa Institute of South Africa.

Spivak, Gayatri. 1998. *In Other Worlds*. Abingdon: Routledge.

Tieku, T. K. 2013. "Exercising African Agency in Burundi via Multilateral Channels: Opportunities and Challenges." *Conflict, Security & Development* 13(5): 513–535.

Vanaik, Achin. 2013. "Capitalist Globalisation and the Problem of Stability: Enter the New Quintet and Other Emerging Powers." *Third World Quarterly* 34(2): 194–213.

Vanaik, Achin. 2014. "Emerging Powers: Rise of the South or Reconfiguration of Elites?" *TNI Working Papers*, September 2014.

Wekesa, Bob. 2017. "New Directions in the Study of Africa–China Media and Communications Engagements." *Journal of African Cultural Studies* 29(1): 11–24.

Zondi, Siphamandla. 2016. "Ubuntu and Sumak Kaway: The Inter-Parliamentary Union and the Search for a Global South Paradigm of Humanist Development." *South African Journal of International Affairs* 23(1): 107–120.

INDEX